Fractured Communities

Nature, Society, and Culture
Scott Frickel, Series Editor

A sophisticated and wide-ranging sociological literature analyzing nature-society-culture interactions has blossomed in recent decades. This book series provides a platform for showcasing the best of that scholarship: carefully crafted empirical studies of socio-environmental change and the effects such change has on ecosystems, social institutions, historical processes, and cultural practices.

The series aims for topical and theoretical breadth. Anchored in sociological analyses of the environment, Nature, Society, and Culture is home to studies employing a range of disciplinary and interdisciplinary perspectives and investigating the pressing socio-environmental questions of our time—from environmental inequality and risk, to the science and politics of climate change and serial disaster, to the environmental causes and consequences of urbanization and war making, and beyond.

Available titles in the Nature, Society, and Culture series:

Diane C. Bates, *Superstorm Sandy: The Inevitable Destruction and Reconstruction of the Jersey Shore*

Cody Ferguson, *This Is Our Land: Grassroots Environmentalism in the Late Twentieth Century*

Anthony E. Ladd, ed., *Fractured Communities: Risk, Impacts, and Protest against Hydraulic Fracking in U.S. Shale Regions*

Stefano B. Longo, Rebecca Clausen, and Brett Clark, *The Tragedy of the Commodity: Oceans, Fisheries, and Aquaculture*

Stephanie A. Malin, *The Price of Nuclear Power: Uranium Communities and Environmental Justice*

Chelsea Schelly, *Dwelling in Resistance: Living with Alternative Technologies in America*

Diane Sicotte, *From Workshop to Waste Magnet: Environmental Inequality in the Philadelphia Region*

Sainath Suryanarayanan and Daniel Lee Kleinman, *Vanishing Bees: Science, Politics, and Honeybee Health*

Fractured Communities

Risk, Impacts, and Protest against Hydraulic Fracking in U.S. Shale Regions

EDITED BY ANTHONY E. LADD

Rutgers University Press

New Brunswick, Camden, and Newark, New Jersey, and London

Library of Congress Cataloging-in-Publication Data

Names: Ladd, Anthony E., 1953- editor.
Title: Fractured communities : risk, impacts, and protest against hydraulic fracking in U.S.
    Shale regions / edited by Anthony E. Ladd.
Description: New Brunswick : Rutgers University Press, 2018. | Series: Nature, society,
    and culture | Includes bibliographical references and index.
Identifiers: LCCN 2017007448 (print) | LCCN 2017023793 (ebook) | ISBN 9780813587684
    (epub) | ISBN 9780813587691 (WebPDF) | ISBN 9780813594248 (mobi) |
    ISBN 9780813587677 (hardback) | ISBN 9780813587660 (paperback)
Subjects: LCSH: Hydraulic fracturing—Social aspects—United States. | Hydraulic
    fracturing—Risk assessment—United States. | Petroleum industry and trade—United
    States. | Shale gas industry—United States. | BISAC: SCIENCE / Environmental
    Science. | NATURE / Environmental Conservation & Protection. | POLITICAL
    SCIENCE / Public Policy / Environmental Policy. | SOCIAL SCIENCE / Regional
    Studies. | SOCIAL SCIENCE / Sociology / Rural.
Classification: LCC HD9565 (ebook) | LCC HD9565 .F723 2017 (print) |
    DDC 338.2/7280973—dc23
LC record available at https://lccn.loc.gov/2017007448

A British Cataloging-in-Publication record for this book is available from the British Library.

www.rutgersuniversitypress.org

Manufactured in the United States of America

This book is dedicated to all the concerned citizens and activists on the planet who struggle against the Treadmill of Production for a more sustainable future.

Contents

Preface

In 1876, an experienced quarry owner and partner in the Ft. Wayne & Southern Railroad, George Carter, was drilling for coal near Eaton, Indiana when he accidentally discovered a vein of natural gas more than 600 feet below the surface. Panicked by the horrid sulphur smell and two-foot flame that shot out of the ground, Carter feared that he had drilled into the outskirts of hell itself and quickly capped the well, lest he be sucked down into the dark void of the underworld. Ten years later, after a series of new gas field discoveries in Indiana and Ohio had been announced, Carter and two business associates returned to drill again at their original site in Eaton and struck gas at 922 feet. On November 11, 1886, the first natural gas well was drilled in nearby Muncie, joining a string of other gas wells in neighboring towns that had come into production during the decade. The Indiana "natural gas boom" was officially underway (Edmonds and Geelhoed 2001).

The "Trenton Field," as it was known, covered over 5,120 square miles of east-central Indiana and became the largest natural gas field in the world. Towns competed to attract new industries with offers of free gas, land, railway access, and tax credits, which in turn created new banks, financial institutions, and other commercial ventures. Some 200 companies emerged that explored, drilled, distributed, and sold natural gas from over 380 producing wells in the region (American Oil and Gas Historical Society 2015). By 1890, 162 factories had been built, over 10,000 jobs had been created, and Muncie became known as "Mighty Muncie," the "Birmingham of the North," and a working-class city where "people made things" (Edmonds and Geelhoed 2001: 45–46). New tinplate, wire, shovel, box, and glass factories migrated to Muncie, including the Ball Brothers Glass Manufacturing plant from Buffalo, New York. By 1915, due to the phenomenal success of Ball Brothers, Muncie became firmly established as the center of glass jar production and food preservation in America, as famous for glass-making and home canning as Detroit was for automobiles or Akron for rubber tires.

While the gas boom drove rapid population growth and industrialization throughout the region, many communities became complacent about what they assumed was an unlimited supply of natural gas beneath their feet. Calls for conservation from concerned citizens and state inspectors went unheeded, gas pressures at wellheads began to drop, and the resulting saltwater intrusion eroded the capacity of many natural gas wells to function properly. Plant closings followed, thousands of jobs were lost, and many of the manufacturers who had come to Indiana for the ready supply of cheap gas either went out of business or were forced to move out of state. By 1913, Indiana was importing natural gas from West Virginia to meet demand and by 1920, the state was consuming more energy than it was producing. The natural gas boom was effectively over (American Oil and Gas Historical Society 2015).

In the two decades that followed, the town of Muncie, Indiana became famous for more than just its natural gas boom or Ball glass mason jars. Thanks to the pioneering research of husband-and-wife sociologists Robert and Helen Lynd from the University of Michigan, Muncie also became known as "Middletown U.S.A." and the site of the first sociological study of an average or "typical" American community. In both *Middletown* (1929) and *Middletown in Transition* (1937), the Lynds adopted the methodological tools of cultural anthropology to analyze Muncie's dominant institutions, power structure, class cleavages, cultural contradictions, and patterns of social interaction. Among the many facets of ordinary life their study documented, the Lynds were particularly interested in how, in the context of the economic hard times of the Great Depression, the introduction of new technological innovations (like cars and radios) had led to an uneven distribution of social impacts, which they saw as weakening Muncie's community ties, creating fissures between families and friends, and diminishing many people's faith in American democracy. The *Middletown* studies not only helped establish the study of American communities as central to the field of sociology (which led to a third *Middletown* study carried out by the late Theodore Caplow and his associates in the 1970s), but made Muncie the go-to city for national pollsters and market researchers who wanted to take the pulse of the typical American on any given political or consumer issue of the day (Gillette 1983). Moreover, the Lynds' landmark research not only represented the first sociological case study that explored how an American city had developed around a local natural gas boom, but how the community became *fractured* in key respects by many of its social impacts.

As fate would have it, Muncie, Indiana is my hometown and the history of its early gas boom and role in the beginnings of American sociology has always had a special personal and professional appeal to me. Growing up there in the

1950s and 1960s (where I attended primary school with descendants of the Ball brothers), I went on to major in sociology at Ball State University (built on land donated to the city of Muncie by the Ball family) from 1972 to 1976. As a youngster, I heard my grandmother's stories about coming to Muncie from Chicago to live with her aunt during the heyday of the gas boom, finding work in the local glass and auto plants, and then meeting my grandfather who was a traveling salesman for the Indiana Gas Company. Later, during my first introductory sociology course at Ball State, I had the opportunity to read critiques of the Lynds' *Middletown* books and learn more about the gas boom and legacy of social research that had made Muncie nationally famous as the "average American city." In the years that followed, when sociologists at conferences would ask me about my origins, I would always joke that I was a "sociologist from Muncie, the birthplace of American sociology"—an admission that was, for me, sort of like being a Catholic from Rome.

Given my Middletown roots, it has always seemed appropriate to me that I became interested in environmental sociology and the impacts of energy development on society. For the past three decades, my primary area of research has revolved around energy-related environmental controversies, technological disasters, and the ways in which industrial pollution can degrade the social and ecological systems on which communities depend. After seeing the film premiere of Josh Fox's *Gasland* at the 2010 True-False Film Festival in Columbia, Missouri, I returned home to New Orleans only to learn that similar issues surrounding natural gas development and hydraulic fracturing were unfolding in my own state in the Haynesville Shale region of northwest Louisiana. With some modest funding provided by a Marquette Fellowship and Faculty Research Grant from Loyola University, I began conducting qualitative field research and interviews with various stakeholders throughout the Haynesville Shale during the summer of 2012. The initial focus of my research was twofold: first, to document the differential opportunity-threat impacts that stakeholders perceived to be associated with unconventional natural gas development and hydraulic fracking in the Haynesville community; and second, to analyze the collective frame disputes and discursive narratives articulated by citizens regarding the major benefits and risks posed by gas fracking for themselves and the environment. The Haynesville residents that I interviewed and the data that I collected that summer had a profound effect on me. Above all, the experience served to justify my early convictions that the subject of fracking constituted an inherently fascinating and relevant area of inquiry for environmental and rural sociologists, as well as social movement scholars interested in community conflicts and protest mobilization surrounding local environmental problems. Another important outcome of the project was that I began working to identify a network of other sociologists around the country who were similarly engaged in research on the community disputes and impacts surrounding unconventional energy production and fracking in their backyards.

In 2015, I was invited to organize and moderate a special session at the American Sociological Association (ASA) meetings in Chicago titled, "No Fracking Way: Risk, Resistance, and the Mobilization of Protest Against Hydraulic Fracking in U.S. Shale Communities." The session featured four sociologists (Abby Kinchy. Suzanne Staggenborg, Sherry Cable, and myself) who were currently involved in community research on local resistance, as well as adaption, to hydraulic fracking in three prominent shale regions of the country (the Marcellus, Utica, and Haynesville/Tuscaloosa respectively). The session was well received by the 60–70 sociologists who attended and included a number of faculty and graduate students who were engaged in fracking-related research projects of some kind in their regions. To my delight, many audience members approached me to say how much they enjoyed the research presentations and how much they had learned. With so many new fracking disputes heating up around the country, it became clear to me that sociological interest in the topic was rapidly growing. Just a year earlier, for example, at the 2014 ASA meetings in San Francisco, less than a half dozen people were listed on the program as presenting work related to unconventional energy development issues. By 2015, the number had quickly jumped to two to three times as many and a number of graduate students at various universities began writing me that they hoped to conduct their dissertation research on fracking protests and impacts if possible. Given the relative success of the ASA session and the promising work presented by its participants, I began to think more substantively about authoring an edited book that would feature new, cutting-edge sociological research—from both prominent and emerging scholars—dealing with community fracking disputes, impacts, and protests in different shale regions across the country. The more colleagues with whom I discussed my ideas about the book project, the more they encouraged me to pursue it with all deliberate speed.

Due to all the kind words of support I received, both from some of the talented sociologists who contributed their research to this volume, as well as a number of editors from reputable university presses—especially Peter Mickulas and Scott Frickel at Rutgers University Press—my decision to organize and edit *Fractured Communities* was sealed. Certainly, every published book represents a labor of love for its author and this one is no exception; its earliest inspiration comes from my *Middletown* past, as well as my longstanding interests in how energy policy shapes the present. But above all, I hope the research included in this volume can play some kind of positive role in redirecting our collective energy future—one that will move us away from our force-fed addiction to fossil fuels—and enable us to begin the transition to a sustainable energy path on which our survival will ultimately depend.

References

American Oil & Gas Historical Society. 2015. "Indiana Natural Gas Boom." Accessed on November 8, 2015. (http://aoghs.org/states/indiana-natural-gas-boom/).

Edmonds, Anthony O. and E. Bruce Geelhoed. 2001. *Ball State University: An Interpretive History*. Bloomington, IN: Indiana University Press.

Gillette, Howard, Jr. 1983. "Review: Middletown Revisited." *American Quarterly* 35(4): 426–433.

Lynd, Robert S. and Helen Merrell Lynd. 1929. *Middletown: A Study in Modern American Culture*. New York: Harcourt Brace.

Lynd, Robert S. and Helen Merrell Lynd. 1937. *Middletown in Transition*. New York: Harcourt Brace.

Fractured Communities

Introduction

————————————●

Energy Matters

ANTHONY E. LADD

Energy constitutes the lifeblood of daily existence and the scaffolding of our civilization. Without an abundant and continuous supply of cheap and accessible sources of energy, virtually every aspect of contemporary life on which humans rely—manufacturing, transportation, agriculture, construction, heating, lighting, refrigeration, sewerage and water systems, computing and communication—would grind to a halt within a matter of minutes. In the transition from the Stone Age to the Petrochemical Age, modern societies have become dependent on thousands of petroleum-derived substances that make up our food, clothing, cleaning products, furniture, pharmaceuticals, cosmetics, plastic credit cards, and energy system (Misrach and Orff 2012). Indeed, few issues are more important to understanding our political economy, dominant cultural values, housing patterns, transportation system, technological infrastructure, or foreign policy alignments around the world than the treadmill of fossil fuel production (Ladd 2016). While oil, natural gas, and coal reserves provide nearly 85% of the energy driving the $19 trillion U.S. economy, these same hydrocarbons also play a prominent role in creating the world's most pressing environmental problems, including atmospheric climate change, habitat loss, species extinction, air, water, and soil pollution, groundwater depletion, and a range of human health issues from asthma to cancer (Cable 2012; Bell 2014). Paralleling the "veil" of secrecy that corporate agribusiness has erected around issues of food production, factory farming, and animal welfare (Schlosser 2012), transnational energy corporations and their petrochemical allies have similarly worked to divert public attention away from the social, economic, and ecological costs of our reliance on finite supplies of carbonized energy. In every sense of the word, fossil fuels—like a Janus-headed creature beneath the earth's surface—represent a veritable Faustian bargain for contemporary society and the planet.

In their most essential form, fossil fuels represent special forms of stored sunlight, molecular bonds of hydrogen and carbon derived from the organic matter of ancient plants and animals that lived over 300 million years ago in primordial swamps and oceans. As these life forms from microorganisms to dinosaurs died, they decomposed and eventually became buried under multiple layers of mud, rock, and sand thousands of feet beneath the earth. In some regions, these materials were even covered by ancient seas before they dried up and receded. Other sedimentary deposits became formed in places where salt was precipitated out of seawater or seashells settled to the bottoms of lakes and oceans. As millions of years passed, different types of fossil fuels were formed from these organic sources, depending on the combination of animal and plant debris that existed, how long the material was buried, and how much it was compressed and "cooked" by the heat, pressure, and bacteria in the earth. For example, oil and natural gas were created from organisms that lived in water and were buried under the ocean or river sediments, while most coal deposits formed from the dead remnants of trees, ferns, and plants on land (Johnson 2014). The subsurface deposits of these fossilized organic materials, once concentrated, become the *source rocks* from which oil and gas is created. Since these energy forms are lighter than water, they tend to migrate upward until they become trapped in permeable reservoirs of sedimentary rock that cap them from escaping. Common sedimentary rocks include sandstone, limestone (carbonate), and shale. Depending on other geologic factors thousands of feet beneath the surface, these gas and oil deposits—including shale natural gas, shale oil, tar sands, coalbed methane, and other hydrocarbons—can be accessed and released should these low-permeability reservoirs become bent, twisted, or fractured by natural or human processes (Freudenburg and Gramling 2011: 21–24).

More than a century after the boomtown decades of the Indiana natural gas fields, the western oil strikes, and the Appalachian coalmines draw to a close, the age of cheap, plentiful sources of conventional fossil fuels on which our industrial society was built is ending and a new energy era of riskier *unconventional* fossil fuel dependence is dawning (Klare 2013). Due to recent innovations in 3-D seismic mapping, multidirectional drilling, and high-volume hydraulic fracturing (HVHF) techniques—more commonly known as "fracking"—national and global energy companies are rushing to tap into a virtual treasure trove of natural gas and oil reserves 5,000 to 10,000 feet beneath the surface that were previously inaccessible (Gold 2014; Klare 2014;

Yergin 2011). Patented in the late 1940s but only commercially viable in the last decade, fracking refers to a controversial well stimulation and completion technique in which millions of gallons of water, sand, and chemicals (many of them considered toxic) are injected under extreme pressure into deep underground shale deposits to fracture nonporous rock formations and release the trapped gas or oil, sending it up to the surface (Ladd 2014). Flannery (2016) describes the complexity of the fracking process in the following manner:

> The fracking of U.S. shales, as it is most commonly done today, involves drilling a curved borehole a mile or more into the earth, with the lower section of the borehole running horizontally through the shale layer. The borehole is lined with steel and concrete, and a perforator is then lowered into it. This tube-shaped device is effectively a gun filled with bullets. When triggered, the bullets leave holes in the section of the borehole that lies in the shale. Then thousands of gallons of water, often mixed with guar (a gelling agent derived from the same bean used to make chewing gum) and some acid, diesel, or other liquids, combined with sand or ceramic proppants, are forced into the well at a pressure twenty times that of domestic hoses. The mix exits forcefully through the bullet holes in the well casing, fracturing the shale at a distance of many yards. When the pressure in the well is lowered by withdrawing part of the fracking fluid, cracks in the shale open and remain open thanks to the action of the proppants, thus allowing the gas and oil to flow to the surface.

Currently, fracking is used in the extraction of over 90% of all unconventional shale gas and oil resources, some 1.3 million wells have been hydraulically fractured since 2008, and over 25,000 wells are fracked each year. Roughly 81% of fracking operations involve oil extraction, with the remaining activity focused on accessing natural gas (Hauter 2016). In addition, more than 15 million Americans live within a mile or so of a well in their community that has been fracked since 2000 (Gold 2014). As a result of the rapid development of fracking technology and its ancillary infrastructure—well pads, access roads, pipelines, compressor stations, wastewater processing, export facilities, and more—shale gas and oil production has risen 12-fold since 2005, and the United States is projected to be a net exporter of natural gas, if not the world's largest gas and oil producer, by as early as 2020 (Boudet et al. 2014; Ladd 2016). The U.S. Energy Information Administration (USEIA) has estimated that some additional 630,000 new onshore wells may be required to extract all the technically recoverable reserves of oil and gas (Hauter 2016: 7).

While disputes over the impacts of fossil fuel production have increased in recent years, few issues have been as contentious as those surrounding unconventional energy development and hydraulic fracking in shale communities across the United States. Historically, the costs of mining fossil fuels were largely borne by poor communities with little to no control over industry operations. In contrast, the extraction of shale gas and oil deposits is increasingly taking place today in working- or middle-class areas in the United States where the population has greater degrees of social capital and political clout (Gullion 2015). Still, most uranium and coal communities, as well as other energy resource–dependent regions in rural America, are usually characterized by volatile boom and bust economic cycles, chronic poverty, spatial isolation, and an uneasy sense among residents that their air, water, land, and public health have been mortgaged to feed the nation's fossil fuel addiction. Caught between the demands of global energy markets on the one hand and the localized burdens of living in a national sacrifice zone on the other, increasing numbers of "reluctant activists" are bonding together—even in oil- and gas-friendly regions traditionally hostile to environmentalists—to protect their drinking water, families, properties, communities, and civil liberties from the impacts of fossil fuel development (Bell and York 2010; Freudenburg and Frickel 1994; Ferguson 2015; Gullion 2015; Malin 2015; Malin and DeMaster 2016). As a result, new forms of political mobilization are emerging in shale communities across the country based on shared grievances that arise from "impending or actual threats to one's accustomed way of life" (Johnson and Frickel 2011: 306).

Applying a Sociological Lens to the Controversy over Fracking

The discipline of sociology in general—and environmental sociology in particular—offers an important and appropriate intellectual lens for understanding the community risks, impacts, and conflicts associated with unconventional energy development and hydraulic fracturing. As noted in the Preface, the study of local communities has a long tradition in sociology going back to the *Middletown* research of the 1920s and 1930s. Ever since, sociologists have examined how small towns and rural areas are often forced to bear the disproportionate burdens of mineral extraction, resource excavation, and waste disposal over which they have little control. Sociologists are also concerned with other key social processes that are central to the controversy surrounding shale fracking, including patterns of political and economic power, social class and inequality, globalization, social movement mobilization, collective protest, discourse analysis, and cultural sociology, to name just a few of the issues that surround the debates over HVHF techniques.

The field of environmental sociology emerged in the United States in the mid-1970s, sparked not only by sociological interest in the growth of environmental awareness and activism, but more specifically by the energy crisis of 1973–1974 that highlighted the dependence of industrialized societies on fossil fuels, as well as their deleterious impacts on air, water, and human health (Dunlap 2015). While scholars like Catton (1982), Humphrey and Buttel (1982), and Rosa, Machlis, and Keating (1988) were among the first to recognize the critical role of energy resources in societal-environmental inter-actions, environmental sociologists in recent years have increasingly focused on such questions as: What are the social, environmental, and public health consequences of energy production? Which communities are most likely to bear the disproportionate burdens of pollution from energy production? Why do some communities protest energy-related pollution, while others do not? What strategies do energy corporations employ to acquire political power, and how do they use it to attain greater profits and market control? (Bell 2014: 138). In addition, environmental sociologists also work to explore the social construction of environmental and energy problems and how activists, advo-cacy coalitions, scientists, and policy-makers engage in claims-making to bring such issues to the public's attention (Dunlap 2015). Questions concerning cor-porate power, technological risk, disaster, the treadmill of production, public epidemiology, animal welfare, human ecology, rural sociology, social impact assessment, and demographic change, among other issues, are also at the heart of their research gaze. Perhaps the most important insight that environmen-tal sociology brings to the analysis of environmental problems is that they are conceptualized as inherently, by cause and consequence, *social problems*, with deep roots in the political economy, human ecological complex, and domi-nant cultural values of contemporary society. As I often remind my students, ecosystems don't pollute themselves and environmental degradation is not the result of something gone awry in nature.

At the same time that environmental sociology and the wider environ-mental movement were taking off in the 1970s, perceptive social critics like E. F. Schumacher (1973) and Amory Lovins (1977) argued that American energy policy was at a critical historical crossroads defined by two mutu-ally exclusive and structurally distinct sociopolitical futures. One fork in the road would lead humanity down the existing "hard energy path" defined by complex and risky technologies like fossil fuels and nuclear power, while the other fork would allow people to choose a "soft energy path" centered around efficiency, alternative technologies, and renewable energy resources like solar, wind, and biomass. Some 40 years later, we are *still* standing at the same pro-verbial crossroads. Despite the growth of the fossil fuel divestment movement (led by organizations like 350.org) in communities across the country, the rapidly declining costs of solar and wind power generation, and the mount-ing evidence that virtually all of our energy needs could be met through the implementation of various energy-saving measures and green technologies over the next two decades, state-corporate energy interests have worked qui-etly behind the scenes to maintain the hegemony of fossil fuels (Klein 2014; Rauber 2014). For some observers, the ascendancy of neoliberal economic policies, the deregulation of markets and industries, and the rise of the shale energy/fracking revolution signals the emergence of a new energy epoch—a "Third Carbon Era"—that will pose serious obstacles to achieving the goals of environmental equity and sustainability in the 21st century (Klare 2013; Perrow 2015; Ladd 2016).

The growing public debate in the United States over hydraulic fracking has become a flashpoint in the wider environmental frame dispute between proponents who see shale gas and oil as the future energy source of the planet—and opponents who view it as the last gasp of a dying fossil fuel era (Ladd 2014: 306). Communities and neighbors are divided, residents are worried, government officials are anxious, and oil and gas companies are angry with grassroots activists who try to block their efforts to extract energy from lands they have leased. Many forms of economic, political, and ecological capital are at stake in this nascent conflict, including the trillions of dollars in profits that stand to be made—or lost—for the energy-industrial complex, a potential seismic shift in the existing distribution of global economic and political power, and whether the acceleration of greenhouse gases in the atmosphere will unleash irreversible social disruption and widespread environmental disaster.

As the ten case studies in this edited volume of sociological research will illustrate, the risks, impacts, and political conflicts surrounding unconventional energy development go far beyond the partisan stakeholders who are on the front lines of the battles over fossil fuel exploration and extraction. In the coming decade, the outcomes of these environmental disputes will have broader implications for the future of the U.S. economy, foreign policy, increased geopolitical tensions between nations, and the balance of global power. Many critics believe that if American energy policy becomes dependent on a new generation of "extreme energy" resources extracted from increasingly fragile and remote environments, the landscape of risk for our human and biotic communities will be irrevocably altered (Klare 2013, 2014).

A Brief History of Hydraulic Fracturing

The use of rudimentary hydraulic fracturing techniques to facilitate oil and water well production has existed since the 1860s, shortly after the nation's first oil well was drilled in 1859 by Col. Edwin Drake in Pennsylvania. Originally called an "oil mine," the well was actually constructed for the purpose of extracting crude oil to sell as a remedy for rheumatism (Gullion 2015). In the decades that followed, oil was used to manufacture kerosene, which replaced whale oil and candlelight as major forms of lighting spaces. However, the fracking process used in energy development today bears little historical resemblance to its early origins and modest impacts. In 1866, Civil War veteran Col. Edward Roberts was issued U.S. Patent #59,936 for an invention he called the

"Exploding Torpedo." Using iron cylinders filled with gunpowder and lubricated by water to provide "fluid tamping" to focus the concussion, Roberts used a cap and wire detonator to ignite the device and "shoot" the explosives down oil wells. The procedure was so successful—increasing oil production in some wells by as much as 1,200%—that Roberts went on to detonate these explosives in thousands of Pennsylvania oil wells, earning a 15% royalty on all subsequent oil production (Heinberg 2013; Manthos 2013).

In 1891, John Campbell was awarded U.S. patent #459,152 for his invention of a small device that could be used to rotate drilling bits for dental applications, although he also believed that it might have utility someday to rotate drive shafts or cables (Bamberger and Oswald 2014). That same year, the Texas Railroad Commission was established to regulate the development of railroads, and later on, the transportation of oil by rail out of the Texas oil fields. In 1929, the first recorded horizontal oil well was drilled in Texas; but it wasn't until 1947 that Floyd Farris of Stanolind Oil and Gas completed the first conventional hydraulic fracturing operation using a thousand gallons of water, along with some chemicals and sand, to fracture a shallow gas well at the Hugoton gas field in southwestern Kansas. Farris's experiments soon led to the first commercial application of the new extraction process when his company and Haliburton successfully fractured two gas wells in Oklahoma and Texas on the same day (Heinberg 2013). Rather than dropping explosives down the well as the early Pennsylvanian oilmen had done, these drillers used a World War II airplane engine as a pump to pressurize the fracking fluid and hydraulically extract the methane contained in the limestone. In 1949, Haliburton took over Stanolind Oil and patented the hydro-fracking process, becoming the first company to commercially apply the technique to producing natural gas (Bamberger and Oswald 2014; Food & Water Watch 2011; Manthos 2013). Despite these innovations, oil and gas drilling remained brutally hard, dangerous work, often resulting in debilitating injuries and fatal accidents for workers, as well as land, water, and wildlife impacts (Gullion 2015).

In the early 1980s, a Houston-based oil and gas producer named George P. Mitchell had been studying recent geological reports and drilling technologies in hopes of finding new ways to produce more natural gas for the market. While he knew that natural gas could certainly be extracted from deep, hard shale rock formations, Mitchell's experiments in the field proved to be extremely difficult, expensive, and far from commercially viable. With the introduction of 3-D seismic imaging technology alongside Section 29—a provision in the 1980 windfall profits tax bill that granted federal tax credits for unconventional natural gas exploration and removed price controls—Mitchell pushed his engineers and geologists to "crack the code" of the Barnett Shale formation beneath Dallas and Fort Worth, Texas (Yergin 2011: 327). By the 1990s, Mitchell had developed "slick-rock" fracturing methods involving the addition of friction-reducing gels to water to increase the flow of frack fluids into wells. Adding fine sand called "proppants" to prop open the newly created cracks and fissures in the shale rock, as well as a long list of chemicals (many of them toxic) to the highly pressurized fracking fluid—combined with his new horizontal drilling techniques—Mitchell's gas field tests began showing results. In 1997, Mitchell Energy and Development Company successfully completed their first high-volume, hydraulic drilling and fracturing operations, using 800,000 gallons of water and 200,000 tons of "light sand" to extract natural gas from deep subterranean source rocks underlying the Barnett Shale (Heinberg 2013; Yergin 2011).

By the time Mitchell sold his company to Devon Energy in 2001 for $3.1 billion in cash and stock, he was heralded as an iconic "independent" Texas oilman who had become rich from his own ingenuity and hard work. In reality, like the fracking technology he helped pioneer, his wealth and reputation as the "Father of Fracking" had also been made possible by decades of federally funded research, tax breaks, and subsidies that helped create propping agents, drill bits, and chemical formulations for gels and foams to make fracking more cost-effective. Combined with making lavish campaign contributions to Texas judges and railroad commissioners, as well as building powerful political connections to energy firms like Enron, Haliburton, and the National Petroleum Council, oilmen like Mitchell were able to profit handsomely from the Reagan and Bush–era policies favoring greater energy deregulation, natural gas promotion, and increased offshore drilling (Hauter 2016).

The final key component in the evolution of hydraulic fracturing technology was the introduction of multiple-well pads or cluster drilling in 2007. With this advancement, over a dozen wells could be drilled from a single industrial platform, allowing operators to concentrate their rigs and materials in one three-square-acre site to reduce costs and permitting complications (Heinberg 2013). While industry marketing narratives often employ a narrow linguistic interpretation of the term "fracking," referencing only the well drilling and fracturing/stimulation phases of energy development, in fact, the entire shale gas and oil extraction process encompasses over a dozen interrelated steps involving significant environmental risks and impacts to the community at each stage (Bamberger and Oswald 2013; Evensen et al. 2014a; Finkel and Law 2016; Heinberg 2013; Humes 2012: 55; Wilber 2015). These impacts include:

- Leasing and clearing the land, including building access roads into the prospective rural/suburban well site;
- Building a well pad that can accommodate eight or more individual wells;
- Digging containment pits and ponds for drilling and fracking fluids;

- Drilling the vertical portion of each well (below the land and water table), which can be anywhere from 5,000 to 10,000 feet;
- Drilling the well's horizontal leg (through the shale layer), which can be 1–2 miles long;
- Installing casing and cement in the well shaft to inhibit gas and chemicals from flowing freely into soil, streams, and aquifers;
- Trucking or piping in millions of gallons (generally 3–7 million) of water for each well and fracking episode (up to ten or more);
- Transporting hundreds of thousands of pounds of equipment, sand, chemicals, and other supplies to and from the well site with eighteen-wheel diesel trucks (involving thousands of round trips on local roads per well);
- Ringing the well with 12–18 high-pressure diesel pumps on flatbed trucks;
- Fracturing the shale to release the methane/oil by using explosives and then injecting fracking fluids at pressures up to 10,000 or more pounds per square inch (psi), along with sand and ceramic "proppants" to keep the fractures open;
- Capturing and removing or recycling the "flowback" wastewaters (15–80% of the fracking fluids injected down the well), containing brine, hydrocarbons, radioactive substances, and various toxic chemicals;
- Controlling, processing, measuring, pressurizing, and piping the gas/oil away from the wellheads to compressor stations and storage facilities for distribution or export;
- Rehabilitating the land around the well site, including installing several protruding well pipes to vent built-up gas from the ground into a couple of small tanks.

When the interactive effects of these complex and intrusive processes, among others, are taken into account, the significant differences between the elementary fracking techniques of 60 years ago and those employed today become easily apparent. As one observer points out (Manthos 2013), "Modern fracking has as much in common with early fracking as an SR-71 Blackbird spy plane has in common with the Wright Flyer." Furthermore, these production techniques highlight the significant qualitative and quantitative distinctions between *conventional* and *unconventional* forms of energy production (Bamberger and Oswald 2014; Heinberg 2013; Humes 2012; Wilber 2015). Whereas older, conventional forms of development involved single-directional, vertical drilling into shallow gas or oil deposits that rarely penetrated local aquifers and groundwater sources, today's unconventional shale development represents a complex technological and organizational system of energy production that is characterized by significantly greater health hazards and freshwater use, more toxic chemical compounds and produced wastewater, more poorly understood long-term risks and hidden costs, and the absence of effective regulatory oversight over the entire process (Ladd and Perrow 2016).

While the combination of these recent technological and infrastructural developments made it possible to extract a new generation of unconventional fuels, fracking never would have become commercially viable had it not been supported by a host of other economic and policy incentives. Chief among these was the passage of the Energy Policy Act of 2005, an omnibus energy bill created during the Bush administration that included, among other things, tax incentives for producers of "alternative energy sources," broadly defined. Buried in the more than 1,700 sections of the law—engineered by the then–vice president and former CEO of Haliburton, Dick Cheney—was a section of provisions that exempted oil and gas drilling—and its wastewater—from the federal regulatory statutes of the Safe Drinking Water Act (SDWA), Clean Water Act (CWA), Clean Air Act (CAA), Solid Waste Disposal Act (SWDA), Resource Conservation and Recovery Act (RCRA), Comprehensive Environmental Response, Compensation, and Liability Act (CERCLA), Emergency Planning and Community Right to Know Act (EPCRA), the National Environmental Policy Act (NEPA), as well as several EPA regulations related to hazardous waste disposal (Food & Water Watch 2015; Kreuze, Schelly, and Norman 2016). In what became known as the "Haliburton Loophole," one provision specifically exempted the oil and gas industry from having to disclose the chemical contents of their fracking fluids under the guise that the ingredients constituted industrial "proprietary secrets" (Heinberg 2013; Perrow 2015). In addition, the Energy Policy Act also granted the Federal Energy Regulatory Commission (FERC) sweeping new powers to overrule local and state governments in the siting of new pipelines and transmission lines, as well as granted the natural gas industry $13.5 billion in subsidies (Gullion 2015; Hauter 2016).

In the face of these federal exemptions, the responsibility to regulate oil and gas fracking activity has been left almost totally to the states, despite their general lack of resources and unwillingness to confront the industry's political and economic muscle. For example, the state of Michigan has only 30 inspectors to oversee 25,000 active wells, while New York has only 31 inspectors to regulate over 125,000 wells. Between 2009 and 2012, Bureau of Land Management (BLM) records showed that the majority of production wells on public lands in states like South Dakota and Pennsylvania went totally uninspected (Kreuze, Schelly, and Norman 2016: 3).

Due to a number of key supporting factors, including the growth of neoliberal economic policies, rising energy prices, federal legislation and subsidies, state severance tax exemptions, increased pipelines and export facilities, and an orchestrated industry push for "free-market" energy deregulation, the large-scale production of gas and oil from deep shale formations became profitable for the first time. Since 2005, unconventional energy production—made possible by hydraulic fracking—has exploded more than 12-fold and has been trumpeted by politicians, pundits, and some national environmental organizations as nothing less than a new economic and energy "game changer" for the United States (USEIA 2015; Ladd 2014, 2016; York 2015). Natural gas markets were also given a boost with the passage of the Natural Gas Act of 2011, which provided a series of tax breaks to various transportation interests to encourage truck, bus, and automobile fleets to convert from gasoline and diesel liquids to natural gas–fueled vehicles (Wilber 2015).

Major U.S. Shale Regions and the Shale Energy Boom

To understand the phenomenal rise in unconventional fossil fuel production and hydraulic fracturing in recent years, it is necessary to geographically identify the key U.S. shale regions and formations (known as "plays") where this new species of energy development is taking place today. Figure I.1 illustrates the major North American shale plays and geological basins, which lie beneath at least 30 of the lower 48 U.S. states. Presently, dozens of major energy conglomerates, as well as smaller independent companies, are engaged in various stages of exploration, drilling, and fracking on roughly 180 million acres of private, state, and federal lands they have leased across the country—an area far greater than the states of California and Florida combined (Ladd 2014). Within these individual shale plays, geophysicists and geologists (employed by government agencies and private energy companies), work to identify suitable locations in shale formations that have the greatest potential to produce commercial volumes of natural gas and oil. For the most part, these areas are identified using rock core samples and assorted geophysical and seismic technologies to produce three-dimensional (3-D) maps that pinpoint where subsurface hydrocarbon deposits are located (USEIA 2016). A brief geographical description of some of the major U.S. shale regions shown in Figure I.1 follows, accompanied by recent federal drilling activity data illustrating shale gas and oil production for July 2016 (see Table I.1).

Underlying much of the Appalachian Mountain range of the northeast, the largest U.S. shale formation is the *Marcellus Shale* region, a mile-deep repository with as much as 500 trillion cubic feet (tcf) of natural gas stretching 104,000 square miles across portions of New York, Pennsylvania, West Virginia, and Ohio. The Marcellus currently produces almost 40% of U.S. shale gas, represents one of the largest gas fields in the world, and is located in the middle of one of the largest energy markets in the country, incorporating the metropolitan areas of New York, Philadelphia, Pittsburgh, and Boston (Wilber 2015). Over 5,800 wells have been drilled in the region and gas well

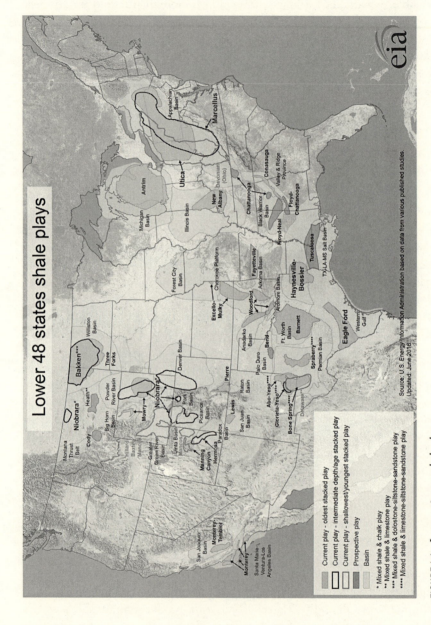

FIGURE I.1. Lower 48 states shale plays.

Table I.1
Natural Gas and Oil Production in Major Shale Regions (as of July 2016)

Region	Gas production (million cubic feet/day)	Oil production (thousand barrels/day)
Marcellus	17,976	40
Permian	6,914	21,980
Eagle Ford	6,014	1,127
Haynesville	5,916	47
Niobrara	4,068	383
Utica	3,663	77
Bakken	1,591	998
Total	46,142	4,652

SOURCE: U.S. Energy Information Administration. 2016. "Drilling Productivity Report." (www.eia.gov/petroleum/drilling/#tabs-summary-2).

production has risen steadily for the most part since 2007. As shown in Table I.1, Marcellus production represents over 17,900 million cubic feet of shale gas per day (Malin and DeMaster 2016; USEIA 2016).

The second largest shale gas play in the United States is the *Haynesville Shale* formation (also referred to as the Haynesville-Bossier Shale) covering a 9,000-square-mile area of northwestern Louisiana and spilling over into a small portion of southwestern Arkansas and eastern Texas. The 150-million-year-old shale formation is estimated to contain 66 trillion cubic feet (tcf) of recoverable natural gas, one of the largest natural gas discoveries in history, and equivalent to a decade's worth of North American consumption (Lavelle 2012). Shale gas production in the Haynesville skyrocketed after 2008 and currently accounts for over 5,900 million cubic feet (mcf) per day (USEIA 2016). Over 2,700 total active wells have been drilled to date, but state officials estimate that some 10,000 wells may eventually be needed to extract all the gas in the formation over the next 20–30 years (Ladd 2013, 2014).

As shown in Table I.1, the third largest producing gas and oil region is the *Eagle Ford Shale*, located in southern Texas below San Antonio and stretching in an arc from the Mexico border eastward some 50 miles wide and 400 miles long. While the shale gas boom there is one of the most recent, emerging largely after 2010, the Eagle Ford is estimated to contain up to 50 trillion cubic feet (tcf) of recoverable shale deposits and currently produces over 6,000 million cubic feet (mcf) of gas per day (Lavelle 2012; USEIA 2016). Moreover, the formation also constitutes the fastest growing and second highest oil production region in the country today, currently generating over 1,100 thousand barrels of oil a day. Since 2008, over 20,000 drilling permits have been issued and more then 3,100 gas and oil wells have been drilled in the Eagle Ford by some of the most active energy companies in the industry, including EOG

Resources, Apache, Petrohawk, Cabot, ExxonMobil, ConocoPhillips, and Chesapeake (Ellis et al. 2016; Heinberg 2013; USEIA 2016).

Northwest of the Eagle Ford in neighboring Arkansas is another key shale region, the *Fayetteville Shale*, a 25,000-square-mile area covering parts of at least eight counties in the central region of the state. Since 2002, over 3,800 wells have been drilled in the formation, which is estimated to hold about 32 trillion cubic feet (tcf) of natural gas (USEIA 2016). More recent production levels, even with increasing drilling, have been essentially flat, suggesting the play is likely past its peak (Heinberg 2013). Additionally, the *Niobrara Shale* region of northern Colorado and southern Wyoming is currently the third largest daily producer of shale gas at over 4,000 million cubic feet of gas per day (USEIA 2016).

As noted earlier in the chapter, the first commercial application of horizontal drilling and hydraulic fracturing to produce natural gas began in 1997 in the *Barnett Shale* formation of north-central Texas, an area covering some 5,000 square miles and at least 18 counties in and around the Dallas–Fort Worth area (Heinberg 2013; Yergin 2011). In 2000, the Barnett Shale was producing 216 million cubic feet (mcf) of natural gas a day, accounting for the vast majority of U.S. shale gas production at the time, which increased to about a half trillion cubic feet a year by 2005. By 2016, shale gas production in the Barnett totaled over 1,600 million cubic feet per day, making it the fifth largest producing shale region in the country (USEIA 2016). Moreover, shale oil production in the Barnett has also been on the rise since 2003, reaching a high of 4,687 barrels of oil a day (Bbl) in 2013, declining in 2015 to 1,885 barrels. Indeed, a recent U.S. Geological Survey estimates that the largest expanses of the Barnett Shale, when taken together, could contain as much as 53 trillion cubic feet (tcf) of recoverable shale gas, 172 million barrels of shale oil, and 176 million barrels of natural gas liquids (Marra et al. 2015). The region is now dotted with over 16,000 horizontal wells, most of which are concentrated near the Fort Worth area (Heinberg 2013). Given that U.S. shale gas production originated in the Barnett formation, it also represents the communities where much of the early sociological research on public perceptions of unconventional energy production and fracking was first conducted (see, for example, Anderson and Theodori 2009; Brasier et al. 2011; Gullion 2015; Theodori 2009, 2013; Wynveen 2011).

To the north of the Barnett Shale in neighboring Oklahoma is the *Woodford Shale*, made up of two sections covering a total of 4,700 square miles in the south and southeastern regions of the state. Led by companies like Newfield Exploration, Devon, and Chesapeake, the Woodford has been producing natural gas since 2005, second only to the Barnett Shale in the region. Current shale production there makes it the sixth largest producing region. Recent estimates suggest that the play may hold as much as 28 trillion cubic feet (tcf) of natural gas, while other observers suggest that the formation has probably peaked and is in decline (Heinberg 2013; Lavelle 2012; USEIA 2016).

North of the Woodford Shale lies the *Bakken Shale* formation covering 200,000 square miles of North Dakota and Montana (as well as portions of Saskatchewan and Manitoba, Canada). The Bakken region represents the seventh most productive shale gas play in the United States, generating around 10 billion cubic feet (bcf) per day (USEIA 2016). However, as a result of its enormous shale oil deposits discovered in the Williston basin of the region, estimated at anywhere from 5 to 500 billion barrels, the Bakken play has become widely known as the fastest growing oil boomtown in the nation and possibly one of the largest oil fields in the world (Brown 2013: 29). Indeed, between 2006 and 2013, oil production in the Bakken increased nearly 150-fold to more than 660,000 barrels of oil a day, increasing again more than four times that volume to around 3 million barrels a day in 2015, making North Dakota the second largest oil-producing state behind Texas (Dobb 2013: 36; USEIA 2016). Despite the hundreds of millions of dollars in state revenues the oil fracking boom has produced for North Dakota, in addition to the lowest levels of unemployment in the nation (under 3%), the industrialization and population increase in the region has created a significant strain on water supplies, sewerage, air and water quality, housing availability, rental costs, and government services. Moreover, the massive influx of some 15,000 workers into the Bakken has also led to increased crime, prostitution, drug and alcohol abuse, suicide, and other undesirable social impacts often associated with oil development boomtowns (Brown 2013; Dobb 2013; Gramling and Brabant 1986).

Three other major shale plays today include the *Antrim Shale* in Michigan, Indiana, and Ohio, with an estimated 22 trillion cubic feet of shale gas, as well as the *Utica Shale* region, located 3,000–7,000 feet beneath an area of the eastern United States larger than the Marcellus region (and Ontario). Because of the depth of the deposits for shale gas and oil in the Utica, as well as its low permeability, most of the leasing and drilling activity there has been confined to eastern Ohio, with current production of shale gas running at more than 3,600 million cubic feet (mcf) of gas a day, as well as about 82,000 barrels of oil a day (USEIA 2016). Finally, the *Permian Shale* region of West Texas accounts for about 6,900 million cubic feet (mcf) of shale gas per day (USEIA 2016). All of the remaining minor shale plays in the United States account for less than 5 billion cubic feet of the natural gas being produced per day and less than 2 million barrels of oil (USEIA 2016). Currently, the United States is estimated to have 610 trillion cubic feet (tcf) of recoverable shale gas resources, ranking it fourth globally after China, Argentina, and Algeria.

Finally, Table I.1 illustrates the most prominent shale plays for shale oil production in the United States, currently totaling around 4.4 million barrels per day (USEIA 2016). Although the Bakken and Eagle Ford shale formations together account for over 80% of current U.S. shale oil production, other plays, such as the Monterey in California, the Niobrara formation in Colorado and Wyoming, and the Tuscaloosa Shale in Louisiana also hold varying degrees of promise (Heinberg 2013; see Ladd, this volume). South Florida and the Florida Everglades also constitute one of the latest frontiers for oil prospecting in the United States, particularly the use of "acid fracking" to stimulate and dissolve limestone bedrock formations that contain oil deposits (see Widener, this volume). Altogether, the United States is estimated to have 59 billion barrels of technically recoverable shale oil resources, ranking it second globally after Russia. While shale gas production is expected to rise significantly to account for over one-half of all U.S. natural gas supplies by 2040, oil production is also likely to surge, depending on whether emerging plays like the Monterey and others are opened up to high-volume fracking operations (USEIA 2016).

Technological Risk and Environmental Controversy over Fracking

The social production of risk has been theoretically positioned as the defining characteristic of the modern age (Beck 1992, 1995; Giddens 1990; Luhmann 1993). For Ulrich Beck and others, the major risks of postindustrial society are no longer naturally occurring hazards and external events beyond human control, but now are largely a function of complex technologies and deliberate human decisions made by unaccountable elites. At its core, the risk society represents the "dark side" of industrial progress, a new era of manufactured uncertainty where human endangerment and the devastation of nature have become the dominant source of social conflict, fear, and anxiety (Rosa, Renn, and McCright 2014; Tierney 2014). In turn, people's perceptions of risk serve to create new forms of social stress, as well as a wider distrust of science, technology, and the power of technical experts. As Nelkin (1984: 10) observed over three decades ago, "controversy seems to erupt over nearly every aspect of science and technology as decisions once defined as technical (within the province of experts) have become intensely political." For Nelkin, the rising tide of citizen protest over technological development reflects a growing sense among ordinary people that their quality of life is deteriorating, that they no longer can shape the technological policies that affect their lives, and that scientific rationality is used by elites to mask their own economic and political objectives. In turn, questions of distributive justice arise as citizens in communities become aware that they must bear the costs—particularly the potential health and environmental hazards—of projects that benefit a different or much broader constituency. Controversy and conflict often become further exacerbated by the sense that landowner rights or other individual freedoms are being denied by corporate or government actors, or that moral frames about what is judged to be equitable, just, or fair are being flouted (Gunter and Kroll-Smith 2007; Nelkin 1984).

Against the backdrop of the Risk Society, perceptions of technological danger and environmental harm reside at the center of the growing divide in communities over the impacts of unconventional energy development. As Gunter and Kroll-Smith (2007: 6–7) remind us, communities are under siege today by a plethora of technological, political, and economic forces that originate beyond their borders, producing acute troubles, difficulties, and conflicts for residents that surface in the wake of real or proposed environmental changes:

> [L]ocal communities in ever-increasing numbers are forced to decide how to act toward and make sense of complicated, value-laden environmental problems. Land-use decisions that could alter the quality of environments for generations to come, toxins in local biospheres, species habitats and human needs, and questions about the use of natural resources for market purposes are among the forced options facing villages, small towns, neighborhoods, and cities throughout the United States and, indeed, worldwide.... Human life and well-being, a house, friendship, a job, a treasured way of life, and justice itself are often endangered when people cannot reach agreement on how to use environments or the kind and degree of danger they pose.

When environmental controversies result in the creation of what Gunter and Kroll-Smith call "volatile places," conflicts can take the form of *conservancy disputes* (struggles over how to define and protect natural areas), *siting disputes* (struggles over proposals to build new technological facilities or modify existing ones), or *exposure disputes* (struggles over the emission of pollutants that threaten human health into air, land, or water supplies (see Gunter and Kroll-Smith 2007: 6–10). Particularly in relatively isolated communities that become unwilling hosts for large-scale extractive industries, preimplementation anxieties and anticipatory social impacts can be experienced by local residents before development even begins (Freudenburg and Gramling 1992, 1993; Ladd and Laska 1991; Willow 2014). Once the new technological risks associated with hydraulic fracking begin to alter the local environment, citizens can become locked in bitter debates with their neighbors over the claims and counterclaims of different stakeholder groups, including scientists, technical experts, industry spokespeople, government regulators, environmentalists, journalists, or other institutional voices in the wider arena.

The title of this book, *Fractured Communities*, is intended to suggest the parallel fault lines and deep divisions that can develop in local shale communities over the differential opportunity-threat impacts associated with unconventional fossil fuel development. The title is also meant to highlight the theoretical linkages between technological risk, the power of complex industrial processes to degrade the environment, their ability to disrupt social systems, harm human relationships, and create political cleavages. In other words, the technologically invasive power of fracking to *fracture* the earth's fragile biophysical landscape for energy extraction also carries with it the power to *fracture* a community's social fabric, its taken-for-granted way of life, as well as its democratic structures of governance (Humes 2012). Unlike the conventional oil and gas boomtowns of the recent past, the scale of today's unconventional horizontal drilling and fracturing operations are much more likely to affect regional ecosystems (particularly groundwater sources, animals, and human health), as well as deplete the social capital and sense of political efficacy of an entire community. Local citizens may view such risks not only as endangering their individual well-being, but also their social well-being, their bonds with the locality they live in, and the ecosystem services on which they depend (Paveglio, Boyd, and Carroll 2016). Tom Wilber (2015: 8), in his account of the fracking conflicts that characterized the development of the Marcellus Shale, for example, conveys a similar insight when he notes: "In all these shale regions, the relationships people have with the land, and with their neighbors, are as complicated and multidimensional as the topographical and geological terrain. Here, too, there are cracks. They are created by forces that sometimes pull in opposite directions, at other times collide with great force, and often are buried from view."

In addition, the introduction of hydraulic drilling and fracturing technologies, supported by a slate of local laws, zoning ordinances, and state legislative initiatives intended to streamline energy development and preempt local opposition, has the potential to dramatically expand the opportunities for political tensions and class conflict, as well as exacerbate existing economic inequalities, in host communities. Bamberger and Oswald (2014: 152), for instance, point to how these cleavages have united unlikely political coalitions, while inflaming tempers and pitting neighbor against neighbor:

> One of the most interesting aspects of this debate is that it cuts across political philosophies and is not a left-versus-right, red-versus-blue issue. We can point to ardent environmentalists and Tea Party Republicans who are bitterly opposed to unconventional fossil fuel extraction and more moderate politicians who think that this process is the best thing that has ever happened for the environment. But irrespective of politics, unconventional fossil fuel extraction is a divisive issue in shale country, unlike anything else except maybe gun control legislation. Perhaps it is easy to see why. The debate involves people who believe they will make a lot of money, and others who see their way of life disintegrating. But caught in the middle are the poor, with little or no prospects of making money from shale gas and little material wealth to protect themselves when their lives are shaken by drilling activities.

In the wake of such economic tensions and conflicts, these volatile communities can splinter along class lines to create what Stephanie Malin (2015: 9) calls "sites of acceptance" and "sites of resistance." In the former case, local citizens accept the burdens of energy development because they believe that natural environments should be used in a utilitarian fashion to create jobs and meet national energy needs; in the latter case, citizens band together to mobilize against industrialized production practices and technological risks out of concerns for their health, safety, and the conviction that the precautionary principle should guide the development process. For Malin, the "toxic ambivalence" that people often experience in these communities reflects their contradictory location between the dictates of extralocal energy market demands on the one hand, and the perception that their sense of place, quality of life, environmental health, and long-term economic stability is in jeopardy on the other (Malin 2015: 3). Depending on the kind and level of perceived ecological threat that energy development poses for a community, particularly its uneven benefits and costs for different stakeholder groups, the potential for citizen mobilization, resistance, or political protest increases (Johnson and Frickel 2011).

Environmental Disputes and Opportunity-Threat Impacts of Oil and Gas Fracking

Heavily promoted by the fossil fuel industry and allied interests as a pathway to energy independence, jobs, royalties, and tax revenues for the economy, fracking in general, and natural gas fracking in particular, has been framed as a safe, abundant, homegrown energy source that offers America (and the world) a solution to climate change, as well as a cleaner energy future. Central to this industry narrative has been to portray the advent of fracking as nothing less than an economic and energy "game changer" that has made the new shale energy revolution possible. For proponents, fracking is viewed as having pioneered an "economic bonanza" for American communities and states, created the potential for natural gas to become the "bridge fuel" to a renewable energy future, and helped reduce the threat of atmospheric climate change by lowering U.S. carbon emissions through the replacement of coal and oil with natural gas (Heinberg 2013; Ladd and Perrow 2016; Wright 2012; Yergin 2011).

At the same time, the rapid development of shale gas and oil reserves through high-volume hydraulic fracturing (HVHF) methods has also spawned a growing antifracking movement that has been gaining political power and scientific support, especially since Josh Fox's Academy Award–nominated documentary *Gasland* was released in 2010 (see also Fox 2013; Vasi et al. 2015). In shale communities across the country like the Marcellus, Barnett, Tuscaloosa, Utica, Monterey, Niobrara, Woodford, and Eagle Ford, citizens have mobilized to express concerns or protest what they perceive as a wide range of negative socioenvironmental impacts associated with unconventional energy development and fracking—particularly its threats to water, air, land, roads, climate, public health, animals, and sustainable economic development. To date, over 475 community bans against fracking have been passed in 24 states, including cities like Pittsburgh, and statewide moratoriums are in effect in New York, Vermont, and Maryland. Additionally, national moratoriums exist in France, Ireland, Germany, Scotland, and Bulgaria, Pope Francis has publicly declared his opposition to fracking, and debates over shale drilling are growing across the European Union, South Africa, South America, and Australia (Ladd 2013, 2014; Food & Water Watch 2015).

More than any other single factor, citizen concerns over the deleterious impacts of gas and oil fracking have revolved around its potential to contaminate local ground and surface waters through methane migration, fracking fluids, wastewater injection, radioactive gases, carcinogenic compounds, corrosive salts, or radioactive elements like cesium and uranium—all leached from rock strata miles underground. Water samples from drilling communities in the Marcellus, for example, have revealed concentrations of methane, barium, bromide, and strontium, along with the chemical fingerprints of many other substances. Fracking fluids, which typically include friction reducers, surfactants, gelling agents, scale inhibitors, acids, corrosion inhibitors, antibacterial agents, and clay stabilizers, among other hazardous compounds, are known to pose threats to plants, fish, aquatic life, and animals (Gullion 2015).

In addition to issues of water quality, issues of water quantity have also been viewed by local residents and officials as key impacts associated with unconventional oil and gas development, particularly the drawing down of local aquifers, surface waters, or public water systems to provide the roughly 2–7 million gallons of water required for each fracking episode. With so many western states like California experiencing severe drought conditions and declining water availability from climate change and related weather events, many critics have assailed the oil and gas industry for its enormous consumption of the scarce water resources needed for drinking, farming, and other daily tasks (Kinchy, Parks, and Jalbert 2015; Ladd 2013). As Hauter (2016: 165) argues, "The competition for water between the oil and gas industry and agriculture will only increase as unconventional oil and gas development expands."

One of the most contentious issues surrounding fracking comes from the additional groundwater contamination risks posed by the underground disposal of its wastewaters through deep-well injection methods. Estimates vary, but between 15% and 80% of the water used in fracking operations is forced back up to the surface of the well each time it is fracked—which can be up to a dozen times over the life of a well. In the United States, 2.4 billion gallons of wastewater are injected daily under high pressure into any of the 187,570 disposal wells that accept oil and gas waste across the country (Concerned Health

Professionals of New York & Physicians for Social Responsibility 2015). Disposal of what the industry calls "produced water" has been difficult to track and can often contain pollutants in concentrations that far exceed those considered safe for drinking water or release into the environment (Bamberger and Oswald 2014). Flowback wastewaters also contain salt brines, heavy metals, radioactive materials, and hydrocarbons, including volatile organic compounds such as benzene, toluene, ethylbenzene, and xylenes, as well as hydrogen sulfide (Food & Water Watch 2015; Gullion 2015; Johnston, Werder, and Sebastian 2016). There are presently no water treatment plants in the United States that can remove all of the toxins and impurities dissolved in the wastewater, so in some states portions of the wastewater only receive partial treatment and then get dumped into rivers, injected back into the ground someplace else, or even are used for road deicing or to suppress dust (Cable 2012). Many northeastern states are trying to ban wastewater injection, while environmental groups in the Ohio Utica region have actively opposed their state becoming a dumping ground for fracking wastes from the Marcellus Shale next door.

In recent years, wastewater injection has also become associated with an alarming increase in induced seismicity and earthquakes in states like Oklahoma, Kansas, Colorado, Texas, Ohio, and Arkansas (Light 2015; Raynes et al. 2016). In Oklahoma especially, regulators and elected officials have failed to calm public fears about the swarms of earthquakes that have occurred in the Woodford Shale region since 2009. During that period, wastewater injection from fracking activity increased by 50%, but the state experienced a 108-fold increase in earthquakes, whose number totaled 5,417 seismic eruptions by 2014 (Hauter 2016: 99; see Mix and Raynes, this volume). Clearly, the centrality of the threat of water contamination from the fracking process, as well as that posed by wastewater injection methods or on-site storage ponds, shows every sign of continuing to be the touchstone issue in the widening debates over shale energy development—as evidenced by the political controversy over the EPA drinking water contamination studies in Wyoming, Texas, and Pennsylvania (Dermansky 2015). Indeed, past research suggests that environmental disputes involving water resources—given their essential value to life, politics, and the economy—tend to be particularly contentious (Krogman 1996; Shriver and Peaden 2009).

The public health impacts of hydraulic fracking have also received considerable attention from sociologists, health officials, journalists, and community activists. As noted, gas and oil fracking involves hundreds of toxic chemical compounds, dozens of which contain known and suspected carcinogens. Research has shown exposure to fracking fluids to be associated with cancer, reproductive disruptions, skin, eye, and respiratory symptoms, impairments of the brain and nervous system, gastrointestinal and liver disease, as well as psychosocial stress (Finkel and Law 2011, 2016; Sangaramoorthy et al. 2016).

Diesel exhaust from engines and heavy trucks, as well as other air pollutants in natural gas production, have been connected to cardiovascular disease, allergies, genetic changes in DNA, increases in childhood illnesses, harm to fetal and neonatal health, as well as increased mortality. Natural gas flaring from oil wells has been shown to produce strong hydrogen sulfide odors (a poisonous gas that smells like rotten eggs), as well as some 60 other air pollutants that travel downwind from flared well sites. Food and soil contamination have also been reported by residents in impacted communities like the Barnett, along with headaches, nosebleeds, thyroid problems, chronic fatigue, and anemia (Gullion 2015). Finally, conflicts over the health impacts of silica sand mining—used as proppants in fracking because of the sand's hardness and shape—have emerged in Iowa, Illinois, Minnesota, and Wisconsin, where dozens of new mines and processing facilities have opened with little state oversight. The tiny silica particles constitute a known carcinogen and can cause respiratory and cardiovascular problems for industry workers or local residents when inhaled (Food and Water Watch 2015; Heinberg 2013).

In states like Pennsylvania, the passage of Act 13 in 2012 had a chilling impact on tracking the public health impacts of fracking and suppressing advocacy by health care professionals. As Hauter (2016: 155) reported:

> Buried in the bill, which ran for more than a thousand pages, was a provision designed to muzzle doctors and nurses, preventing them from warning sick patients that their illness may be related to fracking. Health care professionals who request information about proprietary chemicals must sign a nondisclosure agreement that prohibits them from sharing any information about those chemicals with their patients. At the same time, funding was cut for the registry that tracks respiratory problems, skin conditions, stomach ailments, and other illnesses potentially related to gas drilling.

In 2014, a number of community health professionals came forward and revealed that they had been silenced by the Pennsylvania Department of Health and other state agencies in an attempt to sweep the impacts of fracking in the Marcellus Shale under the rug and minimize the reporting of symptoms of patients who lived near drilling sites (Hauter 2016). As a case in point, research has suggested that children living near shale gas production facilities have experienced throat and nasal irritation, skin rashes, eye burning, nosebleeds, sleep disturbances, headaches, nausea, and back and joint pain (Hauter 2016). In its summary of the more than 150 studies examining the public health impacts of HVHF energy production, one recent independent assessment (Concerned Health Officials of New York & Physicians for Social Responsibility 2015: 4) came to the following conclusion about the nature of these threats: "A growing body of peer-reviewed studies, accident reports, and investigative articles has detailed specific, quantifiable evidence of harm and has revealed fundamental problems with the entire life cycle of operations associated with unconventional drilling and fracking. . . . Despite this emerging body of knowledge, industry secrecy and government inaction continue to thwart scientific inquiry."

A growing body of literature has also pointed to fracking's contribution to global warming and climate change. In recent years, one of the most prominent claims of the fossil fuel industry has been to frame natural gas as "clean" energy, a "bridge fuel" to the future that will effectively reduce CO_2 emissions and the growing dangers of climate change (Ladd 2016). This narrative derives solely from the fact that natural gas, when burned, produces only one-half as much CO_2 as coal and about one-third as much as oil (Nijhuis 2014). Yet, critics suggest that natural gas fracking is not a "climate-friendly alternative" to coal and oil, nor will it help avert the growing threat of global warming and climate change that its advocates claim. For instance, about two-thirds of U.S. climate pollution stems from the oil and gas industry, with nearly 30% attributed to natural gas production alone, which has increased roughly 160% since preindustrial times (Food & Water Watch 2015: 18; Lavelle 2012). Despite its comparatively lower carbon footprint than other fossil fuels, its chief component—methane—can be anywhere from 25 to 87 times more potent than CO_2 as a heat-trapping greenhouse gas, particularly over a 20-year timeframe (Adler 2014; Food & Water Watch 2015; Howarth and Ingraffea 2011). In addition, natural gas still accounts for about 21% of global fossil fuel–generated CO_2, as well as 9% of all greenhouse gas emissions.

When the expected growth (over 150%) of methane emissions from shale gas production is factored in over the next two decades, combined with the average methane loss/leakage to the environment that occurs during most phases of U.S natural gas production (estimated at 6–12%), the overall hidden climate footprint of shale gas fracking is arguably *worse* than coal or oil over its well-to-consumer lifecycle (Howarth and Ingraffea 2011; Light 2014; McKibben 2014). In addition, the U.S. fossil fuel industry is the largest emitter of methane, surpassing livestock and animal manure (Lavelle 2012). Moreover, the deleterious climate impacts of natural gas are even greater when other factors like fugitive emissions from pipelines, gas well flaring, the thawing of arctic permafrost, the operation of export facilities, or the shipping of liquefied gas and condensates abroad are considered (Adler 2015; Blake 2015; Kelly 2014; Light 2014). Indeed, both shale gas and tight-oil production can involve extensive amounts of natural gas flaring (burning off the gas from the wellhead when it is not economical to capture it), or even pure venting in the weeks or months before a well might be fully operational. It also bears noting that methane emissions from fracked gas are up to twice as great as those from conventional gas wells (Hauter 2016).

Finally, many critics view the industry narrative that increased natural gas production will help reduce climate change by displacing the use of fossil fuels as a false scenario. First, the majority of fracking today is aimed at producing shale/tight-oil supplies, not natural gas, which clearly exacerbates greenhouse emissions to the atmosphere. Second, increased natural gas use in the electricity sector tends to displace clean energy solutions such as solar, wind, and efficiency, not just fossil fuels, potentially canceling out the former's benefits. Third, much of the U.S. coal and oil that might be displaced by natural gas is currently slated to be exported and burned in developing countries, offsetting at the planetary level the potential for more significant global reductions in CO_2 (Food & Water Watch 2015: 19; Gelles 2015: 3).

Other key impacts of shale development include increased air pollution, chemical spills, noise, dust, lights, road damage, auto accidents, and transportation congestion associated with the increased daily traffic of the hundreds of heavy diesel trucks used to transport the required water, chemicals, sand, drilling equipment, and other production components to rural well sites (Heinberg 2013; Ladd 2014; Perrow 2015; Theodori 2013; Wilber 2015). For instance, one study estimated that heavy truck traffic in the Marcellus cost the state between $8.5 million to $17 million a year in road damage alone (Hauter 2016: 159). In addition, "fracktivists" and other environmental groups have attacked fracking for its myriad threats to oil and gas workers, well safety, farm communities, pets, wild animals, rural landscapes, property rights, real estate values, tourism, and American democracy itself (Bamberger and Oswald 2014; Brune 2013; Hauter 2016; Heinberg 2013; Ladd 2013; Loki 2015; Pantsios 2014; Seawell 2015; Whitley, this volume).

Finally, fracking has drawn extensive criticism because of its social impacts related to rapid community change, industrialization, cost of living increases, differential economic benefits from signing royalties and leasing contracts, tensions between surface and mineral rights owners, increases in transient workers, crime, prostitution, rape, substance abuse, housing shortages, psychosocial stress, and diminished social capital (Brasier et al. 2011; Ellis et al. 2016; Food and Water Watch 2015; Hauter 2016; Malin 2014; Malin and DeMaster 2016; Sangaramoorthy et al. 2016; Willow 2014). In addition, recent studies suggest that people's perceptions of fracking tend to reflect evolving narratives about the history of a region's extractive practices and sociopolitical interests, as well as lived experiences with sense-of-place connections, trust in government and industry officials, family health problems, local landscapes, water resources, and exposure to alternative frames of media discourse (Ashmoore et al. 2015; Hudgins and Poole 2014; Kinchy and Perry 2012; Mayer 2016; Poole and Hudgins; 2014; Sangaramoorthy et al. 2016; Willow 2014; Willow et al. 2014; Wynveen 2011). In short, the extant literature on the opportunity-threat impacts of unconventional fossil fuel development suggests that hydraulic fracking represents a veritable "double-edged sword" for most host communities. While a majority of residents tend to its welcome its perceived economic benefits, a significant minority are also ambivalent about or directly oppose what they view as a more extensive range of long-term socioenvironmental risks associated with shale energy fracking and waste disposal (Boudet et al. 2016; Cosgrove et al. 2015; Crowe et al. 2015; Eaton and Kinchy 2016; Ladd 2013, 2014; Sarge et al. 2015). To date, nearly 700 peer-reviewed studies have been published providing empirical evidence of the myriad environmental, health, and social impacts of unconventional energy development (Finkel and Law 2016).

Fractured Landscapes, Fractured Communities

Public support or opposition to hydraulic fracking, at both the national and community level, has tended to be associated with the kinds of sociodemographic factors and risk perceptions identified in past research on energy development, technological disputes, or locally unwanted land uses (LULUs). National public opinion polls, for instance, have revealed varying levels of familiarity and knowledge about the issue and sharply divided views. In a 2012 national survey of citizens' views on hydraulic fracturing, Boudet et al. (2014, 2016) found that much of the American populace was largely unaware of and undecided about the issue. Among those who had formed an opinion, respondents were nearly split between support and opposition. Supporters tended to be older men who held a bachelor's degree or higher, were politically conservative, watched TV news more than once a week, and saw fracking as improving the economy or energy independence. Conversely, opponents tended to be women, hold egalitarian worldviews, read newspapers more than once a week, were more familiar with hydraulic fracturing, as well as more concerned about its environmental impacts. Similarly, Sarge et al. (2015) argued that women, Democrats, and urban residents with both proenvironmental views and issue familiarity are more likely to oppose hydraulic fracking.

A 2014 poll by the Pew Research Center found that the percentage of Americans opposed to the increased use of fracking had grown since 2013 from 38% to 47%, particularly among women and younger adults (Pew Research Center 2014). Public opinion has also been found to divide along how the issue is framed or referenced (Evensen et al. 2014b). In a recent national U.S. survey, Clarke et al. (2015) found that people were more supportive of the energy extraction process when it was referred to as *shale oil or gas development* (which was associated with more positive thoughts about economic benefits), but were more opposed when it was referred to as *fracking* (which tended to evoke more negative thoughts about environmental risks). Similarly, a 2012 public opinion survey found that while Louisiana residents were relatively split in their support or opposition to "natural gas extraction" or even "hydraulic fracturing" in the state, the percentage of respondents expressing opposition to "fracking" was significantly higher (Goidel 2012). A 2016 Gallup Poll found that for the first time, a slight majority of Americans, 51%, reported being opposed to fracking (www.gallup.com).

As the ten chapters in this volume illustrate, the debate over the use of hydraulic fracturing in natural gas and oil production has emerged as one of the most contentious environmental issues of our time, provoking serious, if not occasionally rancorous, debate across the country in varying degrees. From the Bakken Shale in North Dakota and Wyoming, to the Barnett and Eagle Ford Shales in Texas, to the Marcellus and Utica Shale in the northeast, grassroots groups in at least a dozen U.S. shale communities have mobilized to protest the threat posed by fracking operations to public health and local water tables through routine emissions, leaking wastewater ponds, injection methods, faulty wells, or fracking fissures—as well as their potential to industrialize the landscape, damage roads, harm property values, and destroy the tranquility of rural community life (Lavelle 2012). Various environmental organizations have organized protests against federal energy policies and agencies, including President Obama for his support of increased natural gas production, as well as the Bureau of Land Management (BLM) for permitting fossil fuel companies to frack near national parks and forests, federal recreational areas, and Native American reservations. Media outlets like National Public Radio (NPR) have come under fire for selling advertising spots to the natural gas industry, while a number of American universities have been criticized for creating "Frackademia" research programs and think tanks on their campuses with corporate funding from oil and gas interests (Horn 2013).

While citizen protests over fracking have mobilized across the country, probably no state has witnessed more conflict over the issue than New York and its "Don't Frack New York" campaign aimed at persuading Governor Andrew Cuomo to extend the nation's first fracking moratorium in 2010. While Cuomo delayed his decision twice to allow state agencies to address a wide range of citizen concerns related to fracking, the issue generated several large protest rallies and concerts throughout New York, one of which was captured in the independent film *Dear Governor Cuomo*. Due to the rising tide of grassroots complaints and protests over well water contamination, aquifer depletion, public health impacts, and rural degradation, among others, Governor Cuomo was forced in 2015 to extend the state's existing moratorium on fracking and shale production (Glick 2012; Wilber 2015).

Local governments in dozens of towns and counties in the Marcellus Shale, largely in Pennsylvania, have also passed community fracking bans in recent years, including Dimock, PA (whose water contamination disputes were featured in Josh Fox's *Gasland* documentary), as well as the city of Pittsburgh (see Staggenborg, this volume). Given the population density of the surrounding region, as well as the fact that some 15 million people draw their drinking water from the Delaware River Basin underlying the geological formation, the rapid growth of natural gas fracking in the Marcellus has generated more social conflict and political protest, as well as more sociological research, than any other shale region in the United States (see, for example, Evensen et al. 2014a; Hudgins and Poole 2014; Jalbert, Kinchey, and Perry 2014; Kinne, Finewood, and Yoxtheimer 2014; Kriesky et al. 2013; Malin and DeMaster 2016; Poole and Hudgins 2014; Weigle 2011; Willow et al. 2014). The neighboring states of Vermont and Maryland, located just to the north and south of the formal geological boundaries of the Marcellus formation, have also joined New York in passing statewide fracking moratoriums since 2012 (Food & Water Watch 2015). In turn, Governor Jerry Brown of California has come under significant pressure from state and national environmental organizations to pass a moratorium against oil and gas fracking in the Monterey Shale region, one of the largest oil and gas formations in the United States (Heinberg 2013).

In response to the Obama administration's domestic push for more widespread natural gas drilling and production (as one policy plank in their "all of the above" energy platform), at least two major national protest rallies against fracking were staged in Washington, D.C. in 2012, along with a coordinated international day of protest in September by activists on five continents dubbed "The Global Frackdown" (Berwyn 2012; Queally 2013). Antifracking groups have also joined ranks with the wider fossil fuel resistance movement (often termed "Blockcadia" by activists) in their common opposition to the XL and Dakota Access pipelines, liquefied natural gas facilities, deepwater offshore oil and gas production, and Arctic drilling—while they work to incorporate issues like fossil fuel divestment, cap-and-trade legislation on carbon emissions, and slowing climate change into their discursive frames and mobilization strategies (Klein 2014). At the December 2015 COP 21 Climate Change talks in Paris, for example, activists illuminated a huge sign under the Eiffel Tower that read: "Climate Leaders—Don't Frack!"

Thanks in part to Josh Fox's 2010 film *Gasland* (Fox 2010; see also Fox 2013), what began as a local controversy over drinking water contamination in rural Pennsylvania has now morphed into a half dozen film documentaries on both sides of the fracking debate (see Vasi, this volume), a 2013 Hollywood movie titled *Promised Land* starring Matt Damon, thousands of media accounts, articles, books, scientific studies, fundraising concerts, artistic exhibits, and other cultural products that have brought national (and international) attention to the issues surrounding hydraulic fracturing, shale gas and oil extraction, and other forms of "extreme energy" development. Given the importance of these issues for society and the environment, the current and original sociological case studies featured in *Fractured Communities* attempt to address some of the key history, risks, socioenvironmental impacts, policy disputes, citizen/leader initiatives, and mobilization of protest taking place today in many of the most prominent shale regions in the United States. As the "fracking wars" heat up across the country, these community studies illustrate how the study of unconventional shale development offers fertile ground for environmental sociologists and other researchers who want to better understand one of the leading sources of social conflict that is fracturing the community, as well as the land and water, beneath our feet.

Overview of the Book

The first three chapters of the book address different components of political economy and cultural conflicts that frame the risks and impacts of shale fracking in the Marcellus Shale, the adjoining Utica Shale of eastern Ohio, the Barnett Shale in north central Texas, or the Bakken Shale in North Dakota. In chapter 1, Sherry Cable ("Natural Gas Fracking on Public Lands: The Trickledown Impacts of Neoliberalism in Ohio's Utica Shale Region") focuses on the decision-making processes, regulatory actions, and citizen mobilization patterns regarding the leasing of public lands for natural gas fracking in four Ohio counties. Analyzing archival data and newspaper accounts from relevant sources, she explores whether citizens took advantage of lessons learned in the Marcellus region to implement proactive safety measures, whether the leasing of public land represented the democratic will of its residents, and whether they mobilized to protest fracking operations in their communities.

In chapter 2, Ion Bogdan Vasi ("This (Gas) Land is Your (Truth) Land? Documentary Films and Cultural Fracturing in Prominent Shale Communities") examines how activists and opposing industry groups used documentary films like *Gasland* and *Truthland* to influence public perceptions about fracking in different shale communities. In particular, Vasi offers a comparative analysis of how these cultural products were used by different nonprofit organizations to mobilize potential constituents in the Marcellus, Barnett, and Bakken Shale regions and the differential impacts of those campaigns.

In chapter 3, Carmel E. Price and James N. Maples ("Disturbing the Dead: Community Concerns over Fracking below a Cemetery in the Utica Shale Region") explore some of the fascinating conflicts surrounding a proposal to horizontally drill and frack for natural gas close to and underneath a historic veterans' cemetery in Youngstown, Ohio in the Utica Shale. Building on theoretical research addressing mountaintop removal coal mining in and around cemetery zones, Price and Maples examine the myriad issues and tensions around land ownership, property and mineral rights, profit motivations, religious beliefs, family rituals, and emotional attachments that have emerged between families of the dead and other stakeholders, particularly those who own and manage cemeteries.

The next two chapters focus on two case studies highlighting different aspects of the conflict surrounding unconventional energy development and hydraulic fracking in the Marcellus Shale, the largest natural gas producing region in the United States. In chapter 4, Suzanne Staggenborg ("Mobilizing against Fracking: Marcellus Shale Protest in Pittsburgh") examines the processes and resource mobilization patterns that explain the growth of the movement against fracking in Pittsburgh, emphasizing how its political success, despite strong opposition, helped create the first moratorium against shale gas development within the limits of a major U.S. city. Building on preexisting movement structures and communities, her research explores the role that movement strategies and tactics can play in mobilizing constituents for collective protest.

In chapter 5, Cameron Thomas Whitley ("Engines, Sentinels, and Objects: Assessing the Impacts of Unconventional Energy Development on Animals in the Marcellus Shale Region") addresses the relatively unexplored past and present history of animals in unconventional energy development, with a particular focus on the Marcellus Shale region. His work specifically examines research on the impacts of the U.S. fracking boom on livestock and wildlife and offers suggestions on what research and policies are needed to support the health and well-being of both humans and animals.

The next three chapters feature case studies that deal with the differential stakeholder perceptions, frame disputes, socioenvironmental impacts, and community conflicts surrounding gas and oil development in the Haynesville, Woodford, and Niobrara Shale regions respectively. In chapter 6, editor Anthony E. Ladd ("Motivational Frame Disputes Surrounding Natural Gas Fracking in the Haynesville Shale") provides a qualitative analysis of the motivational frame disputes of stakeholder groups regarding unconventional natural gas development and fracking in the Haynesville Shale region of Louisiana. On both sides of the Haynesville frame dispute, neither supporters nor opponents articulated a strong sense of urgency, threat severity, or collective agency that called for any conscious efforts to expand or halt shale fracking in the region. Accordingly, Ladd argues that citizens' response to shale development and fracking in the Haynesville to date has been characterized by *quiescence*— the absence of grievance perception in the face of risks—rather than the kind of collective protest, opposition, or political mobilization witnessed in other states and shale communities.

In chapter 7, co-authors Tamara L. Mix and Dakota K. T. Raynes ("Denial, Disinformation, and Delay: Recreancy and Induced Seismicity in Oklahoma's Shale Plays"), explore the various risks and hazards surrounding the exponential increase in seismic activity and earthquakes associated with deep-well wastewater injection methods from oil fracking operations in Oklahoma's major shale regions. Conceptualizing how citizens responded to the exponential increase in earthquake tremors in recent years as a classic case of recreancy and mobilization in an impacted set of communities, the authors examine how these events influenced conceptions of risk, member recruitment, mobilization, and action in a state historically embedded in the culture and landscape of oil and gas development.

In chapter 8, Stephanie A. Malin, Stacia S. Ryder, and Peter M. Hall ("Contested Colorado: Shifting Regulations and Public Responses to Unconventional Oil Production in the Niobrara Shale Region") examine the sociopolitical, environmental, and governance aspects of unconventional shale oil and gas development in the Niobrara Shale region of Colorado's Front Range. Exploring some of the key dynamics surrounding social activism, divisions, and grassroots attempts to oppose fracking at the community, municipal, and state levels, the authors specifically highlight how state leaders provided covert support to the Colorado oil and gas industry to dissuade communities from passing moratoria or permanent bans against fracking in their backyards, as well as pressuring the Colorado Supreme Court to overturn existing bans in Fort Collins and Loveland.

The final chapters of the book conclude with two case studies illustrating some of the contentious issues surrounding fracking and unconventional hydrocarbons that to date have received little attention: protests over "acid fracking" in South Florida and the Everglades, and siting conflicts over liquefied natural gas facilities in Oregon. In chapter 9, Patricia Widener ("Citizen Resistance to Oil Production and Acid Fracking in the Sunshine State") examines the emerging patterns of resistance in South Florida to a new form of extreme energy production—acid stimulation or acid fracking—used to dissolve, rather than break open, limestone bedrock formations that contain oil deposits. In an age when oil and gas development is increasing taking place in deeper, more remote, and previously untapped areas like the Everglades, she documents how citizen groups mobilized against acid fracking not only out of fears of becoming the nation's next energy sacrifice zone, but because of concerns over regional issues like public health, water quality and depletion, the Greater Everglades, Florida panther habitat, sinkhole-prone landscapes, tourism, solar energy development, and climate change. Finally, in chapter 10, Hilary Boudet, Brittany Gaustad, and Trang Tran ("Public Perception and Protest in the Siting of Liquefied Natural Gas Terminals in Oregon") examine how the explosion of shale gas and unconventional energy production in the last decade has shifted the landscape of risk for many coastal communities from the impacts of siting liquefied natural gas (LNG) *import* facilities to those associated with siting LNG *export* facilities. Drawing on insights from the study of social movements, they compare two coastal Oregon communities in terms of how these changes affected citizens' level of support or opposition for such facilities, the natural gas pipeline infrastructure that supplies them, as well as wider attitudes toward shale gas development. They conclude with some implications for understanding disputes over the siting of gas export facilities around the United States and the processes that govern participation in such communities. Taken together, the engaging chapters that follow in *Fractured Communities* offer a comparative analysis of the differential benefits, threats, and impacts associated with unconventional gas and oil development and hydraulic fracking unfolding today across the American landscape.

References

Adler, Ben. 2014. "Keystone Foes Turn Their Fire to Natural Gas Exports." *Grist Magazine*, March 19. Retrieved April 30, 2014. (http://grist.org/climate-energy/keystone-foes -turning-their-fire-to-natu...&utm_medium=email&utm_term=Daily%2520March %252019&utm_campaign=daily).

Adler, Ben 2015. "Climate Hawks Are Not Impressed with Obama's Methane Plan." *Grist Magazine*, January 16. Retrieved January 16, 2015. (http://grist.org/climate-energy /climate-hawks-are-not-impressed-byo...r&utm_medium=email&utm_term=Daily %252olan%252016&utmcampaign=daily).

Anderson, Brooklynn J., and Gene L. Theodori. 2009. "Local Leaders' Perceptions of Energy Development in the Barnett Shale." *Southern Rural Sociology* 24(1): 113–29.

Ashmoore, Olivia, Darrick Evensen, Chris Clarke, Jennifer Krakower, and Jeremy Simon. 2015. "Regional Newspaper Coverage of Shale Gas Development Across Ohio, New York, and Pennsylvania: Similarities, Differences, and Lessons." *Energy Research & Social Science* 11: 119–132.

Bamberger, Michelle, and Robert Oswald. 2014. *The Real Cost of Fracking: How America's Shale Gas Boom Is Threatening Our Families, Pets, and Food*. Boston: Beacon Press.

Beck, Ulrich. 1992. *Risk Society: Towards a New Modernity*. Thousand Oaks, CA: Sage.

Beck, Ulrich. 1995. *Ecological Politics in an Age of Risk*. Cambridge: Polity Press.

Bell, Shannon Elizabeth. 2014. "Energy, Society, and the Environment." Pp. 137–158 in Kenneth A. Gould and Tammy L. Lewis (eds.), *Twenty Lessons in Environmental Sociology* (2nd ed.), London: Oxford University Press.

Bell, Shannon Elizabeth, and Richard York. 2010. "Community Economic Identity: The Coal Industry and Ideology Construction in West Virginia." *Rural Sociology* 75(1): 111–143.

Berwyn, Bob. 2012. "Thousands Rally around the World to Ban Fracking." *Nation of Change*. Retrieved September 24, 2012 (http://wwwnationofchange.org/print/27566).

Blake, Mariah. 2014. "The Chevron Communiques." *Mother Jones* 39: 50–72.

Boudet, Hilary, Christopher Clarke, Dylan Bugden, Edward Maibach, Connie Roser-Renouf, and Anthony Leiserowitz. 2014. "Fracking Controversy and Communication: Using National Survey Data to Understand Public Perceptions of Hydraulic Fracturing." *Energy Policy* 65: 57–67.

Boudet, Hilary, Dylan Bugden, Chad Zanocco, and Edward Maibach. 2016. "The Effect of Industry Activities on Public Support for 'Fracking.'" *Environmental Politics* (http://dx .doi.org/10.1080/09644016.2016.1153771).

Brasier, Kathryn J., Matthew R. Filteau, Diane K. McLaughlin, Jeffrey Jacquet, Richard C. Stedman, Timothy W. Kelsey, and Stephan W. Kelsey. 2011."Residents' Perceptions of Community and Environmental Impacts from Development of Natural Gas in the Marcellus Shale: A Comparison of Pennsylvania and New York Cases." *Journal of Rural Social Sciences* 26(1): 32–61.

Brown, Chip. 2013. "North Dakota Went Boom." *New York Times Magazine*, February 3: 22–31.

Brune, Michael. 2013. "The Opposition's Opening Remarks." *The Economist*, February 5. Retrieved June 2, 2014 (http://www.economist.com/debate/days/view/934/print).

Cable, Sherry. 2012. *Sustainable Failures. Environmental Policy and Democracy in a Petro-Dependent World*. Philadelphia, PA: Temple University Press.

Catton, William R. 1982. *Overshoot: The Ecological Basis of Revolutionary Change*. Urbana, IL: University of Illinois Press.

Clarke, Christopher E., Philip S. Hart, Jonathan P. Schuldt, Darrick T. N. Evensen, Hilary S. Boudet, Jeffrey B. Jacquet, and Richard C. Stedman. 2015. "Public Opinion on Energy Development: The Interplay of Issue Framing, Top-of-Mind Associations, and Political Ideology." *Energy Policy* 81: 131–140.

Concerned Health Professionals of New York & Physicians for Social Responsibility. 2015. *Compendium of Scientific, Medical, and Media Findings Demonstrating Risks and Harms of Fracking* (unconventional gas and oil extraction) (3rd ed). (http://concernedhealthny .org/compendium/).

Cosgrove, Brendan M., Daniel R. LaFave, Sahan T. M. Dissanayake, and Michael R. Doni-hue. 2015. "The Economic Impact of Shale Gas Development: A Natural Experiment along the New York/Pennsylvania Border." *Agricultural and Resource Economics Review* 44(2): 20—39.

Crowe, Jessica, Tony Silva, Ryan G. Ceresola, Amanda Buday, and Charles Leonard. 2015. "Differences in Public Perceptions and Leaders' Perceptions on Hydraulic Fracturing and Shale Development." *Sociological Perspectives* 58(3): 441–463.

Dermansky, Julie. 2015. "Heat on EPA as National Study on Fracking's Risks to Drinking Water Is Challenged." *Truth Out*, December 8. (http://www.truth-out.org/news/item /33943-heat-on-epa-as-national-study...fracking-s-risks-to-drinking-water-is-challenged ?tmpl=component&print=1).

Dobb, Edwin. 2013. "The New Oil Landscape." *National Geographic* 223(3): 28–58.

Dunlap, Riley E. 2015. "Environmental Sociology." Pp. 796–803 in *International Encyclopedia of the Social and Behavioral Sciences* (2nd ed.), Elsevier Publishers.

Eaton, Emily, and Abby J. Kinchy. 2016. "Quiet Voices in the Fracking Debate: Ambivalence, Nonmobilization, and Individual Activism in Two Extractive Communities." *Energy Research and Social Science* 20: 22–30.

Ellis, Colter, Gene L. Theordori, Peggy Petrzelka, Douglas Jackson-Smith, and A. E. Luloff. 2016. "Unconventional Risks: The Experience of Acute Energy Development in the Eagle Ford Shale." *Energy Research & Social Science* (http://dx.doi.org/10.1016/j.erss.2016.05.006).

Evensen, Darrick T., Christopher E. Clarke, and Richard C. Stedman. 2014a. "A New York or Pennsylvania State of Mind? Social Representations in Newspaper Coverage of Gas Development in the Marcellus Shale." *Journal of Environmental Studies and Science* 4(1): 65–77.

Evensen, Darrick T., Jeffrey B. Jacquet, Christopher E. Clarke, and Richard C. Stedman. 2014b. "What's the 'Fracking' Problem? One Word Can't Say It All." *The Extractive Industries and Society* 1: 130–136.

Ferguson, Cody. 2015. *This Is Our Land: Grassroots Environmentalism in the Late Twentieth Century*. New Brunswick, NJ: Rutgers University Press.

Finkel, Madelon L., and Adam Law. 2011. "The Rush to Drill for Natural Gas: A Public Health Cautionary Tale." *American Journal of Public Health* 101(5): 784–785.

Finkel, Madelon L., and Adam Law. 2016. "The Rush to Drill for Natural Gas: A Five-Year Update." *American Journal of Public Heath* 106(10): 1728–1730.

Flannery, Tim. 2016. "Fury over Fracking." *New York Review of Books*, April 21. Accessed April 4, 2016. (http://www.nybooks.com/articles/2016/0421/furv-over-fracking/).

Food & Water Watch. 2011. *The Case for a Ban on Gas Fracking*. Washington, DC: Food & Water Watch.

Food & Water Watch. 2015. *The Urgent Case for a Ban on Fracking*. Retrieved December 6, 2015. (www.foodandwaterwatch.org/sites/default/files/urgent_case_for_ban_on_fracking.pdf).

Fox, Josh. 2010. *Gasland*. Film Documentary. New York: Wow Productions.

Fox, Josh. 2013. *Gasland II*. Film Documentary. New York: Wow Productions.

Freudenburg, William R., and Scott Frickel. 1994. "Digging Deeper: Mining-Dependent Regions in Historical Perspective." *Rural Sociology* 59(2): 266–288.

Freudenburg, William R., and Robert Gramling. 1992. "Community Impacts of Technological Change: Toward a Longitudinal Perspective." *Social Forces* 70(4): 937–55.

Freudenburg, William R., and Robert Gramling. 1993. "Socioenvironmental Factors and Development Policy: Understanding Opposition and Support for Offshore Oil." *Sociological Forum* 8(3): 341–365.

Freudenburg, William R., and Robert Gramling. 2011. *Blowout in the Gulf: The BP Oil Disaster and the Future of Energy in America*. Cambridge, MA: MIT Press.

Giddens, Anthony. 1990. *The Consequences of Modernity*. Cambridge: Polity Press.

Glick, Ted. 2012. "Albany Action a Big Boost to No-fracking Movement." *Reader Supported News*. Retrieved August 12, 2012 (http://readersupportednews.org/opinion2/271-38/13174-albany-action-a-big-boost-to-no-fracking-movement).

Goidel, Kirby. 2012. *2012 Louisiana Survey: Support for Fracking*. Baton Rouge, LA: Reilly Center for Media & Public Affairs, Louisiana State University.

Gold, Russell. 2014. *The Boom: How Fracking Ignited the American Energy Revolution and Changed the World*. New York: Simon & Schuster.

Gramling, Robert, and Sarah Brabant. 1986. "Boomtowns and Offshore Energy Impact Assessment: The Development of a Comprehensive Model." *Sociological Perspectives* 29(2): 177–201.

Gramling, Robert, and William R. Freudenburg. 1992. "Opportunity-Threat, Development, and Adaption: Toward a Comprehensive Framework for Social Impact Assessment." *Rural Sociology* 57(2): 216–234.

Gullion, Jessica Smartt. 2015. *Fracking the Neighborhood: Reluctant Activists and Natural Gas Drilling*. Cambridge, MA: MIT Press.

Gunter, Valarie, and Steve Kroll-Smith. 2007. *Volatile Places: A Sociology of Communities and Environmental Controversies*. Thousand Oaks, CA: Pine Forge Press.

Hauter, Wenonah. 2016. *Frackopoly: The Battle for the Future of Energy and the Environment*. New York: New Press.

Heinberg, Richard. 2013. *Snake Oil: How Fracking's False Promise of Plenty Imperils Our Future*. Santa Rosa, CA: Post-Carbon Institute.

Horn, Steve. 2013. "Frackademia: The People & Money Behind the EDF Methane Emissions Study." *Nation of Change*. Accessed September 17, 2013. (http://www.nationofchange.org/frackademia-people-money-behind-edf-methane-emissions-study-1379425386).

Howarth, Robert W., and Anthony Ingraffea. 2011. "Should Fracking Stop?" *Nature* 477: 271–273.

Hudgins, Anastasia, and Amanda Poole. 2014. "Framing Fracking: Private Property, Common Resources, and Regimes of Governance." *Journal of Political Ecology* 21(1): 303–19.

Humes, Edward. 2012. "Fractured Lives: Detritus of Pennsylvania's Shale Gas Boom." *Mother Jones* 97(4): 52–59.

Humphrey, Craig R., and Frederick R. Buttel. 1982. *Environment, Energy, and Society*. Belmont, CA: Wadsworth.

Jalbert, Kirk, Abby J. Kinchy, and Simona L. Perry. 2014. "Civil Society Research and Marcellus Shale Natural Gas Development: Results of a Survey of Volunteer Water Monitoring Organizations." *Journal of Environmental Studies and Science* 4(1): 78–86.

Johnson, Bob. 2014. *Carbon Nation: Fossil Fuels in the Making of American Culture*. Lawrence, KS: University Press of Kansas.

Johnson, Erik W., and Scott Frickel. 2011. "Ecological Threat and the Founding of U.S. National Environmental Movement Organizations, 1962–1998." *Social Problems* 58(3): 305–29.

Johnston, Jill E., Emily Werder, and Daniel Sebastian. 2016. "Wastewater Well, Fracking, and Environmental Injustice in Southern Texas." *American Journal of Public Health* 106(3): 550–556.

Kelly, Sharon. 2014. "Shale Oil Drillers Deliberately Wasted Nearly $1 Billion in Gas, Harming Climate." *DeSmogBlog*, September 4, 2014. (http://www.dessmogblog.com/print/8463).

Kinchy, Abby J., and Simona L. Perry. 2012. "Can Volunteers Pick Up the Slack? Efforts to Remedy Knowledge Gaps about the Watershed Impacts of Marcellus Shale Gas Development." *Duke Environmental Law & Policy Forum* 22(2): 303–39.

Kinchy, Abby J., Sarah Parks, and Kirk Jalbert. 2015. "Fractured Knowledge: Mapping the Gap in Public and Private Water Monitoring Efforts in Areas Affected by Shale Gas Development." *Environment and Planning C: Government and Policy* (forthcoming).

Kinne, Beth E., Michael H. Finewood, and Davix Yoxtheimer. 2014. "Making Critical Connections through Interdisciplinary Analysis: Exploring the Impacts of Marcellus Shale Development." *Journal of Environmental Studies and Sciences* 4(1): 1–6.

Klare, Michael. 2013. "The Third Carbon Age." *Nation of Change*, August 16. Retrieved August 16, 2014. (http://www.nationofchange.org/print/39644).

Klare, Michael. 2014. "Peak Oil Is Dead." *Nation of Change*, January 10, 2014. Retrieved January 10, 2014. (http://www.nationofchange.org/print/42474).

Klein, Naomi. 2014. *This Changes Everything: Capitalism vs. the Climate*. New York: Simon & Schuster.

Kreuze, Amanda, Chelsea Schelly, and Emma Norman. 2016. *Energy Research and Social Science* (http://dx.doi.org/10.1016/j.erss.2016.05.010).

Kriesky, J., B. D. Goldstein, K. Zell, and S. Beach. 2013. "Differing Opinions about Natural Gas Drilling in Two Adjacent Counties with Different Levels of Drilling Activity." *Energy Policy* 58: 228–36.

Krogman, Naomi T. 1996. "Frame Disputes in Environmental Controversies: The Case of Wetlands Regulations in Louisiana." *Sociological Spectrum* 16: 371–400.

Ladd, Anthony E. 2013. "Stakeholder Perceptions of Socio-environmental Impacts from Unconventional Natural Gas Development and Hydraulic Fracturing in the Haynesville Shale." *Journal of Rural Social Sciences* 28(2): 56–89.

Ladd, Anthony E. 2014. "Environmental Disputes and Opportunity-Threat Impacts Surrounding Natural Gas Fracking in Louisiana." *Social Currents* 1(3): 293–312.

Ladd, Anthony E. 2016. "Meet the New Boss, Same as the Old Boss: The Continuing Hegemony of Fossil Fuels and Hydraulic Fracking in the Third Carbon Era." *Humanity and Society* (http://dx.doi.org/10.1177/0160597616628908).

Ladd, Anthony E., and Shirley Laska. 1991. "Opposition to Solid Waste Incineration: Pre-Implementation Anxieties Surrounding a New Environmental Controversy." *Sociological Inquiry* 61: 299–313.

Ladd, Anthony E., and Charles Perrow. 2016. 'Institutional Dilemmas of Hydraulic Fracking: Economic Bonanza, Renewable Energy Bridge, or Gangplank to Disaster?" Paper presented at the annual meetings of the American Sociological Association, August 20–23, 2016, Seattle, WA.

Lavelle, Marianne. 2012. "Bad Gas, Good Gas." *National Geographic* 222: 90–109.

Light, John. 2014. "Methane Is Leaking Out All Over the Damn Place, Thanks to the Oil and Gas Industry." *Grist Magazine*, December 10. Retrieved on December 11, 2014. (http://grist.org/news/methane-is-leaking-out-all-over-the-damn-plac...r&utm _medium=email&utm_term=Daily%2520Dec%252011&utm_campaign=daily).

Light, John. 2015. "Fracking Is Definitely Causing Earthquakes, Another Study Confirms." *Grist Magazine*, January 7. Retrieved January 7, 2015. (http;//grist.org/news/fracking -is-definitely-causing-earthquakes-anot...tter&utm_medium=email&utm_term=Daily %2520Jan%25207&utm).

Loki, Reynard. 2015. "Fracking May Release Cancer-Causing Radioactive Gas, According to Surprising New Study." *Alternet*, April 10. Retrieved April 10, 2015. (http://www.alternet .org/print/environment/fracking-may-release-cancer-causing-radioactive-gas-according -surprising-new-study).

Lovins, Amory. 1977. *Soft Energy Paths*. New York, NY: Harper and Row.

Luhmann, Niklas. 1993. *Risk: A Sociological Theory*. Trans. Rhodes Barrett. New York, NY: Aldine de Gruyter.

Malin, Stephanie. 2014. "There's No Real Choice But to Sign: Neoliberalization and Normalization of Hydraulic Fracturing on Pennsylvania Farmland." *Journal of Environmental Studies and Science* 4(1): 17–27.

Malin, Stephanie A. 2015. *The Price of Nuclear Power: Uranium Communities and Environmental Justice*. New Brunswick, NJ: Rutgers University Press.

Malin, Stephanie A., and Kathryn Teigen DeMaster. 2016. "A Devil's Bargain: Rural Environmental Injustices and Hydraulic Fracturing on Pennsylvania's Farms." *Journal of Rural Studies* 47: 278–290.

Manthos, David. 2013. "Word Games Are Misleading the American Public about Fracking." *Grist Magazine*, November 17. Retrieved November 18, 2013. (http://grist.org/climate -energy/word-games-are-misleading-the-amer...r&utm_medium=email&utm_term =Daily%2520Nov%252018&utm_campaign=daily).

Marra, K. R., Charpentier, R. R., Schenk, C. J., Lewan, M. D., Leathers-Miller, H. M., Klett, T. R., Gaswirth, S. B., Le, P. A., Mercier, T. J., Pitman, J. K., and Tennyson, M. E. 2015. "Assessment of Undiscovered Shale Gas and Shale Oil Resources in the Mississippian Barnett Shale, Bend Arch–Fort Worth Basin Province, North-Central Texas." *U.S. Geological Survey Fact Sheet 2015–3078*, 2 p. (http://dx.doi.org/10.3133/fs20153078).

Mayer, Adam. 2016. "Risk and Benefit in a Fracking Boom: Evidence from Colorado." *The Extractive Industries and Society* (http://dx.doi.org/10.1016/j.exis.2016.04.006).

McKibben, Bill. 2014. "Obama's Fracking Folly." *Mother Jones* 39: 10–11.

Misrach, Richard, and Kate Orff. 2012. *Petrochemical America*. New York: Aperture Books.

Nelkin, Dorothy (ed.). 1984. *Controversy: Politics of Technical Decisions* (2nd ed.). Beverly Hills, CA: Sage Publications.

Nijhuis, Michelle. 2014. "Can Coal Ever Be Clean?" *National Geographic* (April) 225(4): 28–40.

Pantsios, Anastasia. 2014. "Fracking Linked to Miscarriages, Birth Defects and Infertility." *Nation of Change*, December 9. Retrieved on December 9, 2014. (http:// www.nationofchange.org/2014/12/09/fracking-linked-miscarriages-birth-defects -infertility).

Paveglio, Travis B., Amanda D. Boyd, and Matthew S. Carroll. 2016. "Re-conceptualizing Community in Risk Research." *Journal of Risk Research* (http://dx.doi.org/10.1080/13669877.2015.1121908).

Perrow, Charles. 2015. "Cracks in the Regulatory State." *Social Currents* 2(3): 203–212.

Pew Research Center. 2014. "Views on Increased Use of Fracking Tilt Negative." November 12. Accessed February 13, 2016. (http://www.people-press.org/2014/11/12/little-enthusiasm-familiar-divisions-after-the-gops-big-midterm-victory/).

Poole, Amanda, and Anastasia Hudgins. 2014. "'I Care More about This Place, Because I Fought for It': Exploring the Political Ecology of Fracking in an Ethnographic Field School." *Journal of Environmental Studies and Science* 4(1): 37–46.

Queally, Jon. 2013. "Obama's Possible Frack-friendly Energy Plan a 'Nail in the Coffin' for Climate." *Common Dreams*. Retrieved February 22, 2013. (http://www.commondreams.org/headline/2013/02/22-0?print).

Rauber, Paul. 2014. "Carbon States of America." *Sierra* 98(6): 52–55.

Raynes, Dakota K. T., Tamara L. Mix, Angela Spotts, and Ariel Ross. 2016. "An Emotional Landscape of Place-based Activism: Exploring the Dynamics of Place and Emotion in Antifracking Actions." *Humanity and Society* 40(4): 401–423.

Rosa, Eugene A., G. E. Machlis, and K. M. Keating. 1988. "Energy and Society." *Annual Review of Sociology* 146: 149–172.

Rosa, Eugene A., Ortwin Renn, and Aaron M. McCright. 2014. *The Risk Society Revisited: Social Theory and Governance*. Philadelphia, PA: Temple University Press.

Sangaramoorthy, Thurka, Amelia M. Jamison, Meleah D. Boyle, Devon C. Payne-Sturges, Amir Sapkota, Donald M. Milton, and Sacoby M. Wilson. 2016. "Place-Based Perceptions of the Impacts of Fracking Along the Marcellus Shale." *Social Science & Medicine* 151: 27–37.

Sarge, Melanie A., Matthew S. VanDyke, Andy J. King, and Shawna R. White. 2015. "Selective Perceptions of Hydraulic Fracturing: The Role of Issue Support in the Evaluation of Visual Frames." *Politics and the Life Sciences* 34(1): 57–71.

Schlosser, Eric. 2012. *Fast Food Nation: The Dark Side of the All-American Meal*. New York, NY: Houghton Mifflin.

Schumacher, E. F. 1973. *Small Is Beautiful: Economics As If People Mattered*. New York, NY: Harper and Row.

Seawell, Esther. 2015. "MU Research Team Uncovers Link Between Fracking and Endocrine Disruption." *The Maneater*, April 10, p. 5.

Shriver, Thomas E., and Charles Peaden. 2009. "Frame Disputes in a Natural Resource Controversy: The Case of the Arbuckle Simpson Aquifer in Southcentral Oklahoma." *Society and Natural Resources* 22: 143–57.

Theodori, Gene L. 2009. "Paradoxical Perceptions of Problems Associated with Unconventional Natural Gas Development." *Southern Rural Sociology* 24(3): 97–117.

Theodori, Gene L. 2013. "Perception of the Natural Gas Industry and Engagement in Individual Civic Actions." *Journal of Rural Social Sciences* 28(2): 22–34.

Tierney, Kathleen. 2014. *The Social Roots of Risk: Producing Disasters, Promoting Resilience*. Stanford, CA: Stanford University Press.

United States Energy Information Administration. 2016. *Annual Energy Outlook 2016*. Office of Energy Analysis: U.S. Department of Energy. (www.eia.gov/petroleum/drilling/#tabs- summary-1).

United States Energy Information Administration. 2016. *Energy in Brief: Shale in the United States*. Office of Energy Analysis: U.S. Department of Energy. (www.eia.gov/energy in brief/article/shale in the united states.cfm).

United States Environmental Protection Agency. 2015. *Assessment of the Potential Impacts of Hydraulic Fracturing for Oil and Gas on Drinking Water Resources* (External Review Draft). U.S. Environmental Protection Agency, Washington, DC, EPA/600/R-15/047, 2015.

Vasi, Ion Bogdan, Edward T. Walker, John S. Johnson, and Hui Fen Tan. 2015. "No Fracking Way! Documentary Film, Discursive Opportunity, and Local Opposition against Hydraulic Fracturing in the United States, 2010–2013." *American Sociological Review* 80(5): 934–959.

Weigle, Jason L. 2011. "Resilience, Community, and Perceptions of Marcellus Shale Development in the Pennsylvania Wilds: Reframing the Discussion." *Sociological Viewpoints* 27: 3–14.

Wilber, Tom. 2015. *Under the Surface: Fracking, Fortunes, and the Fate of the Marcellus Shale.* Ithaca, NY: Cornell University Press.

Willow, Anna J. 2014. "The New Politics of Environmental Degradation: Un/Expected Landscapes of Disempowerment and Vulnerability." *Journal of Political Ecology* 21(1): 237–57.

Willow, Anna J., Rebecca Zak, Danielle Vilaplana, and David Sheely. 2014. "The Contested Landscape of Unconventional Energy Development: A Report from Ohio's Shale Gas Country." *Journal of Environmental Studies and Science* 4(1): 56–64.

Wright, Simon. 2012. "An Unconventional Bonanza." *The Economist*, July 12, pp. 3–15.

Wynveen, Brooklynn J. 2011. "A Thematic Analysis of Local Respondents' Perceptions of Barnett Shale Energy Development." *Journal of Rural Social Sciences* 26(1): 8–31.

Yergin, Daniel. 2011. *The Quest: Energy, Security, and the Remaking of the Modern World.* New York: Penguin Press.

York, Richard. 2015. "How Much Can We Expect the Rise in U.S. Domestic Energy Production to Suppress Net Energy Imports?" *Social Currents* 2(3): 222–230.

1

Natural Gas Fracking on Public Lands

The Trickle-down Impacts of
Neoliberalism in Ohio's Utica
Shale Region

SHERRY CABLE

Introduction

The Marcellus Shale formation, home to one of the most substantial natural gas booms of the past decade, has drawn significant attention from sociologists and social scientists researching extractive communities in the United States. Such studies have typically focused on private landowners in the Marcellus region who leased their property to oil and gas companies, examining reasons for signing contracts (Malin 2014), rural environmental injustices (Malin and DeMaster 2016), community concerns about traffic and noise (Metcalfe 2008), the effects of powerlessness on nonmobilization (Eaton and Kinchy 2016), as well as other prominent socioenvironmental issues (Barth 2013; James and Aadland 2011; Korfmacher, Jones, Malone, and Vinci 2013; Krupnick, Gordon, and Olmstead 2013; Papyrakis and Gerlagh 2007; Wang and Krupnick 2013).

This chapter contributes to the sociological fracking literature in three significant ways: it focuses on the Utica Shale formation, rather than the Marcellus

Shale region; it centers on the leasing of public, rather than private lands, for energy development; and it analyzes how the global neoliberal agenda has trickled down to shape local regulation of natural gas fracking in Ohio.

Even though the Utica Shale formation in eastern Ohio is the newest frontier for shale gas development, social scientists have given it little attention (for exceptions, see Price and Maples, this volume; and Willow et al. 2014). Production in the Utica began in August 2011. Natural gas extraction with conventional techniques is largely confined to rural areas that provide extensive acreage for drilling multiple wells. In contrast, unconventional drilling and fracking technologies utilize a single well pad from which multiple wellbores can be drilled, making it feasible to extract gas in more populated urban and suburban areas. Consequently, the fracking industry has set its sights on gaining access to the shale gas deposits beneath tens of thousands of acres of municipal lands. The industry also seeks the shale gas beneath federal and state parks and forests. Both municipal lands and recreational parks and forests are public lands. Management of public lands differs significantly from that of private lands in ways that can provoke conflict.

While private landowners make their own land use decisions, public lands are managed by public officials. Municipalities are governed by local ordinances that manage public lands, such as roads and schoolyards, and that can also affect private properties within municipal boundaries (Ladd 2014). Federal- and state-owned parks and forests fall under the public trust doctrine, a common law principle establishing the right of the public to use certain lands and waters. The doctrine assigns regulatory responsibilities for protecting and regulating these common property resources to federal and state agencies (Rule and Salzman 2006).

The shale gas industry is composed of multiple companies pursuing resources that are spread across both private and public properties. As a consequence, the potential for stakeholder conflict is high among, for example, urban residents, private mineral rights owners, surface rights owners, businesses, conservation organizations, and drilling companies. One crucial function of regulation is to help resolve such conflicts. Yet, as Perrow (2015) argues, since about 1985 the federal government has pulled back from its regulatory responsibility for private sector economic activity, resulting in a much more deregulated state. Deregulation is a hallmark of the larger neoliberal agenda emphasizing privatization, commodification, and marketization that advances economic interests over public interests.

The doctrines that comprise neoliberalism were adopted in exclusive circles of national and international political and economic elites, far from the cities and villages of eastern Ohio. But implementation of environmental regulatory policies inevitably happens on the ground, in real communities, affecting real people. Many scholars emphasize the importance of studying localized, "actually existing neoliberalism" to document its impacts more effectively (Brenner and Theodore 2001: 351). How were land use decisions made at the local level? How did neoliberalism shape local regulatory policies? What conflicts emerged? Were the voices of citizens heard in public land use decisions? These questions are of grave import because the issues associated with unconventional energy development threaten local communities, the environment, and democratic processes.

In the next section, I review literature on neoliberal policies and hydraulic fracking and discuss the analytical focus of my research. Following an account of my research strategy and data, I present my findings from eight conflicts over fracking in the Utica Shale formation. I conclude with a summary and some reflections on the implications of my findings.

Neoliberal Natural Resource Management

Neoliberalism represents a set of economic policies and political projects that reverse the redistributive reforms of the mid-20th century (Harvey 2005). The keystone tenet of neoliberalism is minimal government intervention in the economy. Advocates assert that capital accumulation is best facilitated through decentralization, free-market economic systems, open trade, market-based solutions, and substantially downsized regulation (Bakker 2000; Bridge and Perrault 2009; Castree 201; Mercer, de Rijke, and Dressler 2014). Neoliberal ideologies are "produced and reproduced through institutional forms and political action" (Peck and Tickell 2002: 383). They become embedded within national, regional, and local contexts "defined by the legacies of inherited institutional frameworks, policy regimes, regulatory practices, and political structures" (Brenner and Theodore 2002: 351). Johnson (2011) describes the neoliberal state as "handmaiden to the market" (2011: 304). Neoliberalism and the market mentality are more than just political ideology; in reality they represent a comprehensive cultural orientation, as well as hegemonic discourse, that is commonly accepted and permeates perceptions, worldviews, and political economic practices (Hartwick and Peet 2003; Harvey 2007).

Natural resource management falls under the neoliberal regime, resulting in the commodification and marketization of nature itself. Natural resources are managed "in a way as to favor market-based actors and practices" (Bridge and Perrault 2009: 487). Water, land, air, community, quality of life, and health are yoked to market logic and the quest for maximum profit. Environmental concern has been "ideologically and institutionally incorporated into the global neoliberal hegemony" (Hartwick and Peet 2003: 188) in which natural resource commodification is entrenched. Environmental regulations largely ignore the public interest. For example, Malin's (2014) research among farmers in the Marcellus Shale formation illustrates neoliberalism's penetration of the hegemonic discourse. Malin analyzed how neoliberalism shaped farmers' views of fracking's environmental and economic impacts. Many farmers applied neoliberal logic to normalize fracking, assessing risks via a cost-benefit framework through which fracking was accepted as inevitable and natural.

The literature suggests that neoliberal normalization of fracking occurs frequently. A key finding in a study of regional newspaper representations of natural gas development in eastern Ohio, Pennsylvania, and New York was that newspaper coverage in eastern Ohio, compared to the other states, was more positive in representations of the economic and social impacts of fracking (Ashmoore et al. 2016). At the same time, evidence demonstrates resistance to this hegemonic discourse. Residents of fracking communities and communities sited for fracking are increasingly organizing in opposition to shale gas development, and over 475 communities in 24 states have passed fracking bans (Ladd 2016; Negro 2012). Mobilization represents conflicts among community residents, between residents and industry, and between state and local governments.

I address the issues of conflict and regulation regarding fracking on government-owned lands in the Utica Shale formation. The neoliberal agenda shifted most regulating responsibilities for shale gas development to state and local levels (Malin 2014). But the sudden rise of the fracking industry in eastern Ohio in 2011 caught state officials off guard. They had neither the knowledge nor the infrastructure to regulate this new, unconventional technique for gas extraction. Municipal officials, whose elected positions were frequently part-time, were even less prepared. The neoliberal regime created a regulatory vacuum, ripe for exploitation by the oil and gas industry. In this research, I analyze eight conflicts that emerged within the context of this regulatory vacuum, asking: What was the nature of the conflicts that emerged? Who were the parties in conflict? What role did regulation play in the cause or the resolution of the conflict?

Research Methods and Data

The Utica Shale formation in the Appalachian Basin is located both underneath and adjacent to the Marcellus Shale region, covering portions of Pennsylvania, New York, and West Virginia. The geographical area of the Utica is extensive, underlying much of the northeastern United States. The Utica is 7,000 feet below the Marcellus Shale, but its western edge substantially rises in eastern Ohio where the depth can be less than 3,000 feet below the Marcellus.

Despite the temporary global oil glut and declining market prices for oil and natural gas, Utica Shale gas production continues to increase. The Ohio Department of Natural Resources (ODNR) quarterly report issued in September 2016 shows that shale oil production from the second quarter of 2016 increased approximately 51% over 2015's second quarter (http://oilandgas

.ohiodnr.gov/). Between September 2010 and September 2016, 18 rigs were constructed, 1,800 wells were drilled, and 2,246 permits were granted. The report lists 1,415 currently active horizontal shale wells, of which 1,362 reported production during the second quarter (http://oilandgas.ohiodnr.gov/).

This case study explores regulatory and citizen responses to industry initiatives to frack on public lands in the Utica Shale. The case relies on document analysis, a method in which factual information is gleaned from documents and marshaled to tell a story. Documents included local, regional, and national newspapers and government reports collected with the search phrases "energy and environment" and "horizontal hydraulic fracturing" between August 2011 through September 2016.

New York Times articles provided a larger national context in which to nest the case study, along with the *Columbus Dispatch* published in Ohio's capital city where the General Assembly meets. I relied on four local newspapers: the *Youngstown Vindicator*, the *Athens Messenger*, Lisbon's *Morning Journal*, and Salem's *Salem News*.

Newspaper searches generated a total of approximately 1,100 pages of articles as data, which I arranged in chronological logs. The considerable overlap among the newspapers allowed for corroboration of reported factual information. I integrated the newspaper logs to construct an overall chronology of articles about Utica shale gas and then coded this integrated chronology for conflicts over fracking on public lands. The chronology revealed eight fracking-related conflicts: one over federal land, three over state lands, and four within municipalities. The thematic approach I used was to identify for each the nature of the conflict, the parties in the conflict, and the resolution of the conflict.

Trickle-down from the Neoliberal Regime

Regulation of Federal Lands: The Wayne National Forest

The Bureau of Land Management (BLM) oversees 247 million acres of public lands in addition to 700 million acres of subsurface mineral estate. The Forest Service manages 193 million acres, most of which are in designated national forests. For at least the last 30 years, the federal government's priority in natural resource management has been to reduce the nation's energy dependence by promoting increased exploitation of mineral energy resources on federal lands. Ohio has no national parks and only one national forest, the Wayne National Forest that covers 240,101 acres.

In July 2012, several oil and gas companies expressed to BLM their interest in fracking beneath the Wayne National Forest. Officials accommodated them by announcing a Competitive Lease Sale for December 2015, listing 3,300 acres for auction (Hunt 2012). The announcement drew letters of protest from area residents, the Ohio University president, and environmental groups such as the Buckeye Forest Council. In the meantime, the Forest Service had asked BLM to withdraw the parcels, and the agency consented (DeWitt 2014b). BLM subsequently dismissed the protest letters as moot.

Oil and gas companies again in November 2015 filed documents with the BLM, formally requesting approval to frack 31,900 acres of the national forest (Arenschield 2015b). A spokeswoman for the industry group Energy in Depth complained that BLM's inaction was blocking potential royalty payments for the private owners and lessors of lands adjacent to the Wayne (Arenschield 2015b). BLM held three public meetings in November and invited comments up to January 22, 2016. Elected officials from both political parties and several local unions submitted letters expressing their support for fracking and the jobs it would bring.

Following the close of the comment period, BLM announced its decision to wait until 2017 to pursue drilling in the Wayne. The decision was likely due to an ongoing suit about BLM's authority to regulate fracking on federal lands. In March 2015, BLM announced that the first rules for fracking on federal lands would go into effect in June. The rules would require companies to: disclose on an industry-run website the chemicals used in drilling; use covered storage tanks for fracking wastewater, rather than open pits; conduct tests on the durability of *each* well; and construct strong cement barriers between the wellbore and groundwater (Cockerham 2015).

But enforcement was suspended when industry groups and the states of Wyoming, Colorado, North Dakota, and Utah quickly filed suit in U.S. District Court for the District of Wyoming to block enforcement of the rules, challenging BLM's authority to regulate fracking on federal lands (Gilmer 2015). Fracking opponents countered that not only did BLM have the authority, but that more restrictive fracking regulations were crucial to ensure public safety and environmental protection.

In September 2015, the District Court judge issued a preliminary injunction against BLM's enforcement until the court could consider challenges from states and groups representing the energy industry (Gilmer 2015). The judge's ruling relied on the states' argument that the Safe Drinking Water Act specifically delegated fracking regulation to the Environmental Protection Agency (EPA) and that the 2005 Energy Policy Act had exempted fracking from EPA oversight, removing fracking from all federal oversight. The judge further declared, "The BLM has neither substantiated the existence of a problem this rule is meant to address, identified a gap in existing regulations the final rule will fill, nor described how the final rule will achieve its stated objectives. Rather, the Fracking Rule seems a remedy in search of harm" (Gilmer 2015).

In December 2015, the BLM and other rule supporters appealed to the Tenth Circuit Court on the correctness of the district court's freezing the rule during litigation (Neary 2016). Before the appeals court could rule, the District judge, in June 2016, lifted his preliminary injunction to issue a final ruling, declaring that BLM lacked the authority to set rules for hydraulic fracturing because Congress had not authorized it to do so. The judge wrote that the court's role was not to decide on the merits of hydraulic fracturing, but to interpret whether Congress had given the Department of Interior legal authority to regulate the practice. The litigation with the Tenth Circuit Court will now either be dismissed or rolled into the broader appeal of the decision. In the meantime, no regulations currently govern fracking on federal lands.

Regulation of State Lands

The Utica Shale gas boom that began in late 2011 mushroomed so rapidly that the pace of production increasingly diverged from the pace of scientific knowledge of its impacts (Steinzor, Subra, and Sumi 2013: 56). Indeed, gas wells were producing long before regulators caught up. The ODNR manages more than 590,000 state-owned acres, over half of it in the hill-and-valley region overlying the shale formation. In some cases, the state owns only the surface rights, while the mineral rights are owned either by the federal government or private landowners. The situation originated when the state was given land or bought it at deep discounts because the sellers retained mineral rights (Ludlow 2014). In 2011, the Ohio General Assembly announced plans to allow oil and gas drilling in state parks. Governor John Kasich supported the plan, touting the potential economic windfall (Kovac 2011). The plan stipulated that the governor, by June 2012, appoint a five-member Oil and Gas Commission to oversee the permitting process that included public input. But Kasich changed his mind on the issue and, by default, imposed a fracking moratorium in state parks by neglecting to appoint a commission. I discuss three conflicts that emerged over fracking on state lands.

Blue Rock State Forest. In the early decades of the 20th century, an area in Muskingum County was so extensively logged that it was deemed unsuitable for farming. The U.S. Resettlement Administration bought the land in the 1930s and transferred surface ownership of 4,578 acres to Ohio but retained mineral rights. The state-owned surface land was established as Blue Rock State Forest and is managed by ODNR. BLM manages the underlying federally owned minerals.

In late July 2013, a staff attorney for the nonprofit Buckeye Forest Council browsed the BLM website and read of the agency's proposal to open "state-owned surface/federally owned minerals" for bids from oil and gas companies (Hunt 2013a). Following a link for more details, he discovered the agency's plan for a December auction on 4,525 acres of mineral rights beneath the state forest (Hunt 2013a). The 30-day comment period had already expired without public response (Hunt 2013b). The Buckeye Forest Council and a coalition of environmental groups sent a protest letter to BLM and ODNR, asserting that the public was not given adequate notice of the planned auction and requesting an additional comment period (Hunt 2013b). ODNR spokeswoman Bethany McCorkle responded, declaring that ODNR officials knew that BLM officials were conducting an environmental assessment for drilling in the forest, but were unaware that an auction had been scheduled. McCorkle emphasized that BLM was not legally bound to consult with ODNR in the decision because of BLM's sole authority over mineral rights (Hunt 2013a). In August 2013, BLM indefinitely suspended the auction of mineral rights beneath Blue Rock State Forest (Hunt 2013c).

Public Relations Plan to Promote Fracking. An internal ODNR memo dated August 2012 was released by the agency in February 2014. The document revealed that ODNR had developed, but never implemented, a detailed public relations campaign to promote public support for fracking on state lands (Rowland 2014a). The memo disclosed government views of the public, the fracking industry, and opponents of fracking on public lands. Regarding the public, the memo's author recognized that the "public-relations initiative could blur the public perception of ODNR's regulatory role" and would consequently require "precise messaging and coordination" (Rowland 2014a). The fracking industry was depicted as allies to be enlisted to "minimize public concern" about drilling and convince the public of its benefits, including millions of dollars for park improvements (Rowland 2014a). The memo showed that JobsOhio, the state's privatized economic development agency, was assigned a key role in promoting the job-creation angle of fracking.

Opponents of fracking on public lands were also addressed in the memo, with warnings of the need to overwhelm "zealous resistance by environmental-activist opponents who are skilled propagandists" (Rowland 2014a). Activists would "attempt to create public panic" about possible health risks and "attempt to legally and physically disrupt or halt the drilling projects, including staging dangerous protests" (Rowland 2014a). The latter would necessitate the state's "sustained legal countermeasures and crisis readiness" (Rowland 2014a). Two legislators were explicitly named as opponents. When the memo was released, a Kasich spokesman insisted to media representatives that the governor had never seen the plan. But a Sierra Club activist filed a public records request and found an email from a Kasich adviser to other high-level aides revealing that aides were not only aware of the plan but involved in its development (Rowland 2014b). ODNR spokeswoman McCorkle defended the aides' actions: "Any responsible organization plans in advance what it is going to do, especially when it knows it is going to face fierce opposition to progress. The fact that these secretly funded extremist groups are attacking us today validates the wisdom of anticipating the attack and planning for it" (Rowland 2014b).

Charging that the public relations plan demonstrated the administration's active collaboration with the oil and gas industry, the Ohio Democratic Party on February 19 requested that the Republican House Speaker formally investigate the marketing plan (Rowland and Drabold 2014). On the same day, Kasich announced his new opposition to fracking on state lands, and his office divulged that the governor had deliberately ignored the state law requiring that he appoint by June 2012 a commission to oversee the leasing of park and forest mineral rights (Rowland 2014c).

Brush Creek and Egypt Valley Wildlife Areas. In 2000, ODNR denied a permit to a private mineral rights owner for strip mining coal in the Brush Creek Wildlife Areas (Ludlow 2014). The rights owner filed a complaint against ODNR and the State of Ohio in Jefferson County Common Pleas Court, claiming that strip mining was the only way to reach the coal that he legally owned. The state counter-argued that, because strip mining would destroy surface land, it should not be permitted unless clearly stated otherwise in the 1944 deed. In 2012, the court ruled in the state's favor. The plaintiff appealed to the Seventh District Court of Appeals, whose justices also ruled in the state's favor. The plaintiff then appealed to the Ohio Supreme Court. In 2014, the court ruled 6–1 in the plaintiff's favor: "There is reason to believe that the signatories to the original contract understood that 'reasonable surface right privileges' included the right to strip mine, and there is no reason to believe that the signatories intended to exclude strip-mining" (Phillips 2014). Consequently, the mineral rights owned by non-state parties were permitted to mine or drill in parts of 18 state forests, 24 state parks, and 53 natural areas (Ludlow 2014). All told, parcels of more than 117,000 acres, about 40% of state-owned park and forestland, could be mined (Arenschield 2015a).

The mineral rights owner did not subsequently apply for a strip-mining permit. Instead, Oxford Mining Co. was the first party to take advantage of the ruling, receiving a five-year ODNR permit in July 2015 to strip-mine more than 900 acres of the Egypt Valley Wildlife Area (Ludlow 2015a; Williams 2015). About 80% of the 18,011 acres of the wildlife area is reclaimed strip-mined land purchased and transformed by the state and conservancy groups (Ludlow 2015b; Williams 2015). Oxford Mining leased the mineral rights from the owner, Marietta Coal Co. The Ohio Environmental Council and other advocacy groups contended that ODNR should have done more to protect the area, but ODNR officials said that the 2014 Ohio Supreme Court ruling left them no choice but to issue the permit (Arenschield 2015a). Although the court ruling dealt with surface mining of coal, fracking would likely also be permitted on state-owned lands since strip mining destroys more surface land than does a gas well pad.

Regulation of Municipal Lands

Fracking in urban areas can trigger a regulatory conundrum if private land is leased within a city whose residents vote to ban fracking (see Staggenborg, this volume). In this scenario, two deeply entrenched American values come into conflict: the will of the majority versus private property rights. I explore four such conflicts in cities where fracking within city limits became contested.

Athens. Athens is a small college town in mostly rural southeastern Ohio. In July 2013, the Bill of Rights Committee (BORC) presented the city auditor with a petition requesting that their initiative, the Athens Community Bill of Rights and Water Supply Protection Ordinance, appear on the November ballot. If passed, the initiative would ban fracking activities inside the town and within its jurisdiction, including privately held lands and residences, city parks, and Ohio University's campus (DeWitt 2013a). The petition's reference to "jurisdiction" matched a section of the Ohio Revised Code that authorized a municipality to prosecute polluters of the water supply "within 20 miles of the municipal corporation limits" (DeWitt 2013a). BORC's petition was validated, and the city council passed an ordinance ordering the Athens County Board of Elections to submit the initiative to electors in November.

On August 2, a local attorney formally filed a protest against the initiative on behalf of seven Athens residents (Phillips 2013). The attorney, who worked for a law firm that negotiated oil and gas leases for local property owners, charged that passage of the initiative would likely generate suits against the city either by a supporter suing for enforcement or by a private landowner claiming an unconstitutional "taking" of the value of his property. Board members denied BORC's petition without explanation (DeWitt 2013c). BORC requested that the elections board release the transcript of their proceedings and provide written explanation for their decision (DeWitt 2013d). The board declined, writing: "We are aware of no legal obligation to respond to your request in the manner in which you have specified" (DeWitt 2013e). BORC filed Freedom of Information Act (FOIA) requests for all relevant documentation (DeWitt 2013f), which revealed that the Board of Elections' legal basis for denying their petition was the initiative's intent to impact lands outside the town's limits.

In January 2014, BORC members submitted a petition, revised to delete references to "jurisdiction," requesting that their initiative be placed on the May ballot (DeWitt 2014a). This proposal would ban fracking activities only inside city limits. The initiative further stated that the ordinance "invalidates the state's pre-emptive laws granting special privileges to for-profit corporations over the rights of people to a healthy, clean and safe environment" (DeWitt 2014a). The initiative appeared in November and Athens residents voted for it 79% to 21% (Arenschield 2014).

Youngstown. In April 2013, a citizens group, Frack Free Youngstown, presented a petition to the Mahoning County Board of Elections requesting that their initiative for a city fracking ban be placed on the May ballot. A legal challenge to the initiative was filed by an emergent group, the Mahoning Valley Coalition for Job Growth and Investment, a coalition of business, labor, and several political officials. The group's attorney chaired his firm's oil and gas leasing negotiations (Vindy.com 2013a). But the coalition rescinded its challenge when it was revealed that several coalition members themselves served on the elections board (Davis-Cohen 2015). The fracking ban appeared on the May ballot and was defeated 57% to 43%, a margin of 14 percentage points (Skolnick 2014).

In August, a citizens group, the Youngstown Community Bill of Rights Committee, submitted a petition to the board of elections for an initiative banning fracking operations within city limits (Vindy.com 2013b). The Mahoning Valley Coalition for Job Growth and Investment reunited to campaign against the initiative (Speakman 2013). In November, the fracking ban was defeated again, by a margin of 9.7 percentage points.

The Youngstown Community Bill of Rights Committee persisted, and the initiative was placed on the May 2014 ballot. Voters rejected it for a third time, but by only 8.5 percentage points. The initiative appeared a fourth time on the November 2014 ballot. City voters again rejected the fracking ban, this time by over 15 percentage points, the largest margin of defeat (Skolnick 2014). As the citizens group circulated a petition to put the fracking ban on the November 2015 ballot, a group emerged in opposition. Composed of local business owners and labor leaders, including the Mahoning-Trumbull AFL-CIO Labor Council, the new group, Voters for Ballot Integrity, urged city officials and the board of elections not to place the initiative on the November 2015 ballot (Skolnick 2014).

The county prosecutor advised board members of their mandatory duty to place the initiative on the ballot. But election board members refused, contending that the initiative was unconstitutional because state law invested ODNR with sole authority to regulate Ohio's oil and gas industry (Vindy .com 2015). When the county prosecutor refused to defend the board's decision, the board retained an oil and gas industry attorney at taxpayer expense (Davis-Cohen 2015). The City of Youngstown contended that the board of elections acted illegally in their refusal and requested that the Ohio Supreme Court compel the board to place the initiative on the November 2015 ballot (Skolnick 2015).

Court filings supporting the board of elections were submitted by the Ohio Chamber of Commerce, the Youngstown/Warren Regional Chamber, the Ohio Oil and Gas Association, the Mahoning-Trumbull AFL-CIO Labor Council, and 17 additional local labor unions (Skolnick 2015). In a public statement, the mayor advised that the city's complaint should not be interpreted as an endorsement of the fracking ban. Rather, their intention was to affirm citizens' right to petition the government in accordance with Youngstown's Home Rule Charter and the constitutions of the State of Ohio and the United States.

In September 2015, the Ohio Supreme Court ruled that the Mahoning County Board of Elections "does not have authority to sit as arbiters of the legality or constitutionality of a ballot measure's substantive terms. An unconstitutional amendment may be a proper item for referendum or initiative. Such an amendment becomes void and unenforceable only when declared unconstitutional by a court of competent jurisdiction" (Skolnick 2015). The fracking ban appeared on the November ballot, with Youngstown citizens rejecting it for a fifth time, but with the closest vote yet: 51.5% versus 48.5% (Davis-Cohen 2015).

Broadview Heights. Residents of Cleveland suburb Broadview Heights voted 2–1 in favor of a Community Bill of Rights that banned fracking in November 2012. Broadview Heights' was the first Ohio fracking ban to be subjected to a legal review. In June 2014, Bass Energy Co. Inc. and Ohio Valley Energy sued the city over the ban in Cuyahoga County Common Pleas Court. The companies contended that only ODNR had the authority to regulate fracking (Sandrick 2014). Bass had received an ODNR permit to drill on church property, and Ohio Valley had signed leases with residential landowners to drill on their properties. The companies claimed that the ban wrongly denied them the use of their property (Sandrick 2014).

In March 2015, the Cuyahoga County Common Pleas Court overturned the city's fracking ban, declaring that state law gave ODNR "sole and exclusive authority" to permit, locate, space, and regulate oil and gas wells. The judge ruled that state code prohibited local governments from exercising authority that "discriminates against, unfairly impedes or obstructs oil and gas activities and operations." He viewed Broadview Heights' fracking ban as an attempt to "strip corporate entities of their rights under the U.S. and Ohio constitutions" (Sandrick 2015).

Munroe Falls. Early in 2011, Beck Energy secured a ODNR permit to drill on leased private property within Munroe Falls, a relatively prosperous Akron suburb (Cocklin 2013). When Beck began drilling, the city filed in the Summit County Court of Common Pleas for an injunction to stop the firm from proceeding. The city objected to ODNR's issuing of a permit when city law prohibited drilling in residential-zoned areas, charging that Beck failed to comply with city ordinances that required the firm: (1) to obtain a city drilling permit, a zoning certificate, and the city's planning commission's approval; (2) to pay fees and post a performance bond; and (3) to secure rights-of-way construction permits and pay associated fees (Cocklin 2013). Beck countered that it need not comply with city ordinances because it was already compliant with Ohio Revised Code placing sole responsibility for oil and gas drilling with ODNR.

The common pleas court ruled in the city's favor in 2012 and issued an injunction ordering Beck Energy to cease drilling operations until it complied with city ordinances (Phillips 2013). The court cited the Ohio Constitution's guarantee of "home rule," which confers on cities the power of local self-government and regulatory authority for the protection of public health, safety, morals, and the general welfare of society. The aim of home rule is to facilitate local control and minimize state intervention into municipal affairs on the premise that local authorities are best suited to decide whether drilling would create noise, lights, traffic, and pollution that would harm other property owners (Ludlow 2015a). The court ruled that Munroe Falls possessed the authority to regulate drilling within its right-of-way and that state law did not authorize drilling companies to ignore local regulations, even when a state permit was granted.

Beck Energy appealed to Ohio's Ninth District Court of Appeals (Cocklin 2013). In February 2013, the court overturned most of the lower court's ruling, stating that the appropriate legal issue in the appeal was whether Munroe Falls' ordinances conflicted with Ohio Revised Code 1509. If so, state regulations prevailed, and city requirements such as permits, performance bonds, and public hearings were invalid (Phillips 2013). The City of Munroe Falls asked the Ohio Supreme Court to reverse the appellate court ruling, arguing that the case merited review "both as a matter of great public interest and as a substantial constitutional question" (Ludlow 2015a). The cities of Broadview Heights, Euclid, Heath, North Royalton, and Mansfield filed friends-of-the-court briefs in support of Munroe Falls. Expressions of support for Beck Energy came from the American Petroleum Institute, the Ohio Oil and Gas Association, and the Ohio Chamber of Commerce (Cocklin 2013).

The City of Munroe Falls argued in court that Ohio Revised Code 1509 directly violated Ohio's constitution under the home rule provision (Ludlow 2015a). The appropriate legal question of the case was whether local zoning ordinances were wholly displaced by the authority granted to the ODNR to regulate wells. The city claimed that the regulatory powers of the state and the city were concurrent. But Beck's attorneys challenged the court's jurisdiction, arguing that the appropriate legal question was whether Munroe Falls' local laws conflicted with state law. They contended that the language in the "home rule" section of the constitution prohibited local zoning laws that conflicted with statewide laws and that the city's ordinance was therefore invalid.

The Ohio Supreme Court set a crucial precedent in February 2015, ruling against Munroe Falls. The court held that home rule did not allow a municipality "to discriminate against, unfairly impede, or obstruct oil and gas activities and operations that the state has permitted under R.C. Chapter 1509" (Ludlow 2015a). Munroe Falls' ordinances created a licensing conflict between the state and the municipality, the court stated, and Ohio Revised Code 1509 "explicitly reserves for the state, to the exclusion of local governments, the right to regulate all aspects of the location, drilling, and operation of oil and gas wells, including permitting relating to those activities" (Ludlow 2015a). Beck was free to drill. In July 2015, over 100 local Ohio officials—mayors, county commissioners, and city councilors—wrote a letter to Governor Kasich, petitioning for the right to ban or regulate fracking within their own borders (Bernd 2015). The letter read, in part:

> Fracking imposes particular burdens on local communities, from strained services to ruined roads. As with other extractive booms, the arrival and expansion of fracking operations has been correlated with a wide range of social problems, including increases in domestic violence, drug use, traffic accidents and civil disturbances. Other local impacts, such as water and soil contamination and reduced property values, are likely to persist long after the boom is gone, and we will be the ones left to pick up the pieces.

As it currently stands in the State of Ohio, city, county, and township officials have no authority to ban fracking or to regulate it in their jurisdictions.

The Future of Fracking Bans?

Just a month after the Ohio Supreme Court ruling, the State of Texas became the first state to outlaw local fracking moratoriums, thus prohibiting cities and towns from passing ordinances to ban fracking operations (*Newsweek* June 4, 2015). The Texas governor characterized the law as a defense of "private property rights" and a move to limit government bureaucracy and overregulation. The very next month, the State of Oklahoma outlawed fracking bans (Henry 2015). The Oklahoma governor signed the bill over the objections of municipalities and environmental groups, arguing that the law was intended to provide uniform state standards for oil and gas drilling, rather than a mix of local rules.

In May 2016, the Colorado Supreme Court rejected the City of Longmont's fracking ban and Fort Collins' fracking moratorium, ruling that state power to promote industry eclipsed municipal bans, which were "invalid and unenforceable" (Finley 2016; see Malin, Ryder, and Hall, this volume). The court declared that any measures taken by a municipality that conflict with state law are preempted. The pattern is clear: fracking bans are currently prohibited in five of the major shale gas producing states of the United States, including Ohio, Texas, Oklahoma, Louisiana, and Colorado.

Analytical Summary of the Findings

My analysis of eight conflicts over fracking in the Utica Shale region over a five-year period revealed multiple levels of conflict among federal, state, and municipal sectors. At the federal level, the controversy over fracking in the Wayne National Forest morphed into a legal conflict over BLM's authority to regulate shale gas extraction on federal lands when a coalition of industry groups and four western states sued BLM. Rather than attack the rules, attorneys for the plaintiffs argued that BLM lacked authority to regulate fracking on federal lands because the practice was exempted from federal oversight. The judge sided with the plaintiffs in granting a preliminary injunction against enforcement of the new rules, declaring that BLM failed to demonstrate risks that the rules were intended to minimize. The conflict remains unresolved.

Three conflicts over fracking arose at the state level. The Blue Rock State Forest conflict concerned BLM's auctioning of federal rights to minerals underlying the state-owned surface. Parties to the conflict were BLM versus multiple, statewide environmental organizations. The issue was resolved, at least temporarily, with Governor Kasich's de facto moratorium on fracking on state lands.

The second conflict was over ODNR's plan for a campaign to persuade the public that fracking on state lands was economically beneficial. Conflict ensued even though the campaign plan was never implemented. Parties in the conflict were ODNR versus multiple environmental groups. This conflict forced the governor's office to acknowledge that Kasich's failure to appoint a regulatory commission was deliberate; he had grown to oppose fracking on state lands. The controversy also revealed the trap in which ODNR was caught, as are many regulatory agencies: their contradictory tasks of both regulating and promoting industry. The tasks serve different interests and contribute to the "revolving door" between industry and regulators. For example, the state senator who sponsored Ohio's law investing ODNR with sole regulatory authority was found to have received campaign contributions from the fracking industry. One year after leaving office, he became a registered lobbyist for an oil and gas trade industry group and British Petroleum America (Bernd 2015).

The third conflict related to strip mining for coal in the Brush Creek Wildlife Area. It concerned privately owned rights to minerals beneath state land. Parties to the conflict were ODNR and the private mineral rights owner. The conflict was resolved by a court ruling in favor of the mineral rights owner.

Four conflicts emerged at the municipal level. These battles over fracking bans involved the principle of home rule. The conflict in Athens was whether to ban fracking in the city, spearheaded by a pro-fracking citizens group and opposed by a business group. The battle played out through the county board of elections over whether to place the anti-fracking initiative on the ballot. The conflict was seemingly resolved when the initiative was placed on the ballot and residents overwhelmingly voted for it. The Youngstown conflict over a fracking ban was similar to the Athens case in that the parties were a business group and a citizens group, and the battleground was the county board of elections. From here, the story diverged. The Youngstown conflict went through five cycles in which the anti-fracking initiative appeared on the ballot five times in three years. The dispute's fifth cycle was particularly contentious. The board of elections refused to place the initiative on the ballot. The case went to the Ohio Supreme Court, which ruled for the city, and the initiative appeared on the ballot a fifth time. It was rejected.

The conflict in Broadview Heights occurred after residents voted to ban fracking. The ban was the first in the state to undergo legal review. Pitting corporate rights against majority rule, the lawsuit involved the city versus two fracking companies, who argued that only ODNR could regulate fracking. The conflict was resolved when the court ruled in favor of the companies, overturning the fracking ban. The fourth conflict, in Munroe Falls, concerned whether the city could regulate fracking activities through local ordinances, such as zoning laws. The dispute played out in court when a fracking company with leases on private lands sued the city. A lower court ruled in the city's favor, based on the constitution's guarantee of municipal home rule. On appeal, the court ruled for the fracking company, citing ODNR's sole authority for regulating fracking. The city appealed to the Ohio Supreme Court, and the justices ruled for the fracking company, declaring that the company's ODNR permit was sufficient. This consequential decision not only overruled voter-endorsed fracking bans in Athens, Broadview Heights, and Munroe Falls, it set a powerful precedent in the state by signaling that state laws preempt city laws.

Five of the eight conflicts were or will be resolved in the courts. Three final rulings negated the state's home rule guarantee, favoring either business interests in general or the fracking industry specifically. The fourth final ruling allowed extraction in wildlife areas. The fifth legal conflict remains unresolved, but the temporary ruling suggests the judiciary's tendency to support the oil and gas industry. A sixth conflict was resolved with a majority vote against a city fracking ban.

Two of the eight conflicts were suspended because of Kasich's de facto moratorium. His neglect in appointing a regulatory commission for fracking on state lands was seemingly inconsistent with his otherwise neoliberal ideology. Indeed, he initially preached on the economic benefits of drilling state lands. The inconsistency dissolves with a tracking of Kasich's continuing battle with the Republican-controlled General Assembly over raising severance tax rates on oil and gas extraction.

Ohio is one of 31 state governments that tax oil and gas producers and private landowners receiving royalties. Based on either the volume or the value of production, severance taxes are intended to ensure that costs associated with extraction are not externalized to the public. Kasich's aim in raising severance tax rates was to offset his planned reductions in income taxes, but legislators contended the increase would hurt industry (Siegel 2015; Vardon and Siegel 2013). Between 2012 and 2016, seven proposals for increased severance tax rates were introduced in the General Assembly, but none passed (Downing 2016).

A 2016 *Columbus Dispatch* editorial chided the legislature for its inaction, claiming that many legislators received "generous political donations from oil and gas industries" (Editorial Board 2015). The state government's predilection for business was also displayed in the lack of regulation of the fracking process for public and environmental protection. In June 2014, ODNR announced that the formulation of 20 such rules was in progress. Two years later, only one rule had been implemented (Arenschield 2016). The upshot for fracking in Ohio? Frack, baby, frack!

Conclusion

This research contributes to the sociological literature on hydraulic fracking and the role that neoliberalism plays in unconventional energy development. Focusing on the Utica Shale formation brings this newest fracking front into view and helps to set the foundation for future comparative studies across different shale formations. The federal regulatory vacuum on fracking provides the largest context for this discussion in the United States, but state laws vary. How well do this study's findings apply to other prominent energy regions like Pennsylvania's Marcellus Shale, Texas's Barnett Shale formation, or North Dakota's Bakken Shale play? Additionally, how does the research focus here on the development of shale gas on public lands in the Utica further contribute to such comparisons?

The examination of conflicts centered on the leasing of public lands adds a new dimension to fracking research that has more commonly focused on conflicts between energy companies and private landowners and the privileging of mineral rights over surface rights. With few exceptions, existing studies point to how citizen concerns over environmental quality and public health tend to be discounted by state regulatory officials in favor of industry proposals to control the pace and scale of shale development. The official justifications typically offered in support of such measures are that gas development will contribute to jobs, rural revitalization, energy independence, climate stability, and serve as a "bridge fuel" to a renewable energy future.

However, Ladd (2016) argues that the evidence points less to the realization of these goals and more to the economic profits behind state–corporate agendas premised on maintaining the hegemony of fossil fuels well into the 21st century. Such rationales are also contradicted by the development of new uses for natural gas in areas like transportation and export. The 2011 Natural Gas Act provides tax breaks to transportation interests that transition from gasoline and diesel to natural gas, which can now be observed in municipal bus fleets, UPS trucks, and U.S. Postal Service vehicles (Cardwell and Krauss 2013). Moreover, the energy-industrial complex is busy building dozens of natural gas storage export facilities across the country, whose purpose is not to store these shale gas resources for domestic use, but to make the United States a gas exporter and global market power by 2020 (United States Energy Information Administration 2016).

My focus on regulatory measures yielded somewhat surprising results in that conflicts in the Utica centered on conflicts over civil society and democratic governance, rather than on fracking's environmental and health risks. Who has the right to decide whether fracking occurs on public lands and state lands? The public trust doctrine holds that government-owned resources be maintained for the public's use and in the public's best interest. Who has the right to decide about fracking within city limits: citizens or the state government? In theory at least, home rule states like Ohio and Colorado guarantee municipal self-government with minimal state intervention in local matters—and yet citizens in a growing number of states are finding themselves increasingly powerless to stop the oil and gas industry from constructing fracking and wastewater injection sites in their backyards.

Finally, the findings here also contribute to the critical literature on neoliberalism by revealing that the public trust doctrine and the home rule principle have been trumped by the neoliberal agenda of economic expansion and profits, corporate power, and externalized costs. As Tierney writes in her discussion of disaster resilience and neoliberalism: "what appears on the surface to be a decentralization and localization of policies and programs under neoliberalization, which is discursively constructed as a desirable alternative to centralized state control, is actually the result of public-private collaboration and 'policy steering' at the state [i.e., federal] level" (2015: 1334).

Willow et al.'s (2014) research on the Utica also underscores one of the key outcomes of neoliberal state policies, as many residents experience feelings of disempowerment and vulnerability because they are blocked from determining for themselves the forms of production to be tolerated in their communities. She reports that many residents believed that "the activities of a powerful industry are infringing on fundamental rights and undermining core democratic values" (2014: 247).

The trickle-down impacts of the neoliberal economic agenda not only fracture the earth beneath citizens' feet, but also produce fissures in the democratic processes on which local governance rests. Given these realities, a key question facing the nation today is whether citizens will be increasingly compelled to trade their fragile democratic rights for the kinds of short-term economic gains associated with the boom and bust cycles of unconventional oil and gas development. Indeed, the spatial and temporal disparities between local community costs and national energy benefits constitute one of the most important inequities in need of greater sociological attention.

References

Arenschield, Laura. 2014. "Athens Votes to Ban Fracking." *Columbus Dispatch*, November 6.

Arenschield, Laura. 2015a. "ODNR Criticized for Ok'ing Mining in Ohio's State Parks." *Columbus Dispatch*, September 21.

Arenschield, Laura. 2015b. "Feds Consider Fracking Plan in Wayne Forest." *Columbus Dispatch*, November 4.

Arenschield, Laura. 2016. "Ohio Has Yet to Write Rules for Fracking Industry." *Columbus Dispatch*, January 24.

Ashmoore, Olivia, Darrick Evensen, Chris Clarke, Jennifer Krakower, and Jeremy Simon. 2016. "Regional Newspaper Coverage of Shale Gas Development across Ohio, New York, and Pennsylvania: Similarities, Differences, and Lessons." *Energy Research & Social Science* 11: 119–132.

Bakker, Karen. 2000. "Privatizing Water, Producing Scarcity: The Yorkshire Drought of 1995." *Economic Geography* 76(1): 4–27.

Barth, Jannette M. 2013. "The Economic Impact of Shale Gas Development on State and Local Economies: Benefits, Costs, and Uncertainties." *New Solutions* 23(1): 85–101.

Bernd, Candice. 2015. "Local Ohio Officials Petition Governor for Return of Their Right to Protect Citizens from Fracking." *Truthout*, July 15. Retrieved May 27, 2016. (http://www.truth-out.org/news/item/31808-local-ohio-officials-petition-governor-for-return-of-their-right-to-protect-citizens-from-fracking).

Brenner, Neil, and Nik Theodore. 2001. "Neoliberalism and the Regulation of the 'Environment.'" Pp. 153–159 in *Neoliberal Environments: False Promises and Unnatural Consequences*, edited by N. Heynen, J. McCarthy, S. Prudham, and P. Robbins. New York: Routledge.

Bridge, Gavin, and Thomas Perreault. 2009. "Environmental Governance" In *A Companion to Environmental Geography,* edited by N. Castree, D. Demeritt, D. Liverman and B. Rhoads. Oxford, UK: Wiley-Blackwell.

Cardwell, Diane, and Clifford Krauss. 2013. "Trucking Industry Is Set to Expand Its Use of Natural Gas." *New York Times*, April 22.

Castree, Noel. 2010. "Neoliberalism and the Biophysical Environment A Synthesis and Evaluation of the Research." *Environment and Society: Advances in Research* 1: 5–45. June 17, 2017. (http://www.envirosociety.org/wp-content/uploads/2014/11/Neoliberalism-and-the-Biophysical-Featured-Article-02-ARES-5-45.pdf).

Cockerham, Sean. 2015. "Interior Department Sets Rules for Fracking on Federal Lands." *Tribune NewsService*, March 20.

Cocklin, Jamison. 2013. "Oil, Gas Case Heading to Ohio Top Court Will Spotlight Local Control." *The Vindicator*, September 20.

Davis-Cohen, Simon. 2015. "Government and Gas Industry Team Up Against Local Fracking Ban Initiatives in Ohio." *Earth Island Journal*, September 21. Retrieved May 27, 2016. (http://www.earthisland.org/journal/index.php/elist/eListRead/state_and_gas_industry_team_up_against_ohio_fracking_ban_initiatives/).

DeWitt, David. 2013a. "Group Files Petitions for November Anti-fracking Vote." *Athens News*, July 7.

DeWitt, David. 2013c. "Elections Board Mum on Reasoning for Zapping Fracking Ban from Ballot." *Athens News*, August 18

DeWitt, David. 2013d. "Group Wants Elections Board to Explain Killing of Frack Measure." *Athens News*, August 28.

DeWitt, David. 2013e. "Athens Won't Have Anti-frack Measure on Ballot This Year." *Athens News*, September 22.

DeWitt, David. 2013f. "Group Plans to Try Anti-fracking Ballot Measure Again Next Year." *Athens News*, October 9.

DeWitt, David. 2014a. "Local Anti-fracking Group comes back with 2nd Attempt at Ballot Measure." *Athens News*, January 5.

DeWitt, David. 2014b. "Protest Dismissed, But No Drilling in Wayne Anyway." *Athens News*, August 17.

Downing, Bob. 2016. "Ohio Severance Tax on Natural Gas, Oil Produced Windfall of $21.3 Million in 2014–2015, Total Likely to Top $30 Million in 2015–2016." *Columbus Dispatch*, January 29.

Eaton, Emily, and Abby Kinchy. 2016. "Quiet Voices in the Fracking Debate: Ambivalence, Nonmobilization, and Individual Action in Two Extractive Communities (Saskatchewan and Pennsylvania)." *Energy Research & Social Sciences* 20: 22–30.

Editorial Board. 2015. "Plenty of Room to Hike Frack Tax." *Columbus Dispatch*, November 1.

Finley, Bruce. 2016. "Colorado Supreme Court Rules State Law Trumps Local Bans on Fracking." *Denver Post*, May 2.

Gilmer, Ellen M. 2015. "Courtroom Slugfest Nears as Drilling Creeps toward Ancient Chaco World." Energywire. Retrieved June 17, 2017. (https://www.eenews.net/stories/1060021627).

Hartwick, Elaine, and Richard Peet. 2003. "Neoliberalism and Nature: The Case of the WTO." *Annals of the American Academy of Political and Social Science* 590, Rethinking Sustainable Development: 188–211. Retrieved May 27, 2016. (http://ann.sagepub.com/content/590/1/188.abstract).

Harvey, David. 2005. *A Brief History of Neoliberalism*. New York: Oxford University Press.

Harvey, David. 2007. Neoliberalism as Creative Destruction. Annals of the American Academy of Political and Social Science (610) NAFTA and Beyond: *Alternative Perspectives in the Study of Global Trade and Development*: 22–44.

Henry, Devin. 2015. "Oklahoma Blocks Local Fracking Bans." *The Hill*, June 1. Retrieved August 10, 2015. (http://thehill.com/policy/energy-environment/243645-oklahoma-blocks-local-fracking-bans).

Hunt, Spencer. 2012. "Feds Say 'Fracking' OK in Wayne National Forest." *Columbus Dispatch*, August 28.

Hunt, Spencer. 2013a. "State Forest Proposed as Fracking Site." *The Columbus Dispatch*, July 30.

Hunt, Spencer. 2013b. "State Forest Fracking Plan Halted, for Now." *The Columbus Dispatch* August 1.

Hunter, Spencer. 2013c. "Groups Blast State Forest Fracking Plan" *The Columbus Dispatch*, August 1.

James, Alexander, and David Aadland. 2011. "The Curse of Natural Resources: An Empirical Investigation of U.S. Counties." *Resource and Energy Economics* 33(2):440–

Johnson, Cedric. 2011. *The Neoliberal Deluge: Hurricane Katrina, Late Capitalism and the Remaking of New Orleans*. Minneapolis: University of Minnesota Press.

Korfmacher, Katrina Smith, Walter A. Jones, Samantha L. Malone, and Leon F. Vinci. 2013. "Public Health and High Volume Hydraulic Fracturing." *New Solutions* 23(1): 13–31.

Kovac, Marc. 2011. "Fracking Foes Fight to Block Drilling." *The Vindicator*, April 27.

Krupnick, Alan, Hal Gordon, and Sheila Olmstead. 2013. "Pathways to Dialogue: What the Experts Say about the Environmental Risks of Shale Gas Development." Pp. 1–81. Washington, D.C.: Resources for the Future. Retrieved May 27, 2016. (http://www.rff.org/files/sharepoint/Documents/RFF-Rpt-PathwaystoDialogue_Overview.pdf).

Ladd, Anthony E. 2014. "Environmental Disputes and Opportunity-threat Impacts Surrounding Natural Gas Fracking in Louisiana." *Social Currents* 1(3): 293–312.

Ladd, Anthony E. 2016. "Meet the New Boss, Same as the Old Boss: The Continuing Hegemony of Fossil Fuels and Hydraulic Fracking in the Third Carbon Era." *Humanity & Society* (http://dx.doi.org/10.1177/0160597616628908).

Ludlow, Randy. 2014. "Land Records Show Many State Parks Could See Mining, Drilling." *Columbus Dispatch*, October 19.

Ludlow, Randy. 2015a. "Local Governments Cannot Regulate Fracking, Ohio Supreme Court Rules." *Columbus Dispatch*, February 18.

Ludlow, Randy. 2015b. "Environmentalists Fear Launch of Strip Mining in Ohio State Wildlife Area." *Columbus Dispatch*, August 10.

Malin, Stephanie. 2014. "There's No Real Choice but to Sign: Neoliberalization and Normalization of Hydraulic Fracturing on Pennsylvania Farmland." *Journal of Environmental Studies and Science* 4: 17–27.

Malin, Stephanie A., and Kathryn Teigen DeMaster. 2016. "A Devil's Bargain: Rural Environmental Injustices and Hydraulic Fracturing on Pennsylvania's Farms." *Journal of Rural Studies* 47: 278–290.

Mercer, Alexandra, Kim de Rijke, and Wolfram Dressler. 2014. "Silences in the Boom Coal Seam Gas, Neoliberalizing Discourse, and the Future of Regional Australia." *Journal of Political Ecology* 21: 222–348.

Metcalfe, John. 2008. "On the Web, Drillers Want Their Say, and Call It News." *New York Times*, September 14. Retrieved June 17, 2017. (http://www.nytimes.com/2008/09/15/business/media/15shale.html?mcubz=1)

Neary, Ben. 2016. "The Future of Federal Rules Aimed at Protecting Land, Water and Wildlife from Energy-Production Practices Including Hydraulic Fracturing Now Rests with a Judge in Wyoming." *Associated Press*, April 20.

Negro, Sorell E. 2012. "Fracking Wars: Federal, State and Local Conflicts over the Regulation of Natural Gas Activities." *Zoning and Planning Law Report*, February 35(2): 1–16. Retrieved May 23, 2016. (http://www.ourenergypolicy.org/wp-content/uploads/2013/07/Fracking-Wars.pdf).

Ohio Department of Natural Resources. 2016. "Ohio's Utica Shale Second Quarter Production Totals Released." Oil & Gas Division, September 24. Retrieved September 28, 2016. (oilandgas.ohiodnr.gov/ . . . /ohios-utica-shale-second-quarter-production-totals-released).

Papyrakis, E., and R. Gerlagh. 2007. "Resource Abundance and Economic Growth in the U.S." *European Economic Review* 5(4): 1011–1039.

Peck, J., and A. Tickell. 2002. "Neoliberalising Space." *Antipode* 34: 380–404.

Perrow, Charles. 2015. "Cracks in the 'Regulatory State.'" *Social Currents* 2(3): 203–212.

Phillips, Ari. 2014. "Ohio Supreme Court: It's OK to Strip Mine State Wildlife Areas." *Athens News*, September 19.

Phillips, Jim. 2013. "Protest Filed against Anti-fracking Ballot Measure." *Athens News*, August 1.

Rowland, Darrel. 2014a. "State Made Plan to Promote Fracking While Regulating It." *Columbus Dispatch*, February 15.

Rowland, Darrel. 2014b. "Kasich Aides Knew of Plan for Fracking in State Parks." *Columbus Dispatch*, February 18.

Rowland, Darrel. 2014c. "Fitzgerald Calls for Probe of Fracking Plan." *Columbus Dispatch*, February 27.

Rowland, Darrel, and Will Drabold. 2014. "Kasich Reverses on Fracking in State Parks." *Columbus Dispatch*, February 19.

Rule, J. B., and James Salzman. 2006. "Ecosystem Services and the Public Trust Doctrine: Working Change from Within." *Southeastern Environmental Law Journal* 15(1): 223–239.

Sandrick, Bob. 2014. "Two Drilling Companies Sue Broadview Heights over Ban on Oil and Gas Wells." Cleveland.com, July 17. Retrieved May 27, 2016. (http://www.cleveland.com /broadview-heights/index.ssf/2014/07/two_drilling_companies_sue_bro.html).

Sandrick, Bob. 2015. "Judge Shoots Down Broadview Heights Ban on Future Oil and Gas Wells." Cleveland.com, March 16. Retrieved May 27, 2016. (http://www.cleveland.com /broadview-heights/index.ssf/2015/03/judge_shoots_down_broadview_he.html).

Siegel, Jim. 2015. "Fracking Tax Won't Be Part of Final Budget Deal, Ohio House Insists." *Columbus Dispatch*, June 11.

Skolnick, David. 2014. "Voters Loudly Said No for 4th Time to Ban Fracking in Youngstown." *The Vindicator*, November 2014.

Skolnick, David. 2015. "Ohio Supreme Court Orders Frack Ban onto Nov. Ballot." *The Vindicator*, September 18.

Speakman, Burton. 2013. "Activists Seek to Put Anti-fracking Charter Amendment on Ballot." *Vindy.com*, July 20.

Steinzor, Nadia, Wilma Subra, and Lisa Sumi. 2013. "Investigating Links between Shale Gas Development and Health Impacts Through a Community Survey Project in Pennsylvania." *New Solutions* 23(1): 55–83.

Tierney, Kathleen. 2015. "Resilience and the Neoliberal Project: Discourses, Critiques, Practices—And Katrina." *American Behavioral Scientist* 59(10): 1327–1342.

United States Energy Information Administration. 2016. "Annual Energy Outlook. Early Release: Annotated Summary of Two Cases." Retrieved May 17, 2016. http://www.eia .gov/forecasts/aeo/er/pdf/0383er(2016).pdf. May 25, 2016).

Vardon, Joe, and Jim Siegel. 2013. "Tax Proposals Central to Kasich Budget." *The Columbus Dispatch*, February 5.

Vindy.com. 2013a. "Anti-fracking Issue on May 7 Ballot Triggers Controversy: Can It Be Enforced?" April 28.

Vindy.com. 2013b. "Anti-fracking Amendment Gets Enough Signatures to Be on Ballot Again." August, 12.

Vindy.com. 2015. "Group Seeks to Keep Anti-fracking Item Off Ballot." July 8.

Wang, Zhongmin, and Alan Krupnick. 2013. "A Retrospective Review of Shale Gas Development in the United States: What Led to the Boom?" *Resources for the Future* Discussion Paper 13–12. Retrieved June 17, 2017. (http://www.rff.org/files/sharepoint/WorkImages /Download/RFF-DP-13-12.pdf).

Williams, Geoff. 2015. "Coal Country Unruffled by Strip Mining in Flourishing Egypt Valley." *Aljazeera America*, October 26. Retrieved May 27, 2016. (http://america.aljazeera .com/articles/2015/10/26/coal-country-strip-mining-egypt-valley.html).

Willow, Anna. Rebecca Zak, Danielle Vilaplana, and David Sheeley. 2014. "The Contested Landscape of Unconventional Energy Development: A Report from Ohio's Shale Gas Country." *Journal of Environmental Studies Science* 4(1): 56–64.

2

This (Gas) Land Is Your (Truth) Land?

Documentary Films and
Cultural Fracturing in
Prominent Shale Communities

ION BOGDAN VASI

Introduction

Human civilizations have often measured their achievements in terms of the sheer size of their construction projects, and the ultimate proof of size is visibility from outer space. So far, few human-made structures have achieved this milestone. During the day, only a few bridges and other gargantuan structures are visible from miles above the earth, but during the night, light pollution makes it possible to see very large cities from distances far beyond the planet's orbit. Recently, however, astronauts could see evidence of human activity even in places where very few people live, for example, in the great plains of North Dakota. This is because hydraulic fracturing—popularly known as "fracking"—has created an oil and gas boom so large that almost 30% of the natural gas being generated at drilling sites in North Dakota has to be flared off. As Krulwich (2013) notes, "There are now so many gas wells burning fires in the North Dakota night, the fracking fields can be seen from deep space. [This] is a sign that this region is now on fire . . . to a disturbing degree. Literally."

It is undeniable that hydraulic fracturing has been a major achievement of the oil and gas industry. In 2000, fracked natural gas represented about 1% of the natural gas supply in the United States; by 2013 it represented 40% of total natural gas production and has surpassed production from conventional natural gas wells (Energy Information Administration 2014). Some energy professionals have argued that the shale gas "energy revolution" has transformed the energy marketplace, created new jobs, brought down the cost of natural gas, and made energy independence a topic that is "back in vogue, not as a joke but as a serious topic of political discussion" (Yergin 2013). Others, however, are less sanguine; opponents of fracking have argued that the shale gas energy revolution has potentially dangerous consequences for human and animal life because it pollutes the environment. Echoing Franz Kafka's warning that "every revolution evaporates and leaves behind only the slime (of a new bureaucracy)," environmentalists have argued that this energy revolution will soon evaporate and leave behind a virtual slime of toxic chemicals, polluted water, and fractured landscapes.

The debate over hydraulic fracking has become highly contentious in recent years (Boudet 2011; Boudet et al. 2014; Boudet et al. 2016; Ladd 2013, 2014; Malin 2014; Malin and DeMaster 2016; Vasi et al. 2015). As different chapters of this book document, activists in communities throughout the United State have used tactics such as protests, lawsuits, public gatherings, and citizen monitoring of oil and gas wells to raise awareness about the potential dangers associated with fracking and to prevent oil and gas companies from drilling close to homes, schools, churches, farms, and aquifers. Feeling threatened by activists, the industry has launched a counteroffensive: a well-funded campaign to convince American citizens that hydraulic fracturing is safe and beneficial for local communities. For example, America's Natural Gas Alliance, an industry association that "works with industry, government and customer stakeholders to promote increased demand for and continued availability of our nation's abundant natural gas resource for a cleaner and more secure energy future," has inundated National Public Radio (NPR) and other mass media with its "Think About It" campaign (America's Natural Gas Alliance 2013).

This chapter examines how activists and industry groups in various communities located primarily in the Marcellus, Barnett, and Woodford Shale regions have used cultural products like documentary films to influence public opinion about hydraulic fracturing. The chapter begins with a brief overview of the science and politics of hydraulic fracturing, and then describes how a documentary (*Gasland*) has sent shock waves through the oil and gas industry in the United States. Next, it examines how the industry responded to this perceived threat by sponsoring its own documentary (*Truthland*) and by supporting screenings of another documentary (*FrackNation*). The chapter describes not only the groups that organized public screenings for these documentaries in key states and shale regions across the country, but also how the general public and mass media in different states responded to the films. The conclusion discusses the implications of these findings for theory and research on social movements, mass media, and civil society.

Hydraulic Fracturing: Science, Politics, and Activism

The shale gas revolution has been made possible by technological innovations in oil and gas extraction, as well as key acts of political legislation. While rudimentary forms of hydraulic fracturing have been used in the United States to extract oil and natural gas from wells since the 1940s, the essential elements of the fracking process used today did not emerge until the end of the 1990s and were not economically viable for the industry until the early 2000s (see Ladd, Introduction, this volume). Previously, hydraulic fracturing had been used in some conventional wells involving vertical drilling, but could penetrate only a few tens or hundreds of feet of reservoir rock. During the 1990s, technological improvements paved the way for the introduction of horizontal drilling methods to augment the fracturing process, where operators could drill laterally or horizontally for up to two miles within a shale formation to locate oil or gas deposits that were previously inaccessible. Consequently, compared to vertical drilling, horizontal drilling decreased the total amount of land needed for the drilling platform, but significantly increased the volume of water, chemicals, and sand used to extract oil or gas. The combination of hydraulic fracturing and horizontal drilling made possible the recovery of natural gas deposits in geological formations where it was locked so tightly in the rock that it was technologically and economically difficult to produce. As a result of these technological innovations, among others, unconventional oil and gas development has created energy boomtowns in areas like the Marcellus Shale in the eastern United States, the Barnett Shale in Texas, the Haynesville Shale in Louisiana, the Woodford Shale in Oklahoma, and the Bakken Shale formations in North Dakota and Montana.

The second process that has transformed the energy sector has been political. President George W. Bush followed up on his campaign promise to reduce America's dependency on foreign energy sources by appointing "a cadre of political and policy loyalists to leadership positions within the federal administrative units charged with the management and oversight of domestic energy development" (Forbis 2014: 154). The Bush administration used a three-pronged strategy to expand domestic energy development: use executive power to shift political leadership to administrative agencies; charge them with implementing executive policy directives to change energy policies; and promote the use of hydraulic fracturing. Following the recommendations of the National Energy Policy Development Group, which was chaired by Vice

President Dick Cheney and included only members of the newly elected administration and U.S. energy producing companies, President Bush released executive orders EO 13211 and EO 13212 in 2001. The executive orders charged federal agencies to expedite the means to develop America's domestic energy resources by altering the procedural processes for leasing public lands and speeding the process of approving permits to drill using hydraulic fracturing (Forbis 2014). Moreover, as Forbis (2014: 170) notes, "Republican dominance of congressional committees helped the Bush-Cheney administration achieve the objective of expanding the use of hydraulic fracturing to facilitate increased domestic energy resource development." However, oil and gas fracking never would have become commercially viable had it not been for the increased pipelines, export facilities, favorable energy legislation, tax subsidies, rising energy prices, and other forms of support provided by both the Bush and Obama administrations (Ladd 2014).

The widespread use of hydraulic fracturing has led to a dramatic increase in drilling activities in the United States during the last 25 years. The number of gas and gas condensate wells, for example, increased from 269,790 to 514,786 between 1990 and 2014. However, the drilling activity is concentrated in a number of states with large natural gas deposits, including Texas, Louisiana, Pennsylvania, Kansas, Colorado, Wyoming, and a few others (see Figure 2.1).

The local environmental impacts of increasing drilling activities have been compounded by the growing size of fracking operations. Typical fracking activities during the early 1950s used 750 gallons of fluid (water, mixed with crude oil or gelled kerosene) and 400 pounds of sand; current activities can use up to 7 million gallons of water and 75,000 to 320,000 pounds of sand. Moreover, the hydraulic horsepower (hhp) needed to pump fracking material has increased from an average of about 75 hhp in the early days to an average of more than 1,500 hhp today; big operations require more than 10,000 hhp (MacRae 2012). Given these facts, a growing number of communities around the country have begun to protest against water pollution, the destruction of forests and fields, construction activities involving miles of pipeline, and the truck traffic and compressor stations that burn massive quantities of polluting diesel fuel.

The environmental risks of fracking have been vigorously debated during the last decade (Osborn et al. 2011; Wilber 2015). To date, industry proponents argue that no conclusive research irrefutably proves that hydraulic fracturing is dangerous to human health (but see, for example, Bamberger and Oswald 2014; Concerned Health Professionals of New York et al. 2015). Natural gas industry groups and supporters have argued that the environmental risks of hydraulic fracturing, and in particular the risk of water contamination, are insignificant. The Independent Oil and Gas Association of New York has argued that the chemicals used in fracking are safe and that they are made of

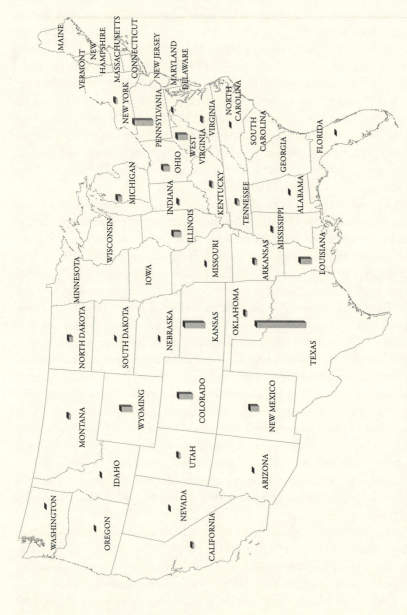

FIGURE 2.1 Total number of oil and gas wells by state in the United States in 2014.
Data source: FrackTracker. See: http://www.fractracker.org/2014/03/active-gas-and-oil-wells-in-us/

"a small amount of diluted, benign additives found in common household products" (Wilber 2015: 117). Moreover, Energy in Depth and other industry groups have tracked numerous studies of health impacts of fracking and systematically attempted to persuade the public that there is no cause for concern. For example, when researchers from the University of Pennsylvania and Columbia University released a study asserting a correlation between natural gas development and an increase in hospitalizations in three Pennsylvania counties, Energy in Depth argued that "the researchers' data show hospitalization rates remaining stable, with high oil and gas developing counties having the lowest hospitalization rates, while the county with no wells had the zip code with one of the highest hospitalization rates."

Naturally, the movement against fracking has been suspicious of industry reports, in particular those that downplayed hydraulic fracturing's impacts on human health and groundwater resources. Hydraulic fracturing has been described as a case of "the health of many versus the wealth of a few," while the opposition to it has been labeled "a movement of the 99 percent—farmers, physicists, journalists, teachers, librarians, innkeepers, brewery owners and engineers" (Cantarow 2012). Environmental activists claim the industry is misleading the public by downplaying the risk of water contamination and other environmental problems, and some civil society groups have been monitoring water quality for evidence of harm (Jalbert et al. 2014). Other environmental groups have claimed that there are almost as many violations as there are wells, with thousands of incidents of spills, contamination, events, and blowouts taking place across the landscape. Given the scientific uncertainties about the health impacts of hydraulic fracturing, the tug-of-war between activists and the oil and gas industry over the environmental consequences of hydraulic fracturing is likely to continue for the foreseeable future. Therefore, it is important to examine how both the antifracking movement and the industry have attempted to persuade the public and policy makers that they are right and their opponents are wrong.

Movements, Industries, and Documentaries

Like most social movements, the antifracking movement has spent considerable resources trying to influence both public perceptions and policy making. Social movement scholars have shown that activists create specific collective action frames that resonate with potential activists and connect them to broader themes in public discourse that can recruit members, strengthen collective identities, and shape policy making (Benford and Snow 2000; Bail 2012; Cress and Snow 2000; McCammon et al. 2007; McCammon et al. 2008). Researchers have argued that contentious collective actors are subject to features of the broader public discourse (Steinberg 1999). In order to have policy effectiveness, social movement frames must not only resonate with potential activists, but also connect with broader themes in public discourse. When movements' ideas are congruent with broader public narratives, activists benefit from "discursive opportunities," or moments when movement ideas align with what the culture at large deems to be sensible, realistic, or legitimate (Koopmans and Statham 1999; McCammon et al. 2007).

Social movements frequently use cultural artifacts to influence public debates and capitalize on discursive opportunities. A number of studies have examined the impact of social movements on the broader cultural context (Berezin 1994; Eyerman and Jamison 1995; Hanson 2008; Lipsitz 2000; McAdam 1994; Steinberg 2004; Zolberg 1997). Research has shown, for example, that abolitionists introduced African American spirituals into white mainstream culture (Cruz 1999), that leftist artists had a major influence on American painting, sculpture, literature, and theater (Hemingway 2002), and that the women's movement of the 1960s and 1970s shaped American poetry (Reed 2005). Studies have also shown that movement participants introduce new ideas through music (Danaher 2010; Eyerman and Jamison 1995; Roscigno and Danaher 2001; Roy 2010) or books (Meyer and Rohlinger 2012). What these accounts often miss, however, is an analysis of how social movements "do culture" and its consequences. This constitutes a relatively neglected issue in the literature because movements that bring people together through poetry reading, singing, and other collective activities can solidify commitment and mobilize actors toward achieving goals (Roy 2010).

The documentary film is one of the most important artifacts used by contemporary social movements. Social scientists have suggested that while films can have substantial effects on public perceptions, the empirical evidence for these arguments is somewhat sparse. Lynch (2012) argued that the documentary *Lovejoy's Nuclear War* acted as a catalyst for the antinuclear protests in the United States during the late 1970s. Whiteman (2003) argued that the 1992 documentary *From the Ground Up*, which examined the environmental impacts of mining in Wisconsin, fostered local opposition against mining in Wisconsin through video screenings at bars, county fairs, churches, and sporting group meetings. Leiserowitz (2004) has shown that individuals who watched *The Day After Tomorrow*, a fictional account of future impacts of climate change, were more likely to perceive global warming as a threat and to be willing to act to address this threat. Jacobsen (2011) has shown that *An Inconvenient Truth*, a documentary in which former vice president Al Gore explained the causes of climate change and its consequences, influenced individuals' decisions to purchase carbon offsets.

Documentaries are important tools for activists because of their potential to overcome quiescence, or the absence of grievance perception, in the face of significant inequality. Studies have shown that while individual grievances about environmental problems may exist in communities where polluting industries are located, these grievances are rarely expressed publicly (Crenson 1971; Gaventa 1982; Roscigno 2011). Indeed, scholars have argued that often residents "endure years of uncertainty in the face of contradictory evidence on the health and environmental risks posed by contaminants" (Aronoff and Gunter 1994: 243). The general public has developed ambiguous and conflicting interpretations of situations because scientists and officials are seldom able to provide definitive answers about risks (Auyero and Swistun 2008; Brown et al. 2000; Brown and Mikkelsen 1990; Elliott and Frickel 2013; Fowlkes and Miller 1982, 1987; Kroll-Smith and Couch 1993; Kroll-Smith et al. 2002). When information is made available, the general public may not understand it because it may be presented using scientific jargon and technical language that is not accessible to many citizens (Freudenburg and Jones 1991). Moreover, as Cable (2012) points out, citizens may distrust information presented by experts because these experts are often embedded in a political system that is more accountable to elites than to the general public. Therefore, documentaries created by artists, activists, or "nonexperts" may present facts in a more accessible manner and may overcome the problem of trust in experts.

The antifracking movement has used documentaries to influence public perceptions and effect change from an early stage. In 2006, Debra Anderson, a film editor and producer whose work has appeared on PBS, National Geographic, and Discovery, began to shoot a documentary about the effects of fracking on human health titled *Split Estate*. Anderson had the idea for this documentary after reading a story in *On Earth Magazine* about a woman from Colorado who developed a tumor after oil and gas companies began drilling in her neighborhood. *Split Estate* premiered in October 2009 on Planet Green, a network of Discovery Communications, and was distributed through Bullfrog Films, the leading U.S. distributor of environmental films. Yet, despite being distributed through a major network for environmental films (or perhaps because of it since the audience for environmental films is rather narrow), this documentary did not reach a wide audience.

The most important documentary of the antifracking movement is *Gasland*, a documentary conceived, directed, and narrated by filmmaker Josh Fox (Fox 2010). The documentary was made in a short time (about eighteen months) by Fox, who began the project as a one-person crew and was later joined by three cameramen. The film begins with Fox receiving a letter from a natural gas company that offered to lease his family's land in Pennsylvania for $100,000 to drill for natural gas. He sets out to see how the process of hydraulic fracturing--or fracking—has impacted communities around the country. Fox interviews citizens who talk about their health problems and attribute them to the contaminated air and water due to natural gas drilling in their communities. He also interviews scientists, politicians, and oil and gas industry representatives and highlights the fact that hydraulic fracturing was exempted from the Safe Drinking Water Act in the Energy Policy Act of 2005 (known as the "Haliburton Loophole").

Gasland has had an extraordinary success in raising awareness about the environmental risks associated with hydraulic fracturing. This is primarily because, in a number of scenes, Josh Fox visits residents living near natural gas wells and witnesses them light their tap water on fire. He attributes this to methane contamination and suggests that this is a major concern for communities affected by hydraulic fracturing. For many viewers, the images of water catching fire evoked the environmental disasters of the 1960s, when the Cuyahoga River in Ohio was so polluted that it caught fire. Like the way in which the burning river energized previous generations of environmental activists, the burning tap water from *Gasland* mobilized numerous activists who have worked to oppose hydraulic fracturing. Indeed, film critics immediately recognized that this documentary could become a powerful mobilizing tool: one critic argued that "*Gasland* may become to the dangers of natural gas drilling what *Silent Spring* was to DDT." Another critic wrote, "*Gasland* is the paragon of first person activist filmmaking done right. . . . By grounding a massive environmental issue in its personal ramifications, Fox turns *Gasland* into a remarkably urgent diary of national concerns" (Variety 2010; IndieWire 2010).

Research shows that *Gasland* had a number of important effects (Vasi et al. 2015). First, it created a discursive opportunity for the antifracking movement by increasing online searches, as well as social media chatter and mass media coverage, about this unconventional method of drilling and extraction. For example, *Gasland* influenced the social media discussion about fracking by focusing attention on problems related to water pollution and, to a lesser degree, to local bans and moratoriums. Second, the documentary contributed to antifracking mobilizations, which, in turn, affected the passage of local fracking moratoriums in Marcellus Shale cities and states. Thus, *Gasland* opened up new discursive opportunities for the movement, while those opportunities supported antifracking mobilization in local communities in the Marcellus Shale region. Because communities with more mobilizations were ultimately more likely to ban fracking, researchers have concluded that the documentary had an indirect effect on the passage of local bans in many communities (see Vasi et al. 2015).

Given its widespread influence, *Gasland* was attacked by the oil and gas industry soon after it was released. The Independent Petroleum Association of America (IPAA) used its Energy in Depth (EID) campaign to "debunk" *Gasland*. The EID website attacked Fox's credentials, noting that he is only "an avant-garde filmmaker and stage director whose previous work has been recognized by the 'Fringe Festival' of New York City." EID listed a number of "the most egregious inaccuracies upon which the film is based"; according to EID, *Gasland* misstated the law, misrepresented the rules, mischaracterized the process, flat-out made stuff up, and recycled discredited points from the past (Energy in Depth 2011).

But the IPAA did not stop at criticizing *Gasland*; it also produced its own documentary called *Truthland: Dispatches from the Real Gasland* (Energy in Depth 2012), which attempted to convince the public that hydraulic fracturing is safe for humans and the environment. As the title implies, *Truthland* is a direct attack on *Gasland*. Jeff Eshelman, vice president of public affairs at the Independent Petroleum Association of America, stated, "This isn't the first time something has been released that sets the record straight on the mountain of misinformation in '*Gasland*.' But it is the first time that these facts have been transmitted in such vivid detail through such a compelling medium" (Energy in Depth 2013). The documentary follows Shelly, a teacher from rural northeast Pennsylvania, who lives with her family on a farm. Having established that Shelly is a well-intentioned mother and credible person, the documentary then presents how she learns various facts that are intended to correct "some of the most pervasive myths that have come to surround the debate over fracturing" (Truthland 2013). For example, she learns that fracking is not a drilling technique but a technology that's used to enhance the flow of energy from a well once the drilling is done, that the fracking fluids are neither a secret nor hazardous, that fracking does not cause water to catch fire, that there are no loopholes in the legislation regulating oil and gas companies, that there is no proof that fracking causes earthquakes, and that natural gas from shale is cleaner than coal. *Truthland*, however, had relatively little influence on the general public, due in part to the fact that the film was funded by the oil and gas industry and thus was not deemed to have any environmental or scientific veracity.

Another documentary that attacked *Gasland* is *FrackNation: A Journalist's Search for the Fracking Truth* (McAleer 2012). The film is directed by Phelim McAleer and Ann McElhinney, two investigative journalists who previously produced two documentaries that attacked environmentalists: *Mine Your Own Business* and *Not Evil, Just Wrong*. According to the film's website, "In *FrackNation* journalist Phelim McAleer faces threats, cops and bogus lawsuits questioning green extremists for the truth about fracking. McAleer uncovers fracking facts suppressed by environmental activists, and he talks with rural Americans whose livelihoods are at risk if fracking is banned. Emotions run high but the truth runs deep." The film emphasizes that it was funded independently "by 3,305 backers on Kickstarter who generously donated $212,265 to have us investigate the truth about fracking" (FrackNation 2012). *FrackNation* was shown in communities around the United States, as well as Europe, but also failed to attract much public attention.

Screenings and Public and Mass Media Reception of Documentaries

While citizens and journalists have used films to document both the dangers and benefits of shale fracking, activists and industry representatives have used these documentaries as ammunition to battle each other. Screenings of *Gasland* have been organized in many communities around the country, in small independent cinemas, libraries, universities, and public spaces. Between July 2010 and March 2013, approximately 330 screenings took place, and most of those screenings took place in the Marcellus Shale states, particularly New York and Pennsylvania (see Figure 2.2). Screenings of *Gasland* were organized mostly by grassroots groups, as well as by regional and national organizations involved in the coalition Americans against Fracking: Marcellus Protest, New Yorkers Against Fracking, Catskill Citizens For Safe Energy, Clean Water Action, Food and Water Watch, Energy Action Coalition, Rainforest Action Network, and others. Many of the screenings had audiences composed of dozens of people, while some had audiences of hundreds; for example, an outdoor screening in Pittsburgh in August 2010 was estimated to have attracted as many as 600 people (Rooftop Films 2010).

Screenings of *Truthland* and *FrackNation*, on the other hand, were organized in states located in the Marcellus Shale region (Ohio, Pennsylvania, New York), as well as in states located in the Barnett and Eagle Ford shales (Texas), the Bakken Shale (Montana), and the Niobrara Shale regions (Colorado) (see Figure 2.2). The total number of screenings for those two documentaries is lower than the number of screenings for *Gasland*, since between May 2013 and July 2015 there were approximately 271 screenings. Moreover, very different groups organized the screenings. For example, *Truthland* screenings were organized mainly by oil and gas associations from different states (Indiana, Arkansas, Kansas, Louisiana), while *FrackNation* screenings were hosted mainly by local chapters of organizations associated with the Tea Party movement (Americans for Prosperity, The 9/12 Project, Tea Party Patriots), by oil and gas associations from various states, and by groups associated with the Republican Party (North Texas College Republicans, Southeast Republican Club, Metropolitan Republican Club).

Public responses to the respective claims of the documentaries can be measured in different ways. Traditionally, a film's impact has been measured using data about box office profits or audience reviews. *Gasland* has made $30,846 as lifetime gross in theaters and has received a positive review of 97% on the film review website Rotten Tomatoes. In contrast, *FrackNation* has received a 50% positive review on Rotten Tomatoes and *Truthland* has not been reviewed at all; in addition, neither *FrackNation* nor *Truthland* have released any data on profits to date. Nowadays, the impact of these documentaries can

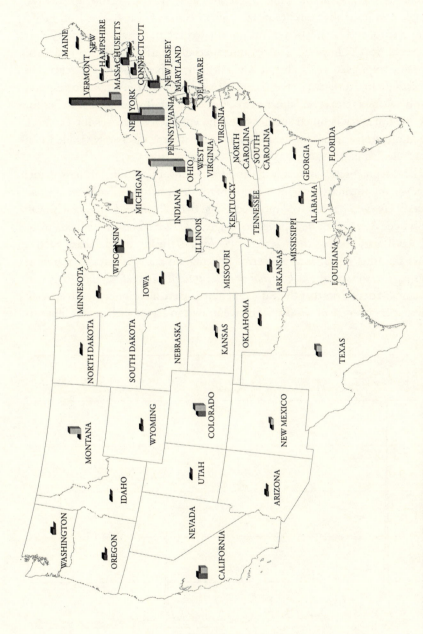

FIGURE 2.2 Screenings of *Gasland* (dark gray) and *Truthland* and *FrackNation* (light gray) in the United States.
Data source: http://www.truthlandmovie.com/category/screenings/

also be measured using data about Internet and YouTube searches. The number of Google searches for *Gasland* is much larger than the number of Google searches for *FrackNation*, while *Truthland* had very few Google searches. Google and YouTube searches peaked when *Gasland* was released on HBO in June 2010, and later, when it was nominated for an Oscar in February 2011 (Vasi et al. 2015). However, Google, and especially YouTube, searches remained relatively high even years after it was released, in part due to the release of *Gasland II* in April 2013 (Fox 2013). Searches for *FrackNation* correspond mainly with its release on January 2013, but the overall volume of searches for this documentary is much smaller than the volume of searches for *Gasland*. The number of searches for *Truthland* is so low that it was barely visible when it was released in June 2012, as shown in Figures 2.3 and 2.4).

The documentaries have also received a significant amount of mass media coverage, but the mass media's reception of the documentaries has been uneven. Overall, *Gasland* has received more coverage than *Truthland* and *FrackNation* combined. A search of newspaper coverage of *Gasland* between January 2010 and April 2013 (before the release of *Gasland II*) using the database Factiva shows that this documentary received coverage in 2,473 articles. A search of newspaper coverage of *FrackNation* and *Truthland* between January 2013 and December 2015 shows that these documentaries were mentioned in 170 articles. Newspaper coverage of *Gasland* is concentrated in regions where the film was

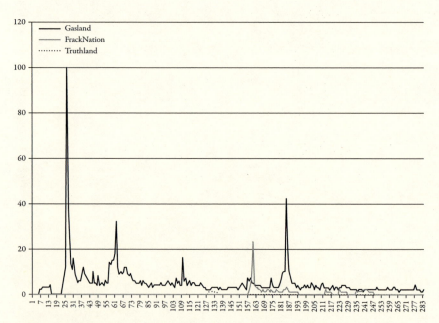

FIGURE 2.3 Weekly Google searches for *Gasland*, *FrackNation*, and *Truthland* between January 2010 and July 2015.

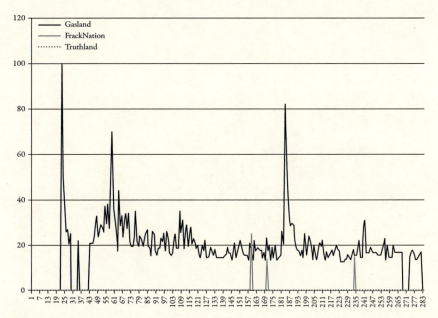

FIGURE 2.4 Weekly YouTube searches for *Gasland*, *FrackNation*, and *Truthland* between January 2010 and July 2015.

screened most often, for example in the Marcellus Shale region (New York, Pennsylvania, Ohio), but states such as California, Texas, and Colorado also had high levels of coverage. Newspaper coverage of *FrackNation* and *Truthland* was also higher in the Marcellus region than in other parts of the country, but the amount of mass media coverage of these documentaries was significantly lower than the amount of coverage of *Gasland*, even in states (like Ohio) with many screenings of *FrackNation* and *Truthland* (see Figure 2.5).

Reel Power: Documentaries and Fracking Policies in the Marcellus, Barnett, and Woodford Shale Regions

The movement against fracking has grown rapidly in the United States and around the world. One outcome of this movement has been the adoption of local and state bans against fracking. A total of 492 towns, cities, and counties had adopted antifracking ordinances in the United States by the beginning of 2016 (Food and Water Watch 2016). Additionally, five states (Connecticut, Maryland, New York, New Jersey, and Vermont) have adopted various antifracking bans or moratoriums. At the same time, as a response to this growing movement, at least four states (Texas, Oklahoma, Colorado, and Ohio) have overruled zoning laws and other restrictions in towns and cities that interfered with oil and gas drilling and adopted state legislation that prohibits local

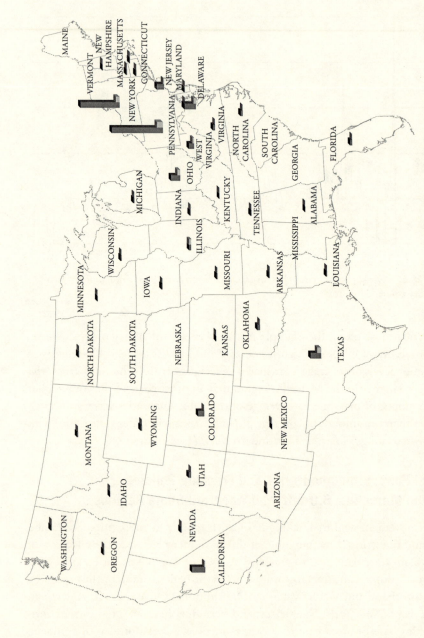

FIGURE 2.5 Mass media coverage of *Gasland* (dark gray) and *Truthland* and *FrackNation* (light gray) in the United States.
Data source: LexisNexis Academic

community bans against fracking (see Figure 2.6). These states' court actions have reenergized the debate about the nature of democratic decisions in America; as one mayor whose municipality adopted a ban that was overridden by a state court declared: "I guess that means the voice of the people doesn't matter. We said 'we want local control,' and then they take it away" (Fulton 2015).

The cases of the Marcellus, Barnett, and Woodford Shale regions are illustrative of the dynamic between the antifracking movement and the oil and gas industry at local and state levels. In the Marcellus Shale region, the most important conflicts between antifracking activists and the industry took place in New York. In this state, the antifracking movement created a broad coalition of "fractivists"—activists concerned about the environmental impacts of hydraulic fracturing on their communities. The main umbrella organization was New Yorkers Against Fracking (NYAF), a coalition that included more than 250 organizations including civic groups, businesses, faith groups, professional associations, labor unions, environmental organizations, social justice groups, and food producers. NYAF acted as a megaphone that "amplified a base of anti-fracking voices that included local groups formed to fight fracking in their communities" (*In These Times* 2015). The organization served an important number of purposes: it provided organizers who mobilized people for action (for example, providing bus trips to the state's capital for protests and demonstrations); it created broad networks of health professionals, businesses, local elected officials, and faith leaders who were able to highlight fracking's impacts on health and the economy; and it enabled collaboration and coordination of tactics and messaging among the different strands of the movement that spanned the state. NYAF also supported screenings of *Gasland* in various public spaces such as small cinemas, outdoor gardens, public libraries, and churches. This was because the main theme in *Gasland*—that fracking poses serious threats to human health due to the possibility of water pollution—had been at the core of NYAF's strategy. As one activist put it, this strategy was based on the realization that "It's really the water that draws people in" (*In These Times* 2015). Consequently, New York not only had the largest number of *Gasland* screenings (more than 80), but also the largest number of local bans (more than 215). In turn, the state adopted a statewide ban against fracking in 2015.

Another important case is that of the Barnett Shale, which is located in northern Texas, a state that has large oil and gas deposits and the largest number of oil and gas wells in the country. In contrast to New York, Texas had few screenings of *Gasland*, only four towns and cities have to date adopted antifracking bans, and screenings of *Truthland* and *FrackNation* were more numerous than screenings of *Gasland*. One of the antifracking bans was adopted in November 2014 in Denton, a city with approximately 120,000 people located north of Dallas. As one activist stated: "We didn't start out wanting a ban.

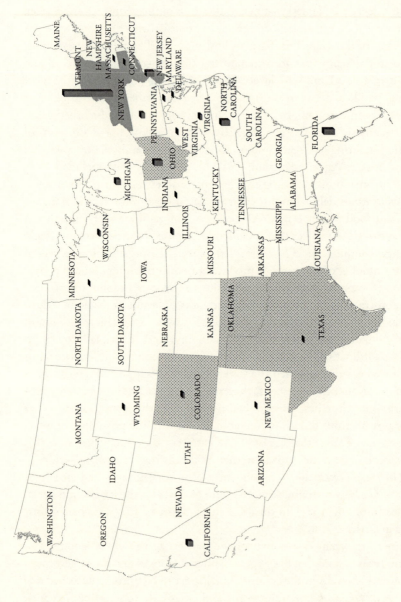

FIGURE 2.6 The adoption of local (city and county) bans against fracking (black bars), state bans against fracking (gray area), and state bans against local bans against fracking (checkered area).

Data collected by the author from Food & Water Watch, and from municipalities' websites.

We're Texans—we are obviously used to drilling and seeing rigs around town. For several years, we tried to work with the industry, but it was creeping into densely populated areas, and they were unwilling to compromise. That's when we voted for the ban" (*Newsweek* 2015). However, given the tremendous influence of the oil and gas industry in the state's economy and politics, Denton's ban against fracking was perceived as an unacceptable challenge by oil and gas companies. These companies lobbied for legislation that would prohibit cities and towns from passing ordinances to ban fracking and regulate oil and gas activity. The anti-anti-fracking legislation was adopted in May 2015 and was a major win for the oil and gas industry. Indeed, Governor Greg Abbott claimed that this law was "a defense of 'private property rights' and a move to limit government bureaucracy and overregulation" (*Newsweek* 2015).

Yet another important case is that of the Woodford Shale, which is located in Oklahoma. Like Texas, Oklahoma has a long history of oil and gas drilling and an economy bolstered by energy-related services. It is estimated, for example, that approximately one-quarter of all jobs in Oklahoma are tied to the energy industry, either directly or indirectly (StateImpact NPR 2015). The statewide hydraulic fracturing boom has created huge amounts of wastewater that is injected deep underground; the practice of injecting wastewater in deep wells is supposed to prevent pollution of freshwater, which is closer to the surface. Yet, the practice is also suspected of contributing to the big increase in the number of seismic events and earthquakes that the state has experienced in recent years, from three earthquakes at or above a magnitude of 3 to 807 earthquakes in 2015 (see Mix and Raynes, this volume). While Oklahoma had very few screenings of *Gasland* and very few communities that mobilized against hydraulic fracturing, public opposition against fracking has recently increased due to the growing seismic activity and mounting scientific evidence about the connection between disposal wells and quakes (StateImpact NPR 2015). In response, Oklahoma governor Mary Fallin signed a bill that banned local ordinances against hydraulic fracturing or other oil and gas drilling operations. According to Governor Fallin, this legislation was necessary to protect the oil and gas industry, the largest employer and taxpayer in the state: "Corporation Commissioners are elected by the people of Oklahoma to regulate the oil and gas industry. They are best equipped to make decisions about drilling and its [e]ffect on seismic activity, the environment and other sensitive issues. We need to let these experts do their jobs. The alternative is to pursue a patchwork of regulations that, in some cases, could arbitrarily ban energy exploration and damage the state's largest industry, largest employers, and largest taxpayers" (Henry 2016).

To understand how the antifracking movement was capable of growing rapidly from small-scale grassroots activism into a movement that threatened one of the most powerful industries in the United States, it is important to examine how the movement used "reel power"—in other words, how it was able to use the documentary *Gasland* to mobilize concerned citizens and attract sympathizers. The documentary was a powerful organizing tool because it used vivid images of burning tap water, sick people, and animals to create awareness about the potential health effects of hydraulic fracturing. Public screenings of *Gasland* were often followed by public debates and discussions that often included the filmmaker Josh Fox and elected officials or celebrities, such as musicians Yoko Ono and Sean Lennon, or actors like Mark Ruffalo and Alec Baldwin. Indeed, Josh Fox was present at many protests and used his connections to celebrities to bring the attention of the mass media to antifracking events. Moreover, Fox and antifracking activists have been very skilled in using social media and the Internet to mobilize the public. *Gasland*'s website, for example, encourages individuals to take action by contacting their elected officials, participating in a demonstration, sharing their stories, and volunteering for local organizations. Similarly, the sequel *Gasland II* (Fox 2013) has a list of tips for hosting a screening that includes suggestions such as: "Create a Facebook event (& join our Facebook fan page), tweet about it, email your friends, or use Evite to invite guests"; "arrange for a guest speaker to join your house party"; "Have a computer accessible so your guests can immediately take action after the film" (Gaslandthemovie.com 2014).

The movement's creative use of *Gasland* was so effective that the natural gas industry tried to counterattack with its own documentary, *Truthland*, yet that film failed to convince the public that it was not a tool of the oil and gas industry and was screened only in a few communities. Moreover, as the previous section shows, the mass media coverage was very limited and the documentary failed to attract interest on the Internet and YouTube. Another documentary, *FrackNation,* claimed that it was independent from the oil and gas industry and managed to attract a bit more attention: it was screened in a number of communities with support from Tea Party groups and oil and gas industry associations. *FrackNation* attracted more media attention than *Truthland*, yet it too had less success than *Gasland*.

A closer look at screenings of documentaries, mass media coverage of documentaries, and oil and gas industry presence in a state revealed a few interesting relationships. Mass media coverage of documentaries appears to have been influenced by two main factors. One factor was the proximity between mass media and screenings, where the mass media in states with many screenings covered the documentaries more often than mass media in states with few screenings. Correlation analysis shows that the association between the number of screenings and the amount of media coverage was strong (approximately 0.9) for *Gasland* and somewhat strong (approximately 0.68) for *Truthland* and *FrackNation*. Another important factor was the strength of the oil and gas industry; in states with numerous oil and gas drilling operations, mass

Table 2.1

Correlation between state-level number of oil and gas wells, *Gasland* screenings, *Truthland* and *FrackNation* screenings, mass media coverage of *Gasland*, mass media coverage of *Truthland* and *FrackNation*, and local bans against fracking

	Oil and gas wells	*Gasland* screenings	*TL* and *FN* screenings	Media coverage of *Gasland*	Media coverage of *TL* and *FN*	Local bans against fracking
Oil and gas wells	1					
Gasland screenings	0.163	1				
TL and *FN* screenings	0.323	0.509	1			
Media coverage of *Gasland*	0.318	0.899	0.593	1		
Media coverage of *TL* and *FN*	0.553	0.758	0.680	0.902	1	
Local bans against fracking	−0.007	0.826	0.304	0.622	0.506	1

media coverage of the documentaries was higher than in states with few operations. Moreover, in states with significant oil and gas industry activity, the mass media covered documentaries that presented fracking in a positive light more frequently than documentaries that presented fracking in a negative light. The correlation between mass media coverage of *Truthland* and *FrackNation* and the number of oil and gas wells was 0.55, while the correlation between the mass media coverage of *Gasland* and the number of oil and gas wells was 0.32. Interestingly, mass media coverage of *Truthland* and *FrackNation* was associated more closely with screenings of *Gasland* (0.76) than with screenings of *Truthland* and *FrackNation* (0.68) (see Table 2.1). Moreover, regarding the adoption of local bans against fracking, there was a strong correlation (0.82) between screenings of *Gasland* and antifracking bans, but a weak (0.3) correlation between the adoption of bans and screenings of profracking documentaries. Taken together, these results suggest that the oil and gas industry influenced mass media coverage of documentaries about hydraulic fracturing, while the antifracking movement used *Gasland* to influence the adoption of local bans against fracking.

Conclusion

At the end of 2015, the United States lifted its 40-year-old ban on exporting oil, and at the beginning of 2016 begin shipping liquefied natural gas to global markets for the first time ever. These historic milestones represent stunning turnarounds from the energy outlook of a decade ago. In 2005, it was projected that in 2015 the United States would be importing 25% of its daily natural gas use; instead, it is now estimated that it will possibly become a net gas exporter by 2017 (Bordoff 2015). Seen in this context, the shale gas energy revolution seems unstoppable, yet many uncertainties remain, including the geopolitics of oil and gas, the fluctuations of the energy markets, major accidents, environmental disasters, and energy tax subsidies, among other key events. Moreover, the American public is currently at a turning point: in March 2015 opinion polls showed that 40% of Americans supported fracking and 40% opposed it; by March 2016 opinion polls showed that 36% of Americans supported fracking and 51% opposed it (Gallup 2016).

The physical process of fracturing shale formations is responsible for the fracturing of many American communities, including increasing the antagonism between those who oppose fracking and those who support it. Documentaries both for and against fracking, however, have deepened these antagonisms by contributing to the "cultural fracturing" of communities around the country. The implications are profound and the questions will continue to be asked for a long time: Should states prevent local communities from deciding whether or not they want fracking? Should the interests of oil and gas companies—in the name of energy security and independence—come before the interests of local communities—or, whose land is this, anyway? Are documentary films the most likely source of future protest or just another source of influence? What can films in this day and age accomplish politically that they could not accomplish in the past, and what makes films about fracking different?

This chapter has shown that, to a certain extent, the antifracking movement has been able to achieve some of its successes by using the documentary *Gasland* to mobilize awareness and opposition. The documentary did not create the antifracking movement, nor did it lead directly to the adoption of antifracking ordinances or moratoriums. Grassroots activists have used a variety of tactics—from protests and petitions to lawsuits and ordinances—to prevent the oil and gas industry from drilling in their communities. They have also used documentary screenings as very effective rallying and recruitment tools. It is a testimony to the influence of the documentary-screening tactic that the oil and gas industry has responded by sponsoring and disseminating its own documentaries. Indeed, two characteristics of the current hydraulic fracturing controversy distinguish it from past environmental controversies: the degree to which activists have used documentaries to mobilize their supporters and influence public opinion, and the degree to which industry groups have also used documentaries to combat activists and build support for oil and gas development. Beyond the issue of fracking, it is likely that, in the age of the

Internet and social media, social movements, as well as their opponents, will use documentaries more and more frequently to reach people, shape perceptions, influence political agendas, and shape public policy.

References

America's Natural Gas Alliance. 2013. "Think About It." Retrieved June 2, 2013. (http://thinkaboutit.org).

Aronoff, Marilyn, and Valerie Gunter. 1994. "A Pound of Cure: Facilitating Participatory Processes in Technological Hazard Disputes." *Society and Natural Resources* 7: 235–252.

Auyero, Javier, and Debora Swistun. 2008. "The Social Production of Toxic Uncertainty." *American Sociological Review* 73: 357–379.

Bail, Christopher A. 2012. "The Fringe Effect: Civil Society Organizations and the Evolution of Media Discourse about Islam since the September 11th Attacks." *American Sociological Review* 77(6): 855–79.

Bamberger, Michelle, and Robert Oswald. 2014. *The Real Cost of Fracking: How America's Shale Gas Boom Is Threatening Our Families, Pets, and Food.* Boston: Beacon Press.

Benford, Robert D., and David A. Snow. 2000. "Framing Processes and Social Movements: An Overview and Assessment." *Annual Review of Sociology* 26: 611–39.

Berezin, Mabel. 1994. "Cultural Form and Political Meaning: State-Subsidized Theater, Ideology, and the Language of Style in Fascist Italy." *American Journal of Sociology* 99(5): 1237–1286.

Bordoff, Jason. 2015. "How Exporting U.S. Liquefied Natural Gas Will Transform the Politics of Global Energy." *The Wall Street Journal.* Accessed online in March 2016 at: http://blogs.wsj.com/experts/2015/11/17/how-exporting-u-s-liquefied-natural-gas-will-transform-the-politics-of-global-energy/.

Boudet, Hilary. 2011. "From NIMBY to NIABY: Regional Mobilization Against Liquefied Natural Gas in the United States." *Environmental Politics* 20(6): 786–806.

Boudet, Hilary, Dylan Bugden, Chad Zanocco, and Edward Maibach 2016. "The Effect of Industry Activities on Public Support for 'Fracking'." *Environmental Politics* (http://dx.doi.org/10.1080/09644016.2016.11537).

Boudet, Hilary, Christopher Clarke, Dylan Bugden, Edward Mibach, Connie Roser-Renouf, and Anthony Leiserowitz. 2014."'Fracking' Controversy and Communication: Using National Survey Data to Understand Public Perceptions of Hydraulic Fracturing." *Energy Policy* 65: 57–67.

Brown, Phil, Steve Kroll-Smith, and Valerie Gunter. 2000. "Knowledge, Citizens, and Organizations: An Overview of Environments, Diseases, and Social Conflict." Pp. 9–28 in *Illness and the Environment,* edited by Steve Kroll-Smith, Phil Brown, and Valerie Gunter. New York: New York University Press.

Brown, Phil, and Edwin Mikkelsen. 1990. *No Safe Place.* Berkeley: University of California Press.

Cable, Sherry. 2012. *Sustainable Failures.* Philadelphia, PA: Temple University Press.

Cantarow Ellen. 2012. "The Fight Against Fracking." *The Nation.* Accessed online in February 2016 at: http://www.thenation.com/article/fight-against-fracking/

Concerned Health Professionals of New York and Physicians for Social Responsibility. 2015. *Compendium of Scientific, Medical, and Media Findings Demonstrating Risks and Harms*

of Fracking (unconventional gas and oil extraction), 3rd ed. (http://concernedhealthny
.org/compendium/).

Crenson, Matthew. 1971. *The Un-politics of Air Pollution.* Baltimore, MD: Johns Hopkins
Press.

Cress, Daniel M., and David A. Snow. 2000. "The Outcomes of Homeless Mobilization:
The Influence of Organization, Disruption, Political Mediation, and Framing." *American
Journal of Sociology* 105(4): 1063–1104.

Cruz, Jon. 1999. *Culture on the Margins.* Princeton: Princeton University Press.

Danaher, William F. 2010. "Music and Social Movements." *Sociology Compass* 4(9): 811–823.

Elliot, James, and Scott Frickel. 2013. "The Historical Nature of Cities: A Study of Urbanization and Hazardous Waste Accumulation." *American Sociological Review* 78: 521–543.

Energy in Depth. 2011. "Debunking Gasland." Accessed online in February 2016 at: http://
energyindepth.org/wp-content/uploads/2011/11/Debunking-Gasland.pdf

Energy in Depth. 2012. *Truthland: Dispatches from the Real Gasland.* Film Documentary.
Washington, DC: Independent Petroleum Association of America.

Energy in Depth. 2013. "'Truthland,' the Factual Alternative to 'Gasland' Launched Nationwide Today." Accessed online in March 2016 at: http://energyindepth.org/wp-content
/uploads/2013/05/Truthland-National-Release1.pdf

Energy Information Administration. 2014. "Shale Gas Provides Largest Share of U.S. Natural Gas Production in 2013." Accessed online in January 2016 at: http://www.eia.gov
/todayinenergy/detail.cfm?id=18951

Eyerman, Ron, and Andrew Jamison. 1995. "Social Movements and Cultural Transformation: Popular Music in the 1960s." *Media, Culture, and Society* 17(3): 449–468.

Food and Water Watch. 2016. Local Resolutions against Fracking. Accessed online in March
2016 at: http://www.foodandwaterwatch.org/campaign/ban-fracking-everywhere

Forbis, Robert. 2014. "The Political History of Hydraulic Fracturing's Expansion Across the
West." *California Journal of Politics and Policy* 6(1): 153–186.

Fowlkes, Martha, and Patricia Miller. 1982. *Love Canal.* Washington, DC: Federal Emergency Management Agency.

Fowlkes, Martha, and Patricia Miller. 1987. "Chemicals and Community at Love Canal."
Pp. 55–77 in *The Social and Cultural Construction of Risk*, edited by B. B. Johnson and
V. T. Covello. Boston: Reidel.

Fox, Josh. 2010. *Gasland.* Film Documentary. New York: International Wow Productions.

Fox, Josh. 2013. *Gasland II.* Film Documentary. New York: International Wow Productions.

Fulton, Deirdre. 2015. "In 'Tragic' Decision, Top Ohio Court Takes Away Local Power
to Ban Fracking." *Common Dreams.* Accessed online in March 2016 at: http://www
.commondreams.org/news/2015/02/18/tragic-decision-top-ohio-court-takes-away-local
-power-ban-fracking

FrackNation. 2012. "About FrackNation." Retrieved in March 2016 (http://fracknation.com
/about/).

Freudenburg, William, and Timphy Jones. 1991. "Attitudes and Stress in the Presence of
Technological Risk: A Test of the Supreme Court Hypothesis." *Social Forces* 69: 1143–1168.

Gallup. 2016. "Opposition to Fracking Mounts in the U.S." Retrieved in April 2016 (http://
www.gallup.com/poll/190355/opposition-fracking-mounts.aspx)

Gaslandthemovie.com. 2014. Retrieved in March 2016 (http://www.gaslandthemovie.com
/pdf/GaslandScreeningGuideDate.pdf)

Gaventa, John. 1982. *Power and Powerlessness.* Chicago: University of Chicago Press.

Hanson, Michael. 2008. "Suppose James Brown Read Fanon: The Black Arts Movement, Cultural Nationalism and the Failure of Popular Musical Praxis." *Popular Music* 27: 341–365.

Hemingway, Andrew. 2002. *Artists on the Left*. New Haven: Yale University Press.

Henry, Devin. 2016. "Oklahoma Blocks Local Fracking Bans." *The Hill*. Retrieved in March 2016 (http://thehill.com/policy/energy-environment/243645-oklahoma-blocks-local -fracking-bans)

IndieWire. 2010. "The Toxic Avenger: Josh Fox's 'Gasland.'" Kohn, Eric (2010–01–30). Retrieved in January 2016. (http://www.indiewire.com/article/review_the_toxic _avenger_josh_foxs_gasland)

In These Times. 2015. "How New York Activists Banned Fracking." By Eric Weltman. Retrieved in March 2016 (http://inthesetimes.com/article/17697/new_york_fracking_ban1)

Jacobsen, Grant. 2011. "The Al Gore Effect: An Inconvenient Truth and Voluntary Carbon Offsets." *Journal of Environmental Economics and Management* 61: 67–78.

Jalbert, Kirk, Abby J. Kinchy, and Simona L. Perry. 2014. "Civil Society Research and Marcellus Shale Natural Gas Development: Results of a Survey of Volunteer Water Monitoring Organizations." *Journal of Environmental Studies & Sciences* 4(1): 78–86.

Koopmans, Ruud, and Paul Statham. 1999. "Challenging the Liberal Nation-State? Postnationalism, Multiculturalism, and the Collective Claims Making of Migrants and Ethnic Minorities in Britain and Germany." *American Journal of Sociology* 105(3): 652–696.

Kroll-Smith, Stephen, and Stephen Couch. 1993. "Symbols, Ecology, and Contamination: Case Studies in the Ecological-Symbolic Approach to Disaster." *Research in Social Problems and Public Policy* 5: 47–73.

Kroll-Smith, Stephen, Stephen Couch, and Adeline Levine. 2002. "Technological Hazards and Disasters." Pp. 295–328 in *Handbook of Environmental Sociology*, edited by Riley Dunlap and William Michelson. Westport, CT: Greenwood Press.

Krulwich, Robert. 2013. "A Mysterious Patch of Light Shows Up in the North Dakota Dark." National Public Radio. Accessed online in January 2016 at: http://www.npr.org/sections /krulwich/2013/01/16/169511949/a-mysterious-patch-of-light-shows-up-in-the-north -dakota-dark

Ladd, Anthony E. 2013. "Stakeholder Perceptions of Socio-Environmental Impacts from Unconventional Natural Gas Development and Hydraulic Fracturing in the Haynesville Shale." *Journal of Rural Social Sciences* 28(2): 56–89.

Ladd, Anthony E. 2014. "Environmental Disputes and Opportunity-Threat Impacts Surrounding Natural Gas Fracking in Louisiana." *Social Currents* 1(3): 293–312.

Leiserowitz, Anthony. 2004. "Before and after the Day After Tomorrow. A U.S. Study of Climate Change Risk Perception." *Environment* 46: 23–37.

Lipsitz, George. 2000. *Time Passages*. Minneapolis: University of Minnesota Press.

Lynch, Lisa. 2012. "'We Don't Wanna Be Radiated': Documentary Film and the Evolving Rhetoric of Nuclear Energy Activism." *American Literature* 84: 327–351.

MacRae, Michael. 2012. "Fracking: A Look Back." Accessed online in February 2016 at: https://www.asme.org/engineering-topics/articles/fossil-power/fracking-a-look-back

Malin, Stephanie A. 2014. "There's No Real Choice But to Sign: Neoliberalization and Normalization of Hydraulic Fracturing on Pennsylvania Farmland." *Journal of Environmental Studies and Science* 4: 17–27.

Malin, Stephanie A., and Kathryn T. DeMaster. 2016. "A Devil's Bargain: Rural Environmental Injustices and Hydraulic Fracturing on Pennsylvania's Farms." *Journal of Rural Studies* 47: 278–290.

McAdam, Doug. 1994. "Culture and Social Movements." Pp. 36–56 in *New Social Movements*, ed. Enrique Larana et al. Philadelphia: Temple University Press.

McAleer, Phelim. 2012. *Fracknation: A Journalist's Search for the Fracking Truth*. Film Documentary. Marina Del Ray, CA: Hard Boiled Films.

McCammon, Holly J., Soma Chaudhuri, Lyndi N. Hewitt, Courtney Sanders Muse, Harmony D. Newman, Carrie Lee Smith, and Teresa M. Terrell. 2008. "Becoming Full Citizens: The U.S. Women's Jury Rights Campaigns, the Pace of Reform, and Strategic Adaptation." *American Journal of Sociology* 113(4): 1104–1147.

McCammon, Holly J., Courtney Sanders Muse, Harmony D. Newman, and Teresa M. Terrell. 2007. "Movement Framing and Discursive Opportunity Structures: The Political Successes of the U.S. Women's Jury Movements." *American Sociological Review* 72(5): 725–749.

Meyer, David S., and Deana A. Rohlinger. 2012. "Big Books and Social Movements: A Myth of Ideas and Social Change." *Social Problems* 59(1): 136–153.

Newsweek. 2015. "To Quiet Calls for Fracking Curbs, Texas Bans Bans." By Leah McGrath Goodman (6/4/15). Retrieved in March 2016. (http://www.newsweek.com/2015/06/12/quiet-calls-fracking-curbs-texas-bans-bans-339164.html)

Osborn, Stephen G., Avner Vengosh, Nathaniel R. Warner, and Robert B. Jackson. 2011. "Methane Contamination of Drinking Water Accompanying Gas-Well Drilling and Hydraulic Fracturing." *PNAS* 108(20): 8172–8176.

Reed, T. V. 2005. *The Art of Protest*. Minneapolis: University of Minnesota Press.

Rooftop Films. 2010. "'Gasland' Tour Draws Huge Crowds, Lively Debate." By Lela Scott MacNeil. Retrieved in March 2016 (http://rooftopfilms.com/blog/2010/09/gasland-tour-draws-huge-crowds-lively-debate.html).

Roscigno, Vincent. 2011. "Power, Revisited." *Social Forces* 90: 349–373.

Roscigno, Vincent J., and William F. Danaher. 2001. "Media and Mobilization: The Case of Radio and Southern Textile Worker Insurgency, 1929 to 1934." *American Sociological Review* 66(1): 21–48.

Roy, William. 2010. *Reds, Whites, and Blues*. Princeton, NJ: Princeton University Press.

StateImpact NPR. 2015. "Exploring the Link Between Earthquakes and Oil and Gas Disposal Wells". Retrieved March 16, 2016. (https://stateimpact.npr.org/oklahoma/tag/earthquakes/).

Steinberg, Marc W. 1999. "The Talk and Back Talk of Collective Action: A Dialogic Analysis of Repertoires of Discourse among Nineteenth-Century English Cotton Spinners." *American Journal of Sociology* 105(3): 736–780.

Steinberg, Marc. 2004. "When Politics Goes Pop: On the Intersections of Popular and Political Culture and the Case of Serbian Student Protests." *Social Movement Studies* 3: 3–29.

Truthland. 2013. "The Facts on Fracturing (and Other Stuff Too)." Retrieved March 2016: (http://www.truthlandmovie.com/the-facts/)

Variety. 2010. "Gasland Movie Review from the Sundance Film Festival." Koehler, Robert (2010–01–25). Retrieved in January 2016. (http://variety.com/2010/film/markets-festivals/gasland-1117941971/)

Vasi, Ion Bogdan, Edward T. Walker, John S. Johnson, and Hui Fen Tan. 2015. "'No Fracking Way!' Documentary Film, Discursive Opportunity, and Local Opposition against Hydraulic Fracturing in the United States, 2010 to 2013." *American Sociological Review* 80(5): 934–959.

Whiteman, David. 2003. *Reel Impact. Stanford Social Innovation Review* 1: 60–63.

Wilber, Tom. 2015. *Under the Surface: Fracking, Fortunes, and the Fate of the Marcellus Shale*. Ithaca, NY: Cornell University Press.

Yergin, Daniel. 2013. "Congratulations, America. You're (Almost) Energy Independent. Now what?" *Politico Magazine*. Retrieved in January 2016 (http://www.politico.com/magazine/story/2013/11/congratulations-america-youre-almost-energy-independent-now-what-098985).

Zolberg, Vera. 1997. *Outsider Art*. Cambridge University Press.

3

Disturbing the Dead

————————————————●

Community Concerns over
Fracking below a Cemetery
in the Utica Shale Region

CARMEL E. PRICE AND

JAMES N. MAPLES

Introduction

At their simplest level, cemeteries are preferred places to put our dead, yet they
also perform crucial but generally unnoticed social roles in our communities
(Snodgrass 2015; Veit and Nonestied 2008). Cemeteries attach individuals to
families, places, and communities that are both near and far, as well as represent a physical, ritualized point of remembrance and memorialization for those
who have passed away (Baptist 2013; Maples and East 2013; Vanderstraeten
2014). Cemeteries also function as a source of data ranging from epidemiological trends to social class patterns (Miller and Rivera 2006). Increasingly,
certain cemeteries function as a source of economic activity via tourism (del
Puerto and Baptista 2015) and, more recently, energy development involving horizontal drilling and hydraulic fracturing—henceforth referred to as
"fracking" (Fernandez 2012; Pompili 2012). Despite the cultural, historic, and
sacred importance of cemeteries, they are increasingly threatened. Whether
abandoned, overlooked, or ignored, cemeteries are often damaged in the face

of development and are frequently excluded from the community planning process (Basmajian and Coutts 2010; Burg 2008). For example, cemeteries have been damaged by mountaintop removal (MTR) methods of coal mining (Maples and East 2013), highway development (Hopkins 2003), and urbanization (Basmajian and Coutts 2010). Today, fracking underneath cemeteries represents a new socioenvironmental concern for residents living within shale regions (Fernandez 2012; Nunez 2015; Pompili 2012). This reality places cemeteries at the crux of the debates about fracking, environmental destruction, land use practices, and the moral and ethical treatment of the dead.

Although this new trend of fracking underneath cemeteries is gaining both traction and opposition, there is a complete dearth of academic literature exploring the decision to frack underneath cemeteries and its resistance by local citizens. In this chapter, we examine community concerns regarding fracking below the Mount Hope Veterans Memorial Cemetery in Youngstown, Ohio in the Utica Shale region. Land investors purchased this property to lease it to oil and natural gas companies and it remains at risk for fracking. We draw on in-depth interviews with community residents and stakeholders, participant observation at a community event, and a thorough review of documents, videos, news articles, and online comments by readers of news articles to examine the discourse around fracking underneath cemeteries. We ground our analysis in the literature on cemetery destruction and ties to place, as well as the sociohistorical context of Youngstown, Ohio. Essentially, we find that resistance to fracking underneath cemeteries relies heavily on a *moral and ethical* discourse, which represents one of three new discursive frames identified by this project. Two of the frames lie within the anti-cemetery-fracking narratives (*moral and ethical* concerns and *beautification and preservation* as a form of resistance) and one within the pro-cemetery-fracking narratives (fracking is *not disruptive* to the dead).

Literature Review

Purpose and Value of Cemeteries

In early sociological research on cemeteries, Mumford (1961) recognized a strong connection between the cemetery and the community, arguing that cemeteries could be treated as "cities of the dead." Cemeteries allow us to perform a needed purpose (managing the removal of the deceased from everyday spaces), while also maintaining a very close physical tie to the deceased. Early geographical work on necrogeography recognized cemeteries as crucial sources of cultural attachment, while also noting the importance of their spatial location to the living (Francaviglia 1971; Kniffen 1967). These early works helped frame the perspective that cemeteries are far more than just burial sites. Cemeteries also exist to meet our emotional and social needs as places of cultural attachment and highly regimented memorialization.

More recent, multidisciplinary scholarship further delves into cemeteries as sources of cultural attachment to place (Hufford 2007; Lokocz, Ryan, and Sadler 2011; Maples and East 2013). Cemeteries (and the family and friends buried within) create a strong link between individuals and the community in which the cemetery is located. This includes returning to graves for cleaning and decoration (Jabbour and Jabbour 2010) or returning to burial sites to feel a sense of reconnection with those who have passed away (Hopkins 2003). This attachment remains strong and even functions as a force that shapes decisions about where the living wish to be buried after their death (Casal, Aragonés, and Moser 2010). Place attachment can also shape how residents, and former residents, desire the protection and preservation of undeveloped land, including cemeteries, when faced with threats of economic growth and development (Lokocz, Ryan, and Sadler 2011).

As part of the burial process, cemeteries function as a memorialization site to attach the living to the dead (Baptist 2013; Vanderstraeten 2014). At their core, cemeteries are places where we can grieve openly, in contrast to the expectations and norms of other public spaces. The cemetery itself creates a physical place (often with highly ritualized symbols such as crosses) to attach a sense of presence to the absence of the deceased (Baptist 2013). By using commemorative markers (such as benches in memorial gardens), cemetery memorials can even make physically absent bodies seem present for the grieving process (Wylie 2009). Memorials can also be a great source of data for future generations and provide a wealth of data for scientists, genealogists, and historians (Veit and Nonestied 2008). Scientists often use cemeteries to track epidemiological and societal changes over time, and cemetery data (whether collected from headstones, public records, or archaeological examination) reveal a wealth of information on life expectancy and mortality (Basmajian and Coutts 2010; Jabbour and Jabbour 2010). Cemeteries also reflect the social class structure of those living in the community, with wealthier people often having more ornate burials or more desirable burial locations (Miller and Rivera 2006; Redfern and Dewitte 2011). Additionally, cemeteries have also become a source of place-based economic activity dubbed cemetery tourism or sometimes "dark tourism." Cemeteries (such as New Orleans' crypts, cemeteries with famous interments, and even military cemeteries) are an attraction for individuals who are interested in history, culture, and, to some degree, a sense of closeness with death (del Puerto and Baptista 2015).

Recently, driven in part by the costs of maintenance, especially for historic and rural cemeteries that have zero to minimal revenue coming in from new burials (see Fernandez 2012), cemeteries are being purchased and/or leased for the purpose of oil and gas fracking. Although fracking underneath cemeteries is an unexamined topic in scholarly work on unconventional energy development to date, there are multiple cases where fracking is occurring underneath cemeteries and some cases where it has been unsuccessfully attempted. For example, over a dozen cemeteries in the Barnett Shale region around Fort Worth, Texas have worked with Chesapeake Energy to allow drilling below their graves (Fernandez 2012). In Colorado, land containing at least one cemetery (Kanza Cemetery) within the Niobrara Shale region has been leased for oil and gas drilling (Nunez 2015). In Pittsburgh, Pennsylvania, the Catholic Cemeteries Association agreed to lease the mineral rights under several cemeteries in the Marcellus Shale region, which helped to justify a fracking ban in Pittsburgh (Fernandez 2012; Staggenborg, this volume). There is presently no directory of cemeteries threatened by resource extraction or watchdog organizations keeping track of this phenomenon. As such, public announcements of fracking underneath cemeteries generally occur at the local level (e.g., newspapers or community-level activism).

Mount Hope Veterans Memorial Cemetery and Youngstown, Ohio

The Mount Hope Veterans Memorial Cemetery (henceforth, Mount Hope) is located at the intersection of Wardle Avenue and Liberty Road in Youngstown, Mahoning County, Ohio. Although its exact date of creation is unclear in local land records, the first burial in Mount Hope took place in 1911 (Queener 2010). As of 2010, Mount Hope contained 1,009 visible burial markers, with most of the burials occurring between the 1930s and 1980s, and a few recent burials since that period (Queener 2010). Mount Hope's approximately 205 veteran burials include members of historic African American units like the 92nd Infantry Buffalo Division, the 369th Infantry Harlem Hell Fighters, and the 332nd Fighter Squadron Tuskegee Airmen.

Like many cemeteries in necrogeographical research, Mount Hope's past and future are attached to (and influenced by) the sociohistorical forces shaping Youngstown, Ohio. Mount Hope is located in the Youngstown–Warren Boardman metropolitan statistical area and is about ten minutes from downtown Youngstown. Youngstown's population grew amid regional industrialization in the early 19th century. Its physical location near coal deposits, the Erie Canal system, and railroads helped Youngstown gain population quickly, attracting immigrants and African Americans seeking work (Linkon and Russo 2002). The development of steel mills in Youngstown pushed the city to reach its largest Census population in 1930 at 170,002 residents. However, the eventual development of the Rust Belt in the region led to mass layoffs, depopulation, economic decline, and vacant housing in the region. Today, the Youngstown area has an estimated 65,602 residents and is still grappling with the decline of the steel industry. Depopulation has also left the city's outskirts destabilized and has spread out the population so that it often feels more rural than urban (Queener 2010).

Fracking-related job growth in Youngstown has been considered a critical burst of economic activity where little existed (Zremski 2014). Mahoning County is at a sweet spot where the Utica Shale region, which runs deep underneath the Marcellus Shale region, is closer to the surface than it is in Pennsylvania and New York. According to DrillingEdge (2016), there are 4,287 permitted oil and gas wells in Mahoning County with 2,177 of them currently producing. Other employment opportunities in Youngstown include health care and education: for example, Mercy Health and Youngstown State University each employ over 1000 people in the area (Regional Chamber of Commerce 2016).

At the local level, Mount Hope is further shaped by changes occurring within its immediate, smaller communities. Mount Hope is at the center of several communities predating Youngstown's annexation of the surrounding area, including Scienceville, the Sharonline, and Geography Hall, each part of the Coitsville area. Integration patterns in most Youngstown neighborhoods reversed in the 1940s, resulting in African Americans living in the neighborhoods around Mount Hope. A 1963 report on Mount Hope communities noted a 74% increase in nonwhite housing units from previous years. Consequently, many African Americans were buried in Mount Hope and its African American community is currently fighting for the preservation of the cemetery (Queener 2010).

Land ownership issues at Mount Hope opened the way for fracking in the cemetery. Reverend Willie Duke purchased Mount Hope in 1985 and the property was transferred to his wife upon his unexpected death (Queener 2010). Upon her death, the property ownership became unclear. Local residents treated it as a communal property (much like Appalachia's treatment of cemeteries in the commons; see Hopkins 2003 or Hufford 2007). Without a specific local church attached to the cemetery or an owner to keep up the property, the local community organized efforts to maintain the cemetery, remove fallen limbs, and keep the grass mowed (Goodwin 2012). Following a lack of clarity concerning its land ownership, the local government issued tax certificates on the Mount Hope property due to unpaid taxes and Ohio Land Management LLC purchased these certificates. Notably, Youngstown has a long history of political corruption. For example, on Friday, February 26, 2016 Mayor John McNally of Youngstown pleaded guilty to multiple charges of conspiracy, bribery, and perjury involving a property deal (Palmer 2016). A few research participants in the study indicated that political corruption was behind the acquisition of Mount Hope, and several residents of Youngstown spoke of political corruption as commonplace.

On January 24, 2013, a group of concerned citizens formed a committee around the preservation of Mount Hope and met with representatives from

Ohio Land Management LLC. In a brief video clip from this meeting, publicly available on YouTube, a community member directly asked: "What was your purpose to purchase a black cemetery?" In response, a representative for Ohio Land Management LLC stated:

> The purpose was not to buy a cemetery. Not to disturb the cemetery. Not to do any damage. Not to defame it. Not to do anything to the history or the heritage of the cemetery. The purpose was for, the purpose, like everything else in Mahoning County right now, is what's underneath the ground. What's underneath the ground—that's what everybody's doing. Regardless, and regardless of what the property is, whether it's city hall, the court house, people's private residences—ultimately there is a high probability that there is going to be a gas line underneath the property.

The Mount Hope Veterans Memorial Committee later considered applying for nonprofit or 501(c)(3) status to get ownership of the cemetery. However, barriers to nonprofit status, as well as concerns over disinterment and new burials, hindered this process. Today, Ohio Land Management LLC still owns the land but the Mount Hope Veterans Memorial Committee maintains the cemetery.

In 2012, residents of the greater Youngstown, Ohio area organized to resist fracking. With help from the Community Environmental Legal Defense Fund (CELDF) and the establishment of Frackfree Mahoning Valley, residents developed a Community Bill of Rights Charter Amendment and placed the measure on the ballot six times (in May and November 2013, May and November 2014, November 2015, and November 2016). According to CELDF (2016b), the measure would "establish rights to clean air, pure water, and the right to local self-government, banning fracking as a violation of those rights" and has received "fierce opposition." Despite opposition to the Community Bill of Rights Amendment, the measure received support from 49% and 45% of voters in November 2015 and 2016, respectively.

Common Threats to Cemeteries

Weakened place attachment helps explain why Mount Hope is at risk of fracking. Cemeteries are historically rooted in and protected by communities. For example, Appalachian cemeteries were typically located in the commons, a space utilized by all community members (Hufford 2007; Maples and East 2013). As future generations moved away, the community changed and as the generational distance between the living and the dead increased, the focus on cemetery preservation often weakened, leaving cemeteries open to damage. Representatives from the Mount Hope Veterans Memorial Committee also reported this problem. As one committee member explained, "My concern is attrition as we age. We've not been successful in getting young people involved. Young people don't know the people directly [buried there] even though they're related to them."

Cemeteries also fare poorly when surrounding land is developed. The proliferation of suburbs around cities and increased problems with urban sprawl generate development in and around cemeteries. In cities, cemeteries are sometimes relocated (and in rare cases, even encased in the cellars of buildings) to maximize space (Basmajian and Coutts 2010). Cemeteries are frequently forgotten when residential, industrial, or commercial sites encroach on undeveloped areas (Snodgrass 2015; Webster 2013; Woodthorpe 2011). Highway construction connecting suburbanized areas to cities can also cross over lost or forgotten cemeteries, which are sometimes identified only after bulldozers have intruded upon the cemetery's borders (Hopkins 2003). Cemeteries in marginalized communities are especially impacted by development (Burg 2008). For example, African American graves have been overrun by thoroughfares and driveways, and historic mountain cemeteries have been damaged or destroyed by mountaintop removal (MTR) methods of coal mining (Burg 2008; Maples and East 2013).

Cemeteries are also subject to damage from natural forces such as wind, rain, and flooding, which can erode soil, damage monuments, and even lift caskets and burials out of the ground (Elliot 2011; Jackson 2006; Miller and Rivera 2009). Earthquakes similarly disrupt cemeteries both above and below the surface, cracking casket seals underground, while shaking monuments around graves (Cucci and Tertulliani 2011; Collins, Kayen, and Tanaka 2012). Earthquake damage represents a grave threat to cemeteries today and is notably due to a mix of natural causes and human activity, such as MTR blasting for coal extraction (Maples and East 2013). Geologists have directly linked multiple earthquakes in Youngstown, Ohio to fracking and wastewater injection activity (Skoumal, Brudzinski, and Currie 2015). Numerous earthquakes ranging in magnitude from 1.0 to close to 4.0 occurred in Youngstown between 2011 and 2014, including one of the largest fracking-induced earthquakes ever recorded. In March 2014, the Ohio Department of Natural Resources ordered one particular well associated with earthquake activity to be shut down and subsequently seismic activity was reduced (Skoumal, Brudzinski, and Currie 2015). Youngstown, Ohio had never experienced earthquakes prior to fracking activity. Although there have not been documented accounts of fracking-related earthquakes affecting cemeteries (in Youngstown or elsewhere), the possibility of damage to cemeteries is real, as it is for any built structure.

A less visible threat to cemeteries is the issue of land and mineral rights. It is common in areas with natural resources for land ownership to be treated separately from mineral extraction rights. As such, one may own a plot of land but a separate entity can own the rights to extract minerals underneath the land. It should be noted that sometimes mineral rights (and even land ownership) in mineral-rich areas are sold under duress or unclear terms (Barry 2011), leading to serious quandaries over land such as cemeteries. Often treated as part of the commons or public property, cemeteries are frequently considered by local communities as shared territory (Maples and East 2013). However, cemeteries are still owned property, whether belonging to a church, a private corporation, local government, or individuals. State and local laws do address what can and cannot be done in cemeteries (such as West Virginia's laws preventing blasting within 100 feet of a cemetery) but vary by place. In Ohio, for example, all cemeteries (excluding family cemeteries and cemeteries with no burials in the last 25 years) must be registered with the state and land containing burial plots can be sold. If the landowner attempts to relocate the cemetery, however, then he or she must acquire consent from next of kin and provide a public notice of the relocation. Beyond this, there are no laws prohibiting mineral extraction below a cemetery in Ohio.

Fracking Disputes and Weak Ties to Place

Those living within energy development zones often experience weakened ties to place. For example, grassroots antifracking activists in Ohio reported that their connections to place were substantially altered due to the changing landscapes caused by energy development (Willow et al. 2014). Environmental degradation caused by fracking has the potential "to convert formerly positive experiences of place into experiences of profound alienation; even when people are not physically displaced, perceptions of pollution can bring about detrimental psychological separation" (Willow et al. 2014: 61). As shale energy development pushes people to physically and emotionally abandon places of strong attachment, that separation then allows space for additional energy development and a cyclical pattern of increasing environmental degradation and community detachment emerges (Cronon 1995; Willow 2014). Conversely, communities faced with the threat of environmental harm from fracking often developed new connections between residents and their environment (Willow et al. 2014).

Sociological researchers have found that support for fracking at the local level is usually associated with its perceived economic opportunities and benefits, including job creation, landowner income, tax revenues, lower energy prices, and improved exports (Ceresola and Crowe 2015; Jacquet and Stedman 2013; Ladd 2013). Moreover, those who view fracking in economic terms, rather than environmental terms, are more likely to support oil and gas development in their community (Jacquet and Stedman 2013; Kriesky et al. 2013). Opponents, however, tend to view fracking as a threat to the environment, public health, and sustainable economic stability. Weigle (2011), for instance, examined perceptions of fracking in the Marcellus shale region of Pennsylvania and found over 400 different concerns related to four categories: (1) environmental concerns, (2) socioeconomic concerns, (3) public health and safety concerns, and (4) political concerns. Citizens who oppose or are ambiguous about fracking typically cite such threats as groundwater contamination, aquifer depletion, air pollution, induced earthquakes from wastewater injection, road damage, differential signing bonuses and royalties paid by energy companies, harm to farm animals and rural landscapes, as well as chemical exposures, drilling accidents, well blowouts, and methane emissions (Ladd 2013, 2014; Schafft, Borlu, and Glenna 2013; Theodori 2009, 2013; Wynveen 2011).

Data and Methods

To exam the discourse around fracking beneath cemeteries, the data for this study included six in-depth phone interviews with four residents of Youngstown, Ohio and two nearby residents with extensive knowledge of the fight against fracking in Youngstown. The interview participants included activists, a historian, a politician, a reverend, and a geologist. Three of the interview participants were male, three female; four interview participants were white and two were African American. Interviewees were recruited via purposeful and snowball sampling methods. Potential participants were contacted by phone or email and were provided information about the objectives and methods of the study. Each interviewee gave verbal consent to participate in the research. Interviews were unstructured as the topic of fracking underneath cemeteries is relatively unexplored in the literature. The interviews were conducted between September 2015 and March 2016 and the interviews lasted between 30 minutes to 1 hour and 45 minutes. Extensive field notes were taken during and after each interview. Four of the six research participants voluntarily followed up with us via email. They provided additional information such as websites, maps, news articles, and videos, which we used to ground our analysis in the socioeconomic, historical, and political context of Youngstown, Ohio.

As additional qualitative data, we included 14 national and regional news articles that examined fracking underneath cemeteries. Five articles were from the Utica and Marcellus Shale regions: two from *The Vindicator*, the local newspaper in Youngstown, Ohio, and one each from the *Pittsburgh Post-Gazette*, the *Pittsburgh Tribune-Review*, and *Cleveland.com*. One article was from the *Times Picayune*, which is one of two local newspapers in New Orleans, Louisiana. Four articles were from nationally recognized publications: the *New York Times*, the *Huffington Post*, *National Geographic*, and the *Associated Press*. Four articles were from publications that focus specifically on energy development, the environment, science, or technology, a group that included *Motherboard*, *Energy Digital*, *Oil Change International*, and *Eco-watch*. The 14 articles were published between August 18, 2010 and March 13, 2015. We also collected the online comments from readers for each news article, if available, to add to our data set. There were a total of 54 online comments from six of the 14 news articles; eight of the news articles either did not have any comments on the article, did not allow readers to comment, or did not allow us to access the comments without a paid subscription to the news outlet. Five of the 54 online comments came from nationally recognized publication outlets. Forty-nine of the 54 comments came from regional news outlets.

As a final source of qualitative data, a research participant invited us to attend the Black History Celebration hosted by Mount Hope Veterans Memorial Committee on Saturday, February 27, 2016 in Youngstown, Ohio. Participant observation was conducted at the formal event and at the craft fair and fellowship before and after the event. We had conversations with dozens of Youngstown residents during the course of the approximately five hours spent at Reed's Chapel and Mount Hope Cemetery. Altogether, our qualitative data set consisted of six in-depth phone interviews, 14 news articles, 54 online comments, and field notes from our participant observation activities.

Analysis and Discussion

Anti-Cemetery-Fracking Perspectives

The most common sources of resistance to fracking have centered on its perceived threats to surface and groundwater supplies, air quality, roads, property values, rural landscapes, climate change, public health and safety, farm animals, and sustainable economic development (Ladd 2013, 2014). We find that objection to fracking underneath cemeteries, however, takes on a more moral and ethical discursive frame. This moral and ethical discourse is not surprising given the larger function of cemeteries, which serve to meet the emotional, spiritual, and social needs of communities. To be clear, our argument is not that typical antifracking narratives lack a moral and ethical grounding. On the contrary, one could easily argue that antifracking narratives, which are concerned with ecological, public health, and economic equity issues, among other things, are inherently moral concerns. However, those opposed to fracking underneath cemeteries seem to argue that it is purely wrong without the need for further justification.

For example, a cemetery activist we interviewed stated, "Our revolution is our moral duty." Another research participant who had family members (a mother, father, sister, nephews, aunt, and uncle) buried at a cemetery where fracking was proposed said that the cemetery was "sacred land" to her and the community and elaborated by explaining, "Our loved ones are interred there for perpetual rest. The thought of them [the gas company operators] desecrating this sacred ground was alarming to us." There seems to be a gut-level reaction among people that fracking underneath cemeteries is "just wrong."

Concerned residents made statements that highlighted the tension between this concept of cemeteries (e.g., that loved ones who have died are resting in their caskets) and the disruptive and volatile reality of fracking (e.g., using heavy, noisy equipment to drill wells). For example, research participants made statements such as "let our loved ones rest" and "[we] put them there with the thought of them resting in peace." Another interviewee said she was "outraged at even the thought of something so invasive" happening in the cemetery. One person described the moment when residents learned about the possibility of fracking in the cemetery: "There were tears; people were distraught."

Online commenters also expressed moral or ethical concerns about fracking underneath cemeteries. Multiple commenters made reference to the idea that the dead should be able to "rest in peace" in frack-free cemeteries. Likewise, many expressed disgust or outrage at the idea of fracking underneath cemeteries, while others referenced cemeteries as sacred spaces that should not be disturbed. Several argued that there should not be fracking out of respect for the dead and their family members, and some simply stated that fracking underneath cemeteries raised moral and ethical concerns. Online statements included sentiments such as: "I don't like the idea of messing with a burial site," "This is disgusting," "Let our ancestors rest in peace," "Cemeteries are sacred ground," and "This idea does creep me out."

Many of the news articles in our data set also focused on the moral and ethical concerns of fracking underneath cemeteries. For example, two news articles made specific reference to morals in their titles: "Is Fracking Cemeteries Immoral? Plot Owners Have No Say" and "Gas Under Graveyards Raises Moral, Money Questions." Another title made a more subtle reference to morals with this question: "Can Fracking Intrude on Final Sleep?" Moreover, those interviewed in the articles indicated moral and ethical concerns. For example, an environmental science professor was quoted as saying, "I could see how people could be deeply offended by this, even if it didn't cause any problems." Similarly, an Ohio township administrator, was quoted as saying: "You know what it is, it's emotional. A lot of people don't want any type of drilling. There's something about disturbing the sanctuary of a cemetery. We're not talking about dinosaurs now and creatures that roamed the earth millions of years ago. We're talking about loved ones who have died, people we knew." Other people interviewed in the news articles mentioned the need to "preserve the mood" of cemeteries, argued against "disturbing the dead," argued that cemeteries were "hallowed" and "sacred" ground, described fracking underneath cemeteries as "disgusting," "sacrilegious," and "discomfiting on a fundamental level," and simply stated that "it's immoral." Fracking underneath cemeteries seemed to cross boundaries between "right and wrong" that made people uncomfortable. As one public official stated, "I don't particularly think that it's something we should do. It's a cemetery; it's your last resting place."

Although concerns about cemetery fracking extended the antifracking discourse into a new, explicitly moral and ethical realm, anti-*cemetery*-fracking narratives sometimes overlapped with more common antifracking diagnostic frames. Several research participants and online commenters expressed concerns similar to those categorized by Weigle (2011), including environmental concerns, socioeconomic concerns, public health and safety concerns, and concerns about political processes. The environmental concerns expressed by research participants and online commenters focused on fracking-induced earthquakes and disruption of the cemetery grounds by things like noise pollution. For example, online commenters referenced environmental or geological concerns. One commenter specifically called for a geological survey to be done prior to fracking as a way of ensuring that graves would not be disturbed. In addition, a few news articles made reference to environmental concerns by arguing that "noisy, smelly, and unsightly" drilling activity was not welcome on sacred cemetery grounds because funeral attendees and gravesite visitors should not have "drilling operations within sight and earshot."

Furthermore, several cemetery activists expressed concern for the environment and fracking-induced earthquakes. One research participant said that her goal was to "create a sustainable future for all species." Another anti-cemetery-fracking advocate explained that she grew up in Youngstown and had never heard of earthquakes in the area until they started drilling for natural gas. Multiple residents also reported cracks in their walls, floors, and ceilings due to earthquakes and expressed concern for the foundations of their houses. Another resident expressed concern for the cemetery specifically because of the earthquakes, hypothesizing that the plots would eventually shift in the ground. Research participants also reported distrust of the gas and drilling companies because of the earthquakes. As one resident explained: "They publically stated that there was no damage following the earthquakes but people had cracks in their walls and cracks in their floors. They said there would be no damage with fracking in the cemeteries. How are we supposed to believe them now?" Similarly, residents expressed concern over the exploitative nature of the fracking industry. For example, one resident specifically wondered if he was going to have to move because his house was located close to a coveted water source.

Environmental concerns quickly spilled over into socioeconomic concerns as people became worried about exploitation and other issues of social and economic justice. For example, online statements included: "Nothing is sacred anymore except the almighty dollar," "If big business can't respect the deceased in a cemetery, the world is really coming to an end," and "wonder which will win, money or morals?" Many commenters also made reference to exploitation, for example: "Corporations like this sit and wait till the town is ripe for the picking, then come in, even talking jobs, but all they want is to take. They're already under your ground, and under your skin I expect, and now they want to do what to your cemetery? I feel they already owe the city an apology." Other commenters discussed concerns and legal issues related to property rights, mineral rights, and cemetery plot ownership. For example, one commenter suggested: "I guess we are going to have to include mineral rights when we die now so we aren't bothered and can rest in peace." Another commenter drew a juxtaposition between the illegality of trespassing and loitering in cemeteries versus the legal act of drilling, while other citizens concerned with property rights simply questioned: "who has the mineral rights" to gravesites? News articles specifically sought to address this question as they quoted residents who were confused, for example: "I thought you own the entire plot" and "people are given deeds to the cemetery plots once they pay for them." Another article quoted the president of a cemeteries association who explained that "plot owners have no legal claim to the mineral rights at a cemetery. Their agreements are for an indefinite rental of sorts at the surface level—and a promise the site will be maintained."

In addition, concerns about inequality and marginalization also surfaced, which extended the moral and ethical discourse to include the idea that certain cemeteries are subject to more desecration than others. For example, one research participant explained: "African Americans are essentially very spiritual people. Our emotions run deep because we've been through so much. A lot of our ancestors that were buried there had no peace in life and now the same in death, it was too much. To be disenfranchised in life and disenfranchised in death is too much." Online commenters agreed and compared the situation to the desecration of Native American burial sites. Others asked if this would be happening at Arlington National Cemetery and suggested that there is an implicit hierarchy of cemeteries where some receive more respect than others. One online commenter simply stated, "They don't respect the living, why should they respect the dead."

Although the environmental concerns previously referenced (e.g., concern for fracking-induced earthquakes and noise pollution) could also be construed as public health and safety concerns, there were no explicit mentions of public health and safety concerns in the entire data set. Perhaps this is because people in cemeteries are already viewed as dead, meaning that issues like water contamination, for example, which is a major concern for communities experiencing fracking (see Ladd 2013) are not seen as posing the same health concerns for the deceased.

There were concerns, however, for a fair and just political process. Several research participants expressed concern that political corruption or "hanky panky" was involved in Ohio Land Management LLC's acquisition of Mount Hope. Although the news articles in this study did not mention political corruption, online commenters did express concerns for a fair political process.

FIGURE 3.1 Landscaping at Mount Hope Veterans Memorial Cemetery. Picture taken by Carmel Price on February 27, 2016.

Commenters suggested that local politicians "sold out" people in the community and described those in government as "political prostitutes." Political concerns also overlapped with other types of concerns. For example, one commenter mentioned a government agency, opportunity costs, and earthquakes all in one post: "Good luck with the fight to stop this, you will need it. ODNR [Ohio Department of Natural Resources] will find a way to allow it at any cost, they always do. It is unfortunate. Hopefully the 4.0 earthquake was big enough to rattle some sense into them."

Given that a new, explicitly moral and ethical discourse emerges when confronted with fracking beneath cemeteries, resistance to such fracking might also take a different shape. Matching the moral, spiritual, and ethical discourse described above, the Mount Hope Veterans Memorial Committee and their allies in Youngstown, Ohio were focused on the *preservation and beautification* of Mount Hope. For example, they partnered with a local Boy Scout troop, which completed several landscaping projects for the Mount Hope Cemetery (as seen in Figure 3.1), and they host an annual black history event every February. The event began in 2013 after Mount Hope was purchased by Ohio Land Management LLC. The black history event could best be described as a celebration; at the most recent 2016 (or fourth annual) event, there was an opening and closing prayer, music, singing, dancing, a candle-lighting ceremony, speeches, recognition of Mount Hope Veterans Memorial Executive Committee members and their service, and poetry. The event was well attended with easily over 100 people gathering at Reed's Chapel A.M.E. Church in Youngstown, Ohio (less than one mile from Mount Hope). The poetry included the work of historical African American poets and social activists, such as Paul L. Dunbar, Langston Hughes, Countee Cullen, Jupiter Hammon, and James Weldon Johnson, as well as poems written by local high school youth. Many children and youth participated in the program. The keynote speaker, Reverend Yvonne Hobson, read an excerpt from Dr. Martin Luther King Jr.'s "I have a dream" speech and then she articulated her own dream. Reverend Hobson, who served as the treasurer for the Mount Hope Veterans Memorial Executive Committee, said that she dreamed of having a perpetual fund to support the long-term preservation, maintenance, and beautification of the cemetery. Before and after the formal program, a community craft fair took place in the basement of Reed's Chapel A.M.E. Church; vendors were selling handbags, jewelry, belts, vases, photographs, and homemade pies, among other things. There was also food (e.g., chili, pasta salad, bread, cookies, chips, water, and soda) and fellowship.

Although there was an explicit reference to Mount Hope, as well as its preservation and beatification, there was no mention of fracking during the formal black history event. There was also no mention of Youngstown's Community Bill of Rights Charter Amendment, which was to be on the ballot in November 2016. The craft fair was designed purely to support community members and vendors and not as a fundraising avenue for Mount Hope. There was no charge for a vendor table or fee to attend the craft fair. The Mount Hope Veterans Memorial Committee occupied one table where they were selling white T-shirts with an inscription and picture of Mount Hope for ten dollars; there were no other fundraising activities and no political activities (i.e., petitions to sign). Before and after the event, multiple community members, including Mount Hope Veterans Memorial Executive Committee members, spoke about their loved ones buried in the cemetery and their concern for its preservation and beautification. The focus of the day's events was on celebrating and preserving the legacy and history of the community, including those buried in the cemetery, as well as promoting the future of the young people in the community. Conversely, the event was not focused on fracking. Perhaps this is because fracking in Youngstown has slowed, especially compared to the boom of several years ago, and fracking in the cemetery has not yet begun. Some community members we interviewed suggested that fracking has declined in Youngstown because of the amendments that have been repeatedly on the ballot; others argued that fracking has slowed because of larger socioeconomic forces (i.e., the fall in oil prices and a nationwide downturn in fracking). Nevertheless, Mount Hope does not seem to be playing a role in the discourse related to the Community Bill of Rights Amendment in Youngstown.

Furthermore, although there was no mention of preservation or beautification in the online comments, a few news articles did mention the "beauty" of a cemetery as important, in terms of honoring the dead and in terms of justifying a resistance to fracking. For example, one anti-cemetery-fracking activist described a local cemetery as a "beautiful, well-maintained property that honors the veterans and their families, and so oil drilling operations on that site are just not appropriate."

Pro-Cemetery-Fracking Perspectives

Our interview data revealed minimal support for fracking underneath cemeteries. As such, we utilized the 14 news articles and 54 online comments to examine pro-cemetery-fracking frames. Even so, the pro-cemetery-fracking sample was relatively small. Similar to those who view fracking as an economic opportunity for local communities (see Ceresola and Crowe 2015; Jacquet and Stedman 2013), proponents of drilling underneath cemeteries also viewed it as an economic opportunity but not in terms of job creation. The costs of cemetery upkeep played a role in the decision to allow cemetery fracking. For example, one cemetery owner explained that revenue from the mineral leases has helped him to "pave roads, repair fences and make other improvements [to the cemetery grounds] during economic hard times." Another article quoted the same owner as saying "you do this to reap the financial benefits. It's the same reason anyone else would." Similarly, two online commenters mentioned economic opportunities. One commenter asked: "How is getting $140,000 plus 16% for the maintenance of the cemetery selling out the community?" And the other online commenter who focused on economic opportunity said that the community "would be crazy not to take the money."

Interestingly, the most frequent argument for fracking underneath cemeteries was that it was not viewed as being "disruptive to the dead." Many of the news articles referenced the technological aspects of fracking (i.e., the depth of drilling and the horizontal drilling process that allows for fracking underneath cemeteries without having to place the heavy drilling equipment in and among the graveyard) as reasons why the dead were not being harmed or disrespected by fracking activity. For example, several articles quoted oil and gas company executives and cemetery owners as stating: "There is no surface or subsurface disturbance that has or will impact the cemetery location," "There certainly wouldn't be any interference with where active burials are occurring," "The drilling itself is not going to occur in the center of an active cemetery," "You're hundreds of feet below the ground, and it's not disturbing any graves," and "I really see nothing that is going to be hurtful." Online commenters similarly believed that fracking was not disruptive to the dead. A few made reference to the technological innovations of horizontal drilling and a few made reference to drilling depth. For example, one commenter noted that "graves are only 6 ft.

deep" and another asked "what is the danger?" Others accused people who were concerned about fracking underneath cemeteries as having a "lack of education," of letting "scare tactics confuse facts," and of being "a moron." One of our research participants also referenced a lack of knowledge and misunderstanding about the drilling process: "The public largely doesn't understand a lot of things that come its way . . . they're presuming that there's going to be a drill rig in the cemetery and they're going to drill right down through grandpa's grave, right through his forehead with a drill pipe." Three online commenters took this point a step further by literally indicating that the dead are not being disrupted because they are deceased. The statements included: "The idea that dead people are sacred is plain stupid," "I suppose it's callous, but the dead won't care," and "They're dead. They do not care. Move them."

Conclusion

This community case study has contributed to the sociological literature on fracking issues in two key respects: it addresses a new area of conflict concerning local disputes surrounding fracking underneath cemeteries; as well as discusses new discursive frames on both sides of the controversy: anti-cemetery-fracking frames associated with *moral and ethical* concerns and *beautification and preservation*; as well as pro-cemetery-fracking frames in which gas development is not perceived by citizens as being disruptive to the dead. One of our main findings is that resistance to fracking underneath cemeteries relies heavily on a moral and ethical discourse. This is consistent with the idea that cemeteries serve a culturally significant purpose in our society and help to meet the social, spiritual, and emotional needs of community members. Even for those residents who support fracking, drilling in cemeteries is perceived as potentially crossing an important moral or ethical boundary.

Another interesting finding concerned activists who opposed fracking underneath cemeteries but did not seem to oppose fracking in general. One story specifically stands out as evidence of this. A cemetery activist explained that she had family members buried in the cemetery and did not want fracking in the cemetery. She did not understand why they wanted to frack in the cemetery when there is so much other land to use. Her stance was not explicitly antifracking, but rather, "why frack here, why frack in this cemetery?" However, it was not a "not in my back yard" or NIMBY argument either. In fact, the activist went on to say, "They can come to my house and frack; I would rather them frack at my house then frack at the cemetery." When she was challenged on this by another cemetery activist, who took a more purely antifracking and environmental stance, she conceded that fracking at her house would not really be a good idea. Nevertheless, she persisted by saying, "I can move but my loved ones [buried in the cemetery] can't." Similarly, Tom

Ridge, former governor of Pennsylvania and director of Homeland Security under George W. Bush's administration, was quoted in August 2010 (while working as a consultant to the Marcellus Shale Coalition) as saying: "I'd have a tough time putting a rig down next to my tomb or next to anyone I'm related to" (Boren and Conte 2010). One research participant pointed out the hypocrisy of Ridge's stance by arguing that some "care more for the dead than the living."

It might be important, however, for antifracking activists to note this deeply moral and ethical opposition to fracking underneath cemeteries, even among those who are not against fracking elsewhere. For example, a regional politician explained to us how he used the indignation over drilling underneath cemeteries to garner public support against fracking in general: "Sure, I'm going to play that card. If that's what gets you up out of your chair, yeah, I'm going to wave that right in front of your face. Say, look, they're going in and drilling grandma's head. That's awful." Fracking underneath cemeteries seems to evoke a very emotional response by residents; those fighting fracking in their community might want to become aware of cemeteries at risk of fracking and connect to people who are working to protect these cemeteries. Although antifracking activists and anti-*cemetery*-fracking activists approached their arguments from slightly different angles, they often shared similar concerns about fracking in their communities concerning environmental quality, economic growth, public health and safety, and local democratic governance (Weigle 2011). Aligning different interest groups together in the fight against fracking might be important as local municipalities find themselves with reduced autonomy.

In 2015, the Ohio Supreme Court ruled by a 4–3 vote that cities and counties do not have the authority to regulate or ban fracking through zoning laws or other ordinances. Essentially city and town ordinances that attempt to regulate fracking conflict with a 2004 state law, which gives authority to the state of Ohio to regulate "all aspects" of fracking (Ludlow 2015). Despite this, as of June 2016, five communities in Ohio, including Youngstown, presented their local government with enough signed petitions to place Community Bill of Rights Amendments on November 2016 ballots. Each measure includes the rights of residents to local community self-government and the right of the people "to ban activities that will harm their communities," including fracking (CELDF 2016a). Youngstown presented 2,489 signatures to place the Community Bill of Rights Amendment on the ballot for the sixth time; only 1,259 valid signatures were required by an August 2016 deadline (Skolnick 2016). The November 2016 ballot measure failed by a similar margin to the November 2015 vote. However, residents remain undeterred and plan to continue placing the amendment on the ballot. If, in the future, the ballot measure passes then it would be in conflict with the state law and it is unclear how this conflict would be resolved or what the next steps would be for antifracking activists.

Furthermore, the concept of cemetery *beautification and preservation*, as a source of resistance to fracking underneath cemeteries, operated as a powerful tool. Beautification and preservation was a reason for opposition taken by the people fighting for Mount Hope; this type of resistance brought the community together in solidarity and support of the entire community and seemed to bring joy, peace, and fellowship to those who participated. Moreover, if one of the primary justifications for fracking underneath cemeteries is *economic opportunity*, as profits from fracking allow for cemetery maintenance and upkeep, then investing in the beautification and preservation of cemeteries might serve as a protective factor against fracking. Beautification and preservation might also be a form of fracking resistance that translates to larger communities; investing in communities (e.g., investing in a strong public transit system and public schools) might provide a protective effect against place detachment, which increases the vulnerability of communities and cemeteries to exploitation by resource-extractive industries (Willow 2014).

It is also important to understand that those who support fracking underneath cemeteries argue that it is *not disruptive* to the dead. Although this specific frame that addressed the technological aspects of drilling underneath cemeteries is more challenging to extend to living populations, it could provide some insight nevertheless. For example, if fracking advocates frame fracking as not disruptive to communities, or if they actually argue for regulations that minimize such disruption, then fracking might be better tolerated by community members as an acceptable trade-off for its perceived economic benefits. Conversely, framing fracking as inherently disruptive to people's quality of life and everyday community activities—as well as disruptive to the memory of veterans—may serve as a new discursive opportunity for future resistance.

References

Baptist, Karen Wilson. 2013. "Reenchanting Memorial Landscapes: Lessons from the Roadside." *Landscape Journal* 32(10): 35–50.

Barry, Dan. 2011. "As the Mountaintops Fall, a Coal Town Vanishes." *New York Times*, April 12. Retrieved June 6, 2016. (http://www.nytimes.com/2011/04/13/us/13lindytown.html?pagewanted=all).

Basmajian, Carlton, and Christopher Coutts. 2010. "Planning for the Disposal of the Dead." *Journal of the American Planning Association* 76(3): 305–317.

Boren, Jeremy, and Andrew Conte. 2010. "Catholic Cemeteries to Permit Gas Drilling Among the Headstones." *Pittsburgh Tribune-Review*, August 18. Retrieved July 17, 2015. (http://triblive.com/x/pittsburghtrib/news/s_695330.html).

Burg, Steven B. 2008. "'From Troubled Ground to Common Ground': The Locust Grove African-American Cemetery Restoration Project: A Case Study of Service-Learning and Community History." *The Public Historian* 30(2): 51–82.

Casal, Aimée, Juan Ignacio Aragonés, and Gabriel Moser. 2010. "Attachment Forever: Environmental and Social Dimensions, Temporal Perspective, and Choice of One's Last Resting Place." *Environment and Behavior* 42(6): 765–778.

Ceresola, Ryan G., and Jessica Crowe. 2015. "Community Leaders' Perspectives on Shale Development in the New Albany Shale." *Journal of Rural Social Sciences* 30(1): 62–86.

Collins, Brian D., Robert Kayen, and Yasuo Tanaka. 2012. "Spatial Distribution of Landslides Triggered from the 2007 Niigata Chuetsu–Oki Japan Earthquake." *Engineering Geology* 127(24): 14–26.

Community Environmental Legal Defense Fund. 2016a. "Press Release: Ohioans Determined to Stop Fracking through Asserting Rights." Retrieved July 1. (http://celdf.org/2016/06/press-release-ohioans-determined-stop-fracking-asserting-rights/).

Community Environmental Legal Defense Fund. 2016b. "Sign Up Now to Support Youngstown, OH, Fighting for Community Rights." Retrieved Feb. 29. (http://celdf.org/support/support-youngstown-community-rights/).

Cronon, William. 1995. "The Trouble with Wilderness; or, Getting Back to the Wrong Nature." Pp. 60–69 in William Cronon (ed.), *Uncommon Ground: Rethinking the Human Place in Nature*. New York: W. W. Norton.

Cucci, Luigi, and Andrea Tertulliani. 2011. "Clues for a Relation between Rotational Effects Induced by the 2009 Mw 6.3 L'Aquila (Central Italy) Earthquake and Site and Source Effects." *Bulletin of the Seismological Society of America* 101(3): 1109–1120.

del Puerto, C. B., and M.L.C. Baptista. 2015. "Cemetery and Space Tourism: Ambivalence Field of Life and Death." *Revista Iberoamericana de Turismo* 5(1): 42–53.

DrillingEdge. 2016. "Oil & Gas Production in Mahoning County, OH." Retrieved May 20. (http://www.drillingedge.com/ohio/mahoning-county).

Elliot, Marie. 2011. "The Fort St. James Cemetery." *British Columbia History* 44(3): 25–28.

Fernandez, Manny. 2012. "Drilling for Gas under Cemeteries Raises Concerns." *New York Times*, July 8. Retrieved March 1, 2016. (http://www.nytimes.com/2012/07/09/us/drilling-for-natural-gas-under-cemeteries-raises-concerns.html?_r=0).

Francaviglia, Richard V. 1971. "The Cemetery as an Evolving Cultural Landscape." *Annals of the Association of American Geographers* 61(3): 501–509.

Goodwin, John W., Jr. 2012. "Mount Hope Vets Cemetery Gets New Life." *The Vindicator*, March 28. Retrieved May 10, 2016. (http://www.vindy.com/news/2012/mar/28/mount-hope-vets-cemetery-gets-new-life/).

Hopkins, Bruce. 2003. *Spirits in the Field—An Appalachian Family History*. Nicholasville, KY: Wind Publications.

Hufford, Mary. 2007. *Ethnographic Overview and Assessment: New River Gorge National River and Gauley River National Recreation Area*. Boston, MA: Northeast Region Ethnography Program, National Park Service.

Jabbour, Alan, and Karen Slinger Jabbour. 2010. *Decoration Day in the Mountains: Traditions of Cemetery Decoration in the Southern Appalachians*. Chapel Hill, NC: University of North Carolina Press.

Jackson, Rodney D. 2006. "Estimates of Soil Erosion from Perched Cemeteries, Sampson County, North Carolina." *Southeastern Geographer* 46(1): 23–34.

Jacquet, Jeffrey B., and Richard C. Stedman. 2013. "Perceived Impacts from Wind Farm and Natural Gas Development in Northern Pennsylvania." *Rural Sociology* 78: 450–472.

Kniffen, Fred. 1967. "Necrogeography in the United States." *Geographical Review* 57: 426–427.

Kriesky, J., B.D. Goldstein, K. Zell, and S. Beach. 2013. "Differing Opinions about Natural Gas Drilling in Two Adjacent Counties with Different Levels of Drilling Activity." *Energy Policy* 58: 228–236.

Ladd, Anthony E. 2013. "Stakeholder Perceptions of Socioenvironmental Impacts from Unconventional Natural Gas Development and Hydraulic Fracturing in the Haynesville Shale." *Journal of Rural Social Sciences* 28(2): 56–89.

Ladd, Anthony E. 2014. "Environmental Disputes and Opportunity-Threat Impacts Surrounding Natural Gas Fracking in Louisiana." *Social Currents* 1(3): 293–311.

Linkon, Sherry Lee and John Russo. 2002. *Steeltown, U.S.A.: Work and Memory in Youngstown.* Lawrence: University Press of Kansas.

Lokocz, Elizabeth, Robert L. Ryan, and Anna Jarita Sadler. 2011. "Motivations for Land Protection and Stewardship: Exploring Place Attachment and Rural Landscape Character in Massachusetts." *Landscape and Urban Planning* 99(2): 65–76.

Ludlow, Randy. 2015. "Local Governments Cannot Regulate Fracking, Ohio Supreme Court Rules." *Columbus Dispatch*, Retrieved July 1. (http://www.dispatch.com/content/stories /local/2015/02/17/Supreme-Court-rules-fracking.html).

Maples, James N., and Elizabeth A. East. 2013. "Destroying Mountains, Destroying Cemeteries: Historic Mountain Cemeteries in the Coalfields of Boone, Kanawha, and Raleigh Counties, West Virginia." *Journal of Appalachian Studies* 19(1–2): 7–26.

Miller, DeMond Shondell, and Jason David Rivera. 2006. "Hallowed Ground, Place, and Culture: The Cemetery and the Creation of Place." *Space and Culture* 9(4): 334–350.

Miller, DeMond Shondell, and Jason David Rivera. 2009. *Hurricane Katrina and the Redefinition of Landscape.* Lanham, MD: Lexington Books.

Mumford, Lewis. 1961. *The City in History.* New York: Penguin.

Nunez, Christina. 2015 "Fracking Next to a Cemetery? 10 Unlikely Sites Targeted for Drilling." *National Geographic*, March 13. Retrieved March 1, 2016. (http://news .nationalgeographic.com/energy/2015/03/150313-oil-gas-drilling-fracking-cemetery -unlikely-drilling-sites-report/).

Palmer, Kim. 2016. "Ohio Mayor Pleads Guilty to Corruption Charges, Remains in Office." *Reuters.* Retrieved February 26. (http://www.reuters.com/article/us-ohio-mayor -idUSKCN0VZ2RG).

Pompili, Dan. 2012. "Youngstown Rally Targets Fracking." *Tribune Chronicle*. Retrieved September 13. (http://www.tribtoday.com/page/content.detail/id/576541.html).

Queener, Nathan L. 2010. "The People of Mount Hope." Master's thesis, Department of History, Youngstown State University, Youngstown, OH.

Redfern, Rebecca C., and Sharon N. Dewitte. 2011. "Status and Health in Roman Dorset: The Effect of Status on Risk of Mortality in Post-conquest Populations." *American Journal of Physical Anthropology* 146(2): 197–208.

Regional Chamber, Youngstown and Warren. 2016. "Largest Employers of the Youngstown-Warren Area." Retrieved Jan. 21. (http://regionalchamber.com/EconomicDevelopment /FactsFigures/LocalEconomy/LargestEmployers.aspx).

Schafft, Kai A., Yetkin Borlu, and Leland Glenna. 2013. "The Relationship between Marcellus Shale Gas Development in Pennsylvania and Local Perceptions of Risk and Opportunity." *Rural Sociology* 78(2): 143–166.

Skolnick, David. 2016. "An Anti-fracking Charter-Amendment Proposal Could Be in Front of Youngstown Voters for a Sixth Time." *The Vindicator.* Retrieved July 1. (http://www .vindy.com/news/2016/jun/28/group-submits-petitions-for-anti-frackin/).

Skoumal, Robert J., Michael R. Brudzinski, and Brian S. Currie. 2015. "Earthquakes Induced by Hydraulic Fracturing in Poland Township, Ohio." *Bulletin of the Seismological Society of America* 105(1): 189–197.

Snodgrass, Anthony. 2015. "Putting Death in Its Place: The Idea of the Cemetery." Pp. 187–199 in *Death Rituals, Social Order and the Archaeology of Immortality in the Ancient*

World, edited by Colin Renfrew, Michael J. Boyd, and Ian Morley. New York: Cambridge University Press.

Theodori, Gene L. 2009. "Paradoxical Perceptions of Problems Associated with Unconventional Natural Gas Development." *Southern Rural Sociology* 24(3): 97–117.

Theodori, Gene L. 2013. "Perception of the Natural Gas Industry and Engagement in Individual Civic Actions." *Journal of Rural Social Sciences* 28(2): 122–134.

Vanderstraeten, Raf. 2014. "Burying and Remembering the Dead." *Memory Studies* 7(4): 457–471.

Veit, Richard F., and Mark Nonestied. 2008. *New Jersey Cemeteries and Tombstones: History in the Landscape*. Rutgers, NJ: Rivergate Books.

Webster, Richard. 2013. "Iberville Redevelopment Threatens St. Louis Cemeteries Preservationists Contend." *Times Picayune*, March 15. Retrieved March 1, 2016. (http://www.nola.com/politics/index.ssf/2013/03/iberville_redevelopment_threat.html)

Weigle, Jason L. 2011. "Resilience, Community, and Perceptions of Marcellus Shale Development in the Pennsylvania Wilds: Reframing the Discussion." *Sociological Viewpoints* 27: 3–14.

Willow, Anna J. 2014. "The New Politics of Environmental Degradation: Un/Expected Landscapes of Disempowerment and Vulnerability." *Journal of Political Ecology* 21(1): 237–257.

Willow, Anna J., Rebecca Zak, Danielle Vilaplana, and David Sheely. 2014. "The Contested Landscape of Unconventional Energy Development: A Report from Ohio's Shale Gas Country." *Journal of Environmental Studies and Science* 4(1): 56–64.

Woodthorpe, Kate. 2011. "Sustaining the Contemporary Cemetery: Implementing Policy Alongside Conflicting Perspectives and Purpose." *Mortality: Promoting the Interdisciplinary Study of Death and Dying* 16(3): 259–276.

Wylie, John. 2009. "Landscape, Absences, and the Geographies of Love." *Transactions of the Institute of British Geographers* 34(3): 275–289.

Wynveen, Brooklynn J. 2011. "A Thematic Analysis of Local Respondents' Perceptions of Barnett Shale Energy Development." *Journal of Rural Social Sciences* 26(1): 8–31.

Zremski, Jerry. 2014. "Youngstown, Ohio, Is a City Changed by Fracking." *Buffalo News*, May 18. Retrieved March 1, 2016. (http://www.buffalonews.com/city-region/youngstown-ohio-is-a-city-changed-by-fracking-20140518).

4

Mobilizing against Fracking

Marcellus Shale Protest
in Pittsburgh

SUZANNE STAGGENBORG

Introduction

Across the United States, conflicts have emerged in many towns, cities, and states where the process of hydraulic drilling and fracturing, known as "fracking," has created a new era of oil and gas development in major shale formations. In the past decade, a natural "gas rush" took place in Pennsylvania, as advances in drilling techniques spurred companies to invest in extracting shale gas reserves from the Marcellus Shale region, sending thousands of "landmen" to the state to urge residents to sign leases to allow drilling on their property (Wilber 2015). Controversies quickly emerged as thousands of wells were drilled, bringing economic booms to some areas, but also disrupting communities and raising concerns about the health and environmental impacts of shale gas development. Grassroots protests against fracking have also spread in recent years in response to the documentary film *Gasland* (Fox 2010; Vasi et al. 2015), as well as the growth of fossil fuel resistance movements across the globe (Klein 2014). To date, scholars have examined various aspects of the controversy, including how and why perceptions of the impacts of fracking vary

(Boudet et al. 2014; Brasier et al. 2011; Ladd 2013; Malin 2014; Sangaramoorthy et al. 2016; Sarge et al. 2015; Theodori 2013); how the issue has been framed by proponents and opponents of fracking (Hudgins and Poole 2014; Ladd 2014; Weigle 2011; Wright 2013); how civil society organizations and governments have responded to drilling practices (Davis 2012; Jalbert, Kinchy and Perry 2014); and how movements against fracking have emerged in certain communities (Mazur 2016; Simonelli 2014; Vasi et al. 2015). Research shows that local perceptions of the costs and benefits of fracking vary with the extent to which drilling sites are developed and the direct experience of residents living in communities where drilling takes place (Ladd 2013: 64).

In this chapter, I explore how the antifracking movement emerged in Pittsburgh, where activists began organizing in response to the perceived threats posed by unconventional natural gas development to the environment, human health, and quality of life. In November 2010, the Pittsburgh City Council, citing health and environmental concerns, voted unanimously to become the first U.S. city to ban natural gas drilling within its borders (Smydo 2010a). The Pittsburgh ban was unprecedented, but the debate took place in the context of nationwide conflicts over fracking and warnings about the impacts of fracking on communities (Lord 2010). The city ban directly inspired other communities to attempt to enact similar bans (Smydo 2010b), and the Pittsburgh movement continued to organize collective action aimed at exposing the dangers of fracking and banning the practice throughout the state. To get off the ground, the antifracking movement in Pittsburgh had to create an organizational infrastructure that could support ongoing collective action campaigns. I examine how that was accomplished and the challenges facing the movement. How does a new movement create effective organizational structures, and how can a local movement keep up momentum?

To address these questions, I employ participant observation and interview data on the process of mobilization. I focus on the organizational network known as "Marcellus Protest" (MP) that sparked the movement in Pittsburgh and has worked to keep the movement going since 2010. Drawing on social movement theory, I begin by discussing key processes involved in mobilizing movements and organizations. After describing my data and methods, I provide a brief history of the origins of the local movement. I then demonstrate how the Pittsburgh movement mobilized support and maintained collective action around fracking. I conclude with a discussion of challenges facing the movement.

Social Movement Mobilization

The process of mobilizing a social movement is anything but simple or automatic, and few studies provide close-up descriptions of how groups organize, maintain momentum, and engage in ongoing collective action. While a number of studies explain why mobilization occurs when it does by examining factors such as political opportunities, resources, and frames (e.g., Cress and Snow 1996; McCammon 2001), fewer studies examine how mobilization occurs. Some work does, however, provide insight into aspects of the process, which includes creating organizational structures and carrying out various types of collective action. The process of creating organizational structures is a difficult one, influenced by the backgrounds and ideologies of activists and by their experiences and interactions in an emergent movement. Activists create the types of organizations with which they are familiar, and different types of organizational structures, such as bureaucratic or collective ones, appeal to different types of activists (Freeman 1975). Movements that attract diverse constituents face particular challenges due to cultural differences associated with class and race (Lichterman 1995). Thus, one of the challenges in mobilizing a movement is to create organizational structures and decision-making procedures that appeal to diverse constituents.

Research on how social movement organizations (SMOs) are able to develop workable structures so that they can maintain themselves and formulate strategies points to some key processes (Zald and McCarthy 1987). One is the development of "leadership teams" of organizers who regularly discuss strategy and teach skills to new activists (Ganz 2000; Polletta 2002; Reger and Staggenborg 2006). While some groups avoid talk of "leadership" because of its hierarchical connotations, most struggle with issues of how labor is divided, how decisions are made, and how participants can be integrated into the group. The structures of SMOs vary with regard to features that affect mobilization, such as the number and nature of group meetings, the types of "task committees" or "working groups" formed, the ways in which meetings are conducted, the extent of centralization and bureaucratization, and the number of actively involved members (Gamson 1990; Han 2014; Lichterman 1996; McCarthy and Wolfson 1996; Staggenborg 1989). These structural characteristics of organizations are important in part because they allow groups to recruit individual activists, who put significant energy into organizing collective action (Corrigall-Brown 2012; McCarthy and Wolfson 1996).

Since movement growth requires ongoing collective action, activists and organizations need to devise effective strategies and tactics. In doing so, they must frame issues in ways that interpret problems, propose solutions, and motivate people to participate in collective action (Benford and Snow 2000). Movements that survive over a number of years tend to consist of a series of actions and campaigns that serve to attract participants, gain media attention, generate public support, and expand resources, even if they do not have immediate or direct impacts on targets. To sustain collective action, movements need an organizational infrastructure capable of nourishing activists and generating strategies and tactics. In this research, I examine the ability of the antifracking movement in Pittsburgh to mobilize and sustain collective action by looking at efforts to create organizational structures and devise strategies in response to concerns about the health and environmental impacts of shale gas drilling.

Data and Methods

This chapter draws on extensive participant observation research on the local Pittsburgh environmental movement. Since beginning my study in 2010, I have investigated a wide range of groups and events in the local environmental movement community, which includes cultural spaces as well as political organizations, coalitions, and networks (Staggenborg 1998). Long-standing political organizations in the local environmental movement include the Group Against Smog and Pollution (GASP), founded in 1969 to combat Pittsburgh's air pollution problems, and the Allegheny Group of the Sierra Club, founded in 1970. In addition, the local movement includes more recently formed environmental justice groups and an antifracking movement, which has spawned coalitions such as Protect Our Parks (POP), organized in 2013 in response to the announced intention of Allegheny County's chief executive to lease park land for shale gas drilling.

While paying attention to a broad range of organizations and activities in Pittsburgh, I selected some groups and activities for intensive study based on variations in ideological and organizational characteristics. I gained extensive access to these groups by participating in their activities and contributing to them in various ways, while at the same time being completely open about my research intentions. For the purposes of this chapter, I use the portion of my data focused on the antifracking movement, particularly Marcellus Protest, a network of antifracking activists in Southwestern Pennsylvania; the Shadbush Environmental Justice Collective, a radical environmental collective; and the POP coalition, which included the Sierra Club, Marcellus Protest, and Shadbush, as well as other organizational members. I collected extensive data on these groups, including participant observation of meetings, collective actions, and interviews with participants. In the case of Marcellus Protest, I attended 77 meetings and events since 2010 and interviewed 11 activists. I joined the Shadbush collective in 2011, attended 47 meetings and events through 2014, and interviewed 6 participants. (Shadbush disbanded in 2014 to form a new group, Three Rivers Rising Tide.) I attended over 45 meetings and events of the POP coalition, which ceased meeting in 2016. I recorded detailed field notes following all meetings and events, transcribed all interviews, and used a qualitative analysis program to code and analyze the data. These data allowed me to examine a number of key processes involved in movement mobilization and the development of effective strategies and tactics.

Origins of the Pittsburgh Antifracking Movement

The movement against fracking in Pittsburgh and Southwest Pennsylvania originated in late 2009 and 2010 as residents began to learn of threats posed by shale gas drilling in the Marcellus Shale region and across the country. Hundreds of people first saw *Gasland* in Pittsburgh on June 5, 2010 at a special showing attended by Josh Fox before the debut of the film on HBO on June 21. At the same time, neighbors began talking to one another and organizing in their communities after learning about the leasing of land for Marcellus Shale drilling in and around Pittsburgh. A few concerned residents of the Lincoln Place neighborhood called their city councilor, Doug Shields, to express their concerns, leading him to start investigating the issue and to eventually sponsor Pittsburgh's ban on fracking (interview, April 3, 2012). The residents formed neighborhood groups, including the Lincoln Place Action Group and the Lawrenceville Marcellus Action Group, and began meeting to talk about the issue. Some large community meetings were held in Lincoln Place, a part of the city where landmen were active in trying to secure leases from homeowners. Another special screening of *Gasland* in Frick Park with director Josh Fox in August 2010, attended by over 500 people, was particularly important in raising consciousness about the issue among larger numbers of residents.

The emerging movement expressed strong concerns about the dangers of fracking, including possible contamination of drinking water, the depletion of aquifers, the generation of toxic wastewater, the release of volatile organic chemicals and methane into the air, and contributions to climate change. Meanwhile, the Community Environmental Legal Defense Fund (CELDF) drafted an ordinance to ban fracking and approached members of the Pittsburgh city council to drum up interest. Doug Shields ended up spearheading the ordinance in the City Council, and residents became involved in a citywide movement that lobbied for the ban on shale gas drilling in Pittsburgh. The City Council was provided with all kinds of information about the dangers of fracking, which helped to counter the industry's promise of jobs and economic benefits (Lord 2010). Health and environmental concerns led to a unanimous vote in favor of a ban in Pittsburgh; then Pittsburgh City Council President Darlene Harris memorably concluded that most of the job creation would be "for funeral homes and hospitals" and asked "is it worth it?" (Smydo 2010a).

Some of the people who became involved in the early movement were longtime activists, either in other social movements, such as peace movements, or as members of local environmental organizations, such as GASP and the Sierra

Club. Others were ordinary citizens with little or no previous involvement in social movements who became alarmed about fracking and started to investigate the issue after being asked to sell their drilling rights. For example, Mark, a resident of Lincoln Place who did not consider himself an environmentalist at the time, explained that he educated himself using the Internet even before attending community meetings: "I googled it and googled it and googled it. . . . I would just learn, learn, learn about the whole process of hydraulic fracturing, aquifers, fractured earth, fractured bedrocks, radon, argon, every gas, every element in the earth that you don't want to come out, every element that's in what I'll call the shallow earth, things that you can reach with these drill heads two, three, four miles down, that's still shallow earth, all the plutonium, P2–38s, P2–35s. Everything that they're pulling up out of the ground, I knew about before I went to the meeting" (Interview, May 15, 2012). Mark also learned of people in Pennsylvania who had experienced fracking firsthand and were speaking out about the practice and its effects on their land and the health of their families. He watched YouTube videos made by Ron Gulla, a farmer who had leased his land in 2002 and later spoke out about the disrespect shown by the gas company and the damage done to his land by fracking. Mark emailed Gulla and spoke to him on the phone, and he and his wife and three of their neighbors later drove to Gulla's farm in Hickory, Pennsylvania to meet him and see for themselves the damage caused by fracking. They also met other people in the area and talked to them about fracking and the health problems they and their children were facing. Others in Pittsburgh told similar stories of learning about the health and environmental risks of fracking and hearing the very real stories of people harmed by it, even before *Gasland* was widely viewed by large numbers of people.

An ad hoc group of experienced activists and other concerned citizens decided to try to build the movement against fracking by organizing a large demonstration in Pittsburgh on the occasion of a shale gas industry conference at the convention center in downtown Pittsburgh. Originally, the gas industry planned an October 1, 2010 meeting in Pittsburgh and activists began gearing up to protest it; however, the meeting was cancelled, possibly in response to the scheduled protests. A national summit on shale gas drilling was later scheduled by the gas industry for November 3, 2010, which gave the emerging movement more time to organize a protest on that date. While the Marcellus Shale Coalition, an industry trade group, promoted the benefits of shale gas drilling for the region, the emerging movement sought to counter profracking messages. On September 3, 2010, the first of a series of public meetings to plan the protest was held. Meetings were typically attended by about 50 people and a number of working groups were formed, such as Outreach, Media, Logistics, Fundraising, and Art.

The working groups were extremely active, attending to all the details needed for a large demonstration. The Art group created large, colorful banners, puppets, and signs. The Fundraising committee held a number of events, including a very successful "Shale Trail" awareness night and fundraiser at a bar crawl in the Lawrenceville area of Pittsburgh. The Outreach committee worked on making contacts with other groups, such as environmental organizations, and sent out letters asking for endorsements and contributions for the November 3 march. Organizers also worked with the Pittsburgh Student Environmental Coalition (PSEC) to organize students and show *Gasland*, a critical organizing tool, at area colleges. A radical environmental group called the Shadbush Environmental Justice Collective, which had formed earlier the same year in response to issues such as mountaintop removal coal mining, became very involved with the protest organizing. Members of the Shadbush collective were particularly active in organizing door-to-door outreach, using their "What the Frack?" pamphlet, which explained the process of hydraulic fracturing and detailed problems and potential impacts, such as the use of huge amounts of water from nearby waterways, the disposal of wastewater, groundwater and surface water contamination, and other ecological impacts.

It was a busy and exciting time, and the November 3, 2010 demonstration was the payoff for all of the planning. Pittsburgh is a great place for marches because protesters can march across one of its many bridges into downtown Pittsburgh. The march began with a rally at the Allegheny Landing on the North Shore of the Allegheny River, with speakers and performers who revved up the demonstrators. On a beautiful fall day in Pittsburgh, hundreds of activists then proceeded across the Rachel Carson Bridge (named for the environmental pioneer from the Pittsburgh area) to the site of the protest in front of the convention center, where the gas industry was holding its conference. There, the protesters were inspired by more performers and speakers, including *Gasland* producer Josh Fox and city councilor Doug Shields. Stephen Cleghorn, an organic farmer from Jefferson County, spoke movingly about the threat of nearby shale gas drilling to his farm and about how fracking is degrading our quality of life (fieldnotes, November 3, 2010). Participants came from all over Pennsylvania, New York, and other places, as well as Pittsburgh, and spent the afternoon talking and building solidarity with one another in the emerging movement. Soon after the protest, on November 16, 2010, the Pittsburgh City Council voted 9–0 to ban shale gas drilling in the city based on perceived threats to public health and the environment (Smydo 2010a).

Inspired by the success of the demonstration and the victory in Pittsburgh, organizers began to plan the future of the movement. On November 12, 2010 they held an open meeting to ask, "where do we go from here?" after the November 3 protest. At that meeting, many participants expressed their enthusiasm about the large demonstration as a "powerful" event. As one protester said, "Walking over that bridge was an amazing feeling. I was proud to be part of that group. . . . It was awe inspiring" (fieldnotes, November 12, 2010).

Newcomers who were mobilized by the protest joined the protest organizers in brainstorming about the future direction of the movement. Many were from areas outside Pittsburgh, such as Butler County, that were immediately threatened by fracking, and worried about their land and their health. The group became known as Marcellus Protest, taking its name from the protest and website (www.marcellusprotest.org), which was set up in advance of the protest and used to promote antifracking activities. Then came the hard work of building a movement by creating organizational structures and devising ongoing strategies and tactics.

Organizational Structures

The successful November 3 march, as well as the fracking ban in Pittsburgh, generated a great deal of enthusiasm and helped launch Marcellus Protest. Nevertheless, the group faced significant challenges in creating and sustaining an organizational structure. Participants at the November 12, 2010 meeting were diverse with regard to their movement experience and ideology. Approximately half of about 60 people in attendance had not been involved prior to the November 3 protest (fieldnotes, November 12, 2010). Many had never before participated in efforts to build a mass movement, while others were seasoned activists. Some were alarmed by the impacts of fracking on their communities, but new to environmentalism and progressive politics. Others considered themselves radicals, including those with experience dating back to the 1960s, as well as young people influenced by anarchism and radical environmentalism. While participants shared concerns about fracking, they lacked common ideas about how to organize and they disagreed on tactical matters such as the value of electoral politics.

The organizational structures used to mobilize the demonstration provided something of a template for how the participants might proceed. At the November 12 meeting, activists managed to create a basic structure for the group, which included a Coordinating Committee (CC), general assemblies, and working groups. However, it was difficult to work out details of how the group would function at the meeting, at which activists tried to plot a future direction for both an organization and a movement. The role of the CC was the subject of some mild disagreement at the start, which later resurfaced. Initially, the committee was seen by some as an administrative working group that would attend to running the organization, just as the logistics group had operated for the demonstration. Others talked at the start about a steering committee, possibly with elected members, to provide leadership to the organization, but participants who favored more "horizontal" structures rejected this idea. The CC struggled to structure itself and to build MP as an effective organization. While it was supposed to consist of representatives of working groups and member organizations such as Shadbush, it was difficult to get all of these groups to send a representative to every meeting. Despite having committed members who took turns serving on the CC until 2013, when it finally stopped meeting, MP never really developed a leadership team or a way of developing new leadership. Instead, the group ran on individual enterprise, depending on those willing to step up and organize events. The working groups, which included Outreach, Action, Internet, and a few that never really got off the ground, such as a Media group, were intended to drive the organization, but they met with mixed success. Some met only sporadically and were short-lived, while a few met regularly until they fulfilled their missions. The most vital working group to form out of the organizing meeting was a new Outreach committee, with a number of new members from rural areas of Southwestern Pennsylvania, as well as from Pittsburgh and its suburbs. As a result of the work of the Outreach committee, several rural or suburban groups formed outside of Pittsburgh in areas directly affected by shale gas drilling.

After MP's exciting start, there was a great deal of enthusiasm for action, and many people were eager to participate in demonstrations and other tactics. Eventually, however, MP struggled to keep up the momentum, experimenting with various organizational structures and strategies aimed at building the movement and having an impact on shale gas drilling in the region. To some extent, the group was a victim of its own success; once the ban on fracking was passed in Pittsburgh, there was no longer an obvious strategy to pursue within the city. Although neighborhood groups that had formed in Pittsburgh generally did not take a Not in My Back Yard (NIMBY) position, the ban did reduce the immediate threat that residents felt when they thought fracking might take place in the city. However, shale gas drilling continued to expand in Pennsylvania and MP contributed to efforts to combat industry influence by attending demonstrations and providing testimony at hearings across the state. MP activists helped spread the model of the CELDF ordinance, holding workshops with CELDF and providing expertise to other community groups in Pennsylvania. After the Outreach committee spawned several suburban and rural groups, most of their members became less active in MP and the Outreach committee eventually dissolved after participation declined. But ties among groups in the MP network remained strong, and many activists outside Pittsburgh continued to attend special events and meetings called by MP.

While some working groups functioned for a time and then fizzled out, those that formed to complete a specific task probably enjoyed the most success. Generally, it seemed to work better to create groups as they were needed for particular tasks, rather than to form standing committees that might generate new tasks or actions. MP activists followed developments in the battle over fracking in Pennsylvania and other states and often responded to industry strategies and governmental decisions. As an example of a successful working group, participants interested in framing the issue and combating messages from the gas industry formed a strategic messaging group that met regularly with the goal of coming up with effective ways of talking about fracking. After reading materials about messaging and thinking about the likely reception of various frames by different audiences, the group came up with a strategic messaging "toolkit" that was shared on the website (www.marcellusprotest.org /toolkit). The toolkit emphasized the need to promote positive values, such as preservation of the environment for future generations, community self-determination, health and safety, environmental justice, and sustainability. It also identified threats to those values, such as greed and consumerism, the failure of government to represent citizens rather than corporations, and industrial destruction of the environment. In this case, the working group not only had a specific task, but consisted of a congenial group of people who enjoyed meeting together, discussing works on strategic messaging, and thinking about how to be effective. Once its mission was completed, after about seven months of meetings, the working group disbanded. The Outreach group, during the year when it met, was similar in that participants enjoyed one another's company and felt they were achieving goals. When Outreach disbanded, most felt it had fulfilled its purpose of raising consciousness and generating a network of groups and individuals inside and outside of Pittsburgh. In the case of the Action committee, however, there was little agreement about the need to meet; one organizer felt that meetings were a waste of time and that calls for action should simply be put out on the email list when needed. This worked for a while, when individuals organized targeted demonstrations, but did not produce any long-term strategy or firm commitments to the organization.

From 2010 to 2014, MP held a series of general assemblies, some of which attracted newcomers and produced promising strategic ideas. For instance, in May 2011, a meeting attended by about 40 people, including a lot of new faces, lasted for about three hours, but remained energetic for the whole time. Part of the meeting was devoted to small group strategizing, which generated new ideas and seemed to energize participants, who were eager to respond to the various conflicts that were occurring around fracking in the Marcellus Shale. For example, one group discussed a possible boycott of farm producers that had leased land for drilling, but also their concern about those who had leased without understanding the consequences. At another general assembly in January 2012, nearly 50 people attended for a potluck dinner, short films dealing with fracking by a local filmmaker, and discussion of a proposed action around gas pipelines. Participants were concerned about a variety of issues around the expanding network of pipelines in Pennsylvania, including the risk of explosion and other safety concerns, the development of compressor stations associated with pipelines, dangers to wildlife, and the potential for companies to build pipelines over the objections of property owners through "eminent domain." The action around pipelines was planned as a "counter-summit" to a gas company event focused on infrastructure in March 2012. Thus, some strategic planning and actions came out of general assemblies, which helped to build the movement by addressing issues of widespread concern. But MP was largely decentralized with no strong leadership team to direct the organization. The CC planned general assemblies, often as educational events, but it was sometimes difficult to get very many people to come. When attendance was good and strategic ideas were generated, there was often little follow-through. While general assemblies were originally envisioned as the place where strategies would be formulated and decisions made, different people attended them and there was little continuity from one assembly to the next.

Organizationally, MP never solved the problem of relying on individuals to step up and initiate actions, as opposed to having working groups or leadership teams that got things done. For several years, a relatively small number of about 20 committed and talented individuals did much of the work needed to keep Marcellus Protest alive—serving on the Coordinating Committee, organizing educational forums and other special events, attending hearings and demonstrations, putting out calls for participation in various collective actions and events, developing new literature, speaking at public events, putting out a newsletter, and keeping up the website, calendar, and Facebook page. Despite their efforts, MP encountered many challenges in keeping the organization going and devising ongoing strategies and tactics. Eventually, MP became more of a network of activists than an organization, held together by its email lists, website, and a core of committed activists. Nevertheless, as exchanges on the Google groups demonstrate, activists in the MP network remain engaged and aware of continuing developments in the conflict over fracking. The larger movement has survived and MP and its offshoots continue to play an important role in grassroots resistance to fracking in the region.

Strategies and Tactics

The inaugural demonstration on November 3, 2010 illustrated the value of a good tactic for building a movement. It was an exciting time planning for the demonstration, which brought experienced activists together with people new to movement organizing. The various committees had concrete tasks, such as working out the logistics of the march, creating props, raising money, and holding educational events. As in other places around the country, the film *Gasland* was particularly important as an organizing tool (Vasi et al. 2015); many people who attended the organizing meetings for the demonstration had attended a special screening of the documentary in June or August 2010, and activists continued to show the film in a variety of venues to educate and mobilize community members. Planning meetings were energetic as participants showed up to work toward a specific goal—a large demonstration that would help stimulate the movement and send a message to the city and the industry. Individuals came up with various ways to contribute; for example, one of the artists had weekly brunches at his house to raise money and create solidarity among organizers. The artists also designed popular "Don't Frack with Pittsburgh" T-shirts and sweatshirts and held workshops where anyone could go and help paint signs and make props. All of this hard work paid off with an inspiring march that energized the movement and demonstrated support for a ban on fracking in Pittsburgh.

After the big November 3 march, activists were eager to continue to organize smaller scale demonstrations, but their success depended on individuals taking the lead and there was not always consensus on strategy. For example, on the day afterward, November 4, an activist organized a demonstration at the site of a municipal waste facility to protest the disposal of wastewater generated by state fracking operations. There was concern over whether or not the protest would involve civil disobedience, and Marcellus Protest agreed to endorse the action only after assurances were made that it would not involve arrests. The action came off without incident and drew about 30 protesters who were enthusiastic about the demonstration. Its organizers planned to continue with similar protests, but they were never able to produce a sustained campaign around the issues of wastewater disposal and suspected illegal dumping, perhaps because it was very hard for a few people to organize such a campaign without a strong organizational infrastructure.

In early December 2010, activists protested at the Carnegie Museum in Pittsburgh in response to the possibility that the museum would lease land for fracking that it owned outside the city and used for scientific research. In response to the protests, however, the museum soon decided not to do so. A few days later, activists sang antifracking Christmas carols in downtown Pittsburgh on the steps of the City County Building. In January 2011, one activist put a great deal of energy into organizing a bus to go to Harrisburg to join other activists from around the state to protest the inauguration of profracking governor Tom Corbett. Unfortunately, icy weather prevented many from attending, leaving the bus only partly filled, and the person who spent so much time arranging the trip vowed "never again!" For those who traveled to Harrisburg, it was a very uncomfortable demonstration in the freezing cold, but connections were made with activists across the state and others applauded the Pittsburgh participants at a large meeting following the demonstration (fieldnotes, January 18, 2011). Pittsburgh was becoming known as a movement center after staging the large demonstration, launching a highly informative and frequented website, and passing the first U.S. city ban on fracking. Activists from Pittsburgh interacted with those from other regions as they participated in protests and lent their expertise to the movement and as activists from other regions visited Pittsburgh.

In 2011, activists from Pittsburgh attended other rallies in Harrisburg and testified at hearings around the state. Marcellus Protest worked with others in the movement community, raising the issue of fracking along with the Sierra Club, for example, at the annual St. Patrick's Day parade in Pittsburgh in 2011, and causing a stir among parade organizers who withdrew the Sierra Club's invitation to the parade the following year. The Thomas Merton Center, a local peace and justice center, gave its annual "New People Award" to Marcellus Protest in 2011. Activists worked to ban fracking in other municipalities, with some successes, such as in neighboring Wilkinsburg. When Occupy Pittsburgh formed in October 2011, Marcellus Protest members carried their banner in the inaugural march (and a few began to spend more time with Occupy than with MP). Shadbush activists recruited participants from the Occupy camp for a demonstration at a second annual gas industry conference in November 2011. In December, there was more Christmas caroling at the City County Building.

In 2012 and 2013, activists continued to participate in demonstrations, hearings, and supportive actions in other parts of the state as well as in Pittsburgh. The health and environmental impacts of shale gas drilling continued to motivate actions, as did concerns about people directly affected by fracking. On March 20, 2012, Marcellus Protest held a demonstration and educational forum focused on pipelines in response to a gas industry conference on Marcellus Shale infrastructure. The Shadbush collective was very active and organized with an offshoot of the Marcellus Protest Outreach committee in Butler County called Marcellus Outreach Butler (MOB) to occupy the office of their profracking state representative in May 2012. MP and Shadbush both participated in MOB's "Tour de Frack" bicycle ride to Washington DC, putting on a demonstration for the Pittsburgh leg of the tour in July 2012. Shadbush organized several tours of fracking sites in heavily drilled areas, and developed relationships with some of the people affected by fracking, including families in Butler County who had lost use of the water in their homes (see https://sites.google.com/site/waterforwoodlands). To assist them, Shadbush organized the first of two large fundraisers in Pittsburgh in October 2012. Shadbush then organized a well-attended action camp on an organic farm located near a Shell fracking site, demonstrating against the company in November 2012 at the conclusion of the camp. Having developed a strong relationship with the family that owned the farm, Shadbush wanted to help them protest against Shell. In addition, the collective decided to follow up on its action camp to make it more influential than those typically staged by radical groups that culminate in a one-shot action. Consequently, Shadbush also organized a much more involved demonstration held in January 2013 after two months of planning. To symbolize the organic farm and attract attention, the group constructed a giant papier-mâché pig. On the day of the demonstration, the pig was transported to the drilling site—much to the surprise of police—and several activists locked themselves to the pig while others played supporting roles. This creative action received a great deal of media and police attention, as well as acclaim from activists.

Shadbush was earning a reputation as an innovative organizer of direct action and its members played a key role in organizing some high-profile demonstrations. On Earth Day 2013, Shadbush worked with Marcellus Protest on a demonstration at the office of the Pennsylvania Department of Environmental Protection (DEP), along with a statewide coalition that organized demonstrations against the DEP across the state. The DEP was targeted for not adequately protecting communities against fracking, but it was also an attractive geographic target in Pittsburgh. Because the DEP office in Pittsburgh is located on an island, Shadbush activists organized an "Allegheny Armada" of kayaks and canoes (despite the freezing water) to meet other demonstrators who marched across a bridge for a brief occupation of the DEP. Following this memorable demonstration Marcellus Protest joined with a number of groups on an organizing committee to bring Josh Fox and *Gasland II* (Fox 2013) to Pittsburgh in June 2013. As a result, some 1,700 persons attended the showing in the city, the final stop on Fox's nationwide tour before the debut of the documentary on HBO. Beginning in July 2013, MP became heavily involved in the Protect Our Parks campaign against fracking under the Allegheny County parks, which helped to revitalize the group. In October 2013, MP and Shadbush demonstrated in support of POP with Power Shift, a youth climate summit, when it came to Pittsburgh. Both groups continued to be active in POP and other activities. In early 2014, Shadbush held a second fundraiser to benefit people who had lost the use of their domestic water source as a result of fracking.

Thus, antifracking groups remained active, holding many inspiring and memorable demonstrations that helped build the movement. MP continually tried new ideas and coordinated with other groups, including the Shadbush collective. Many ideas were partially explored or generated a few actions, but most activities failed to result in sustained campaigns. One reason for this was that it was not always easy to find clear targets. For example, some members of MP were alarmed that farmers in Pennsylvania were selling their surface or mineral rights for drilling without understanding how fracking might damage their water, animals, and crops. Some participants suggested boycotts of farms that allowed fracking, but others wanted to be careful not to blame farmers, who had a difficult enough time earning a livelihood. MOB, the group in Butler County (about an hour away from Pittsburgh), organized a demonstration at a drilling site on a dairy farm in the summer of 2011, but was careful to blame the gas industry rather than the farm in its messaging. There was some talk of a "frack-free farming campaign" to support farms that did not allow fracking, but the campaign never got off the ground. This was due in part to disagreements over targets, but also because of a lack of entrepreneurs willing to spearhead such a campaign. Issues around fracking created passion among movement constituents, but organization was necessary to mobilize that sentiment; without a strong organizational infrastructure, it was difficult even for highly motivated individuals to initiate activities. Campaigns, which involved a series of collective actions, were even more difficult to orchestrate.

The Protect Our Parks coalition that began in 2013 emerged from a deliberate attempt by activists to create an ongoing campaign. The effort was initiated by Shadbush, after the group decided at a retreat that it should try to spearhead a campaign, rather than continuing to engage mostly in "one-off" actions. Shadbush invited activists from other groups, including MP, to join them in brainstorming campaign ideas. One interest was in public land; activists across Pennsylvania and in other areas of the country were working to protect public parks and forests from fossil fuel extraction. By the second meeting, the Allegheny County Executive, Rich Fitzgerald, had expressed to the media his intent to lease the rights to drill under the county parks, after previously leasing the rights to drill at the Pittsburgh airport. (He said the drilling would not occur in the parks, only under them, with well pads on private property adjacent to the parks.) This sparked outrage and a decision to build a campaign around the county parks, which activists were optimistic could yield a victory. The result was a campaign coalition consisting of individuals and representatives of environmental groups such as the Sierra Club, GASP, PennEnvironment, and Clean Water Action, as well as MP and Shadbush.

Beginning in August, 2013, POP members began testifying during the public comment period at County Council meetings, talking to legislators, and collecting signatures against fracking on petitions in advance of a vote by the council on signing a lease to allow drilling under Deer Lakes Park, one of Allegheny County's nine public parks. They testified about the health and environmental impacts of fracking on air, water, and wildlife and they also expressed concern that county parks, intended as places of retreat, would be marred by fracking sites, even if the well pads were near the parks rather than in them. The campaign attracted numerous participants, including many from Marcellus Protest and other environmental groups such as the Sierra Club. POP members lined up supporters to testify and lobby the County Council, and developed friendly relations with some of the councilors, even bringing a large cake to a meeting before the holidays to thank the council for listening. Nevertheless, the County Council voted to lease the drilling rights to Deer Lakes Park, delivering a stinging defeat to the movement. POP fought back by collecting signatures for the first ever citizens' ordinance (permissible under the Allegheny County charter) that called for a moratorium on further drilling in the parks, which County Council also voted down. Nevertheless, POP continued to devise new tactics and find new targets, such as efforts to force the Allegheny County Health Department to consider the health effects of fracking. Although participation in POP declined to a core of about 10 regulars at meetings, the coalition remained active to fight against drilling under the county's remaining parks, helping to keep the larger movement alive.

The movement also expanded across the state of Pennsylvania and in 2014 a new statewide coalition, Pennsylvanians Against Fracking, organized with the initial goal of winning a wider moratorium on fracking. Pennsylvanians Against Fracking worked with constituencies such as Pennsylvania farmers and faith leaders opposed to fracking (www.paagainstfracking.org). A number of activists from MP were involved in creating the state coalition, as well as other groups, such as Friends of the Harmed, which produced two volumes of *Shalefield Stories*, compilations of the testimonies of people who have been harmed by shale gas drilling. Many contributors spoke of how their peaceful, scenic areas had been destroyed by the noise and traffic associated with shale gas wells and compressor stations. They also described health problems associated with fracking, including severe rashes, breathing difficulties, kidney and liver problems, and stomachaches—in addition to tales of dead animals, contaminated water supplies, and irreversible damage to the land. For activists, these stories provided motivation to continue struggling against fracking. Although a number of the original MP activists went on to work in statewide groups and to help people adversely affected by fracking, they remained part of the network and often showed up for tactics such as testimony at County Council meetings, even if they didn't attend the meetings of groups such as POP. Google groups, which were set up in 2010 to allow for email exchanges, are still being used today to share information and calls for action in the network.

The MP website, Facebook page, newsletter, and calendar, which covered events across the state, have also been important tools for the movement. A few activists maintained these, so there was at least a continuous online presence that allowed new people to find the movement. In fact, people from around the world contacted MP through the website and these contacts helped to connect Pittsburgh activists to others, both nationally and internationally. Soon after Pittsburgh passed its ban on fracking, activists in Mont St. Hilaire, a city in Quebec, initiated contact with Pittsburgh activists that resulted in the two cities becoming "sister cities," with an official proclamation passed by the Pittsburgh City Council on December 6, 2011 (www.marcellusprotest.org/international_proclamation). Two Marcellus Protest members who went to Montreal for a demonstration in 2012 maintained these ties, and in 2016, the Quebec activists asked MP to add its name in support of a protest against renewed threats from the shale gas industry. Delegations from three different countries—Denmark, Australia, and the Netherlands— have contacted MP asking for representatives to meet with them along with government and industry representatives when they visited Pennsylvania. A group of MP activists met with the Danish delegation, for example, and one activist impressed them with a graphic presentation showing a map of Denmark superimposed over a map of Pennsylvania, which was about the same size. He then proceeded to show all the gas wells drilled in Pennsylvania each year and how it would look if they were being drilled in Denmark. Despite what had occurred in Pennsylvania, he told them, they might be able to stop fracking in Denmark—a warning that seemed to make a big impression. The Netherlands delegation came twice and after the second visit, in July 2015, the Minister of Economic Affairs, who was part of the delegation, announced a five-year moratorium on shale gas drilling in the Netherlands.

Conclusion: The Challenges of Movement Building

The history of the Pittsburgh-area antifracking movement demonstrates how an important grassroots movement can be built, as well as how difficult it is to maintain and expand collective action. Concerns about the environmental and health effects of fracking were aroused by successful efforts to alert the public, notably through the documentary film *Gasland*, and by the actions of the gas industry, which became very active in protecting its interests in the Marcellus Shale region. While research shows that people are most likely to be upset about fracking when they have direct experience with its negative impacts in their communities, many Pittsburgh residents lacked such direct experience. Nevertheless, activists did form relationships with others in areas directly impacted by fracking, including those who had suffered serious impacts, and these ties created strong motivations to build the movement against fracking. Many activists visited fracking sites in Pennsylvania and some traveled to different parts of the country. Everyone was informed about developments in the conflict throughout the country and, increasingly, around the world, as information was frequently shared through email lists, websites, blogs, Facebook, and other communications.

As of 2016, the movement against natural gas drilling in the Marcellus Shale remains alive and well, and grassroots activists still care deeply about the issue. Many local organizers have formed strong relationships with one another and they have also made connections to other state, national, and international antifracking movements. Yet, threats and concerns about issues are not enough to build a movement; while networks among activists support the ongoing movement, organizations and resource mobilization are also crucial. The movement has enjoyed much success, but activists have also encountered difficulties in organizing and finding effective strategies, and the battle against fracking is far from won. Local activists continue to confront significant challenges involved in building the movement.

Strategies and tactics are critical to movement growth, and early organizers against fracking in Pittsburgh were wise to kick off the movement with a large demonstration. Yet the experience of Marcellus Protest shows that building a lasting organization is extremely difficult. One challenge for MP was that the group included participants with a wide range of backgrounds. Although largely white, the group included working-class participants, professionals, and people with no prior movement experience, as well as long-time activists. This diversity meant that participants lacked common ideological perspectives and ideas about organizational structures and strategies. Although an organizational structure was created, it was difficult in the long run to maintain participation in the working groups and general assemblies. In contrast to MP, Shadbush was much more cohesive and worked well as a collective, but the group also had difficulty expanding and ultimately disbanded to make way for Three Rivers Rising Tide, a new organization that is part of a national network and potentially able to reach more constituents.

Individual activists played a key role in organizing events and tactics, but MP lacked a means of creating leadership teams and developing new leaders. While energetic individuals can mobilize demonstrations and other events, it is much more difficult to mobilize ongoing campaigns without organizational support. It takes leadership teams to build organizational structures and mobilize campaigns; once participants create organizational structures, they need to find ways to develop new leadership and bring in new participants. Leadership need not be hierarchical, but there must be ways of creating commitment and accountability. The antifracking movement in Pittsburgh currently has a core of hardworking organizers who know and trust one another. The movement also has a much larger circle of supporters who are connected by email lists and social media. MP tried to become a stable organization for several years, but eventually became more of a network than an organization.

Nevertheless, a movement community of environmental activists opposed to fracking remains active in the Marcellus Shale region, and an oppositional culture can be mobilized to act against new threats. Consciousness of the dangers associated with fracking is now widespread, with a recent Gallup poll showing a rise in opposition to fracking from 40% in 2015 to 51% in 2016 (www .gallup.com). In addition to the MP network—with its website, calendar, Google groups and Facebook page—offshoots of the group, such as MOB, continue to engage in regular actions. Three Rivers Rising Tide, the successor to Shadbush, now organizes direct actions in Pittsburgh around fracking and other environmental issues, and links activists—many of them young people—to the national and international Rising Tide networks. Long-standing organizations such as the Sierra Club also remain active and contribute to various campaigns. Members of MP continue to participate in other local groups and events, and some participate in statewide organizations such as Pennsylvanians Against Fracking. Many activists are connected to others in networks across the country. When movement constituents feel strongly about an issue, as supporters of the antifracking movement in Pittsburgh do, they continue to come out for well-organized actions. Although organizational structures are difficult to maintain, the movement continues to build support and raise public consciousness about the environmental and health problems created by shale gas drilling in the Marcellus Shale and beyond.

References

Benford, Robert D., and David A. Snow. 2000. "Framing Processes and Social Movements: An Overview and Assessment." *Annual Review of Sociology* 26: 611–639.

Boudet, Hilary, Christopher Clarke, Dylan Bugden, Edward Maibach, Connie Roser-Renouf, and Anthony Leiserowitz. 2014. "'Fracking' Controversy and Communication: Using National Survey Data to Understand Public Perceptions of Hydraulic Fracturing." *Energy Policy* 65: 57–67.

Brasier, Kathryn J., Matthew R. Filteau, Diane K. McLaughlin, Jeffrey Jacquet, Richard C. Stedman, Timothy W. Kelsey, and Stephan W. Kelsey. 2011. "Residents' Perceptions of Community and Environmental Impacts from Development of Natural Gas in the Marcellus Shale: A Comparison of Pennsylvania and New York Cases." *Journal of Rural Social Sciences* 26(1): 32–61.

Corrigall-Brown, Catherine. 2012. *Patterns of Protest: Trajectories of Participation in Social Movements*. Stanford, CA: Stanford University Press.

Cress, Daniel M., and David A. Snow. 1996. "Mobilization at the Margins: Resources, Benefactors, and the Viability of Homeless Social Movement Organizations." *American Sociological Review* 61(6): 1089–1109.

Davis, Charles. 2012. "The Politics of 'Fracking': Regulating Natural Gas Drilling Practices in Colorado and Texas." *Review of Policy Research* 29(2): 177–191.

Fox, Josh. 2010. *Gasland*. Film Documentary. New York: International Wow Productions.

Fox, Josh. 2013. *Gasland II*. Film Documentary. New York: International Wow Productions.

Freeman, Jo. 1975. *The Politics of Women's Liberation*. New York: Longman.

Gamson, William A. 1990. *The Strategy of Social Protest* (2nd ed.). Belmont, CA: Wadsworth.

Ganz, Marshall. 2000. "Resources and Resourcefulness: Strategic Capacity in the Unionization of California Agriculture, 1959–1966." *American Journal of Sociology* 105(4): 1003–62.

Han, Hahrie. 2014. *How Organizations Develop Activists: Civil Associations and Leadership in the 21st Century*. New York: Oxford University Press.

Hudgins, Anastasia, and Amanda Poole. 2014. "Framing Fracking: Private Property, Common Resources, and Regimes of Governance." *Journal of Political Ecology* 21: 303–319.

Jalbert, Kirk, Abby J. Kinchy, and Simona L Perry. 2014. "Civil Society Research and Marcellus Shale Natural Gas Development: Results of a Survey of Volunteer Water Monitoring Organizations." *Journal of Environmental Studies and Sciences* 4(1): 78–86.

Klein, Naomi. 2014. *This Changes Everything: Capitalism vs. the Climate*. New York: Simon and Schuster.

Ladd, Anthony E. 2013. "Stakeholder Perceptions of Socioenvironmental Impacts from Unconventional Natural Gas Development and Hydraulic Fracturing in the Haynesville Shale." *Journal of Rural Social Sciences* 28(2): 56–89.

Ladd, Anthony E. 2014. "Environmental Disputes and Opportunity-Threat Impacts Surrounding Natural Gas Fracking in Louisiana." *Social Currents* 1(3): 293–311.

Lichterman, Paul. 1995. "Piecing Together Multicultural Community: Cultural Differences in Community Building among Grass-Roots Environmentalists." *Social Problems* 42(4): 513–534.

Lichterman, Paul. 1996. *The Search for Political Community: American Activists Reinventing Commitment*. Cambridge: Cambridge University Press.

Lord, Rich. 2010. "Council Hears Sharp Warnings against Marcellus Shale Drilling." *Pittsburgh Post-Gazette*, October 19: B1-B2.

Malin, Stephanie. 2014. "There's No Real Choice But to Sign: Neoliberalization and Normalization of Hydraulic Fracturing on Pennsylvania Farmland." *Journal of Environmental Studies and Sciences* 4(1): 17–27.

Mazur, Allan. 2016. "How Did the Fracking Controversy Emerge in the Period 2010–2012?" *Public Understanding of Science* 25(2): 207–222.

McCammon, Holly J. 2001. "Stirring Up Suffrage Sentiment: The Formation of the State Women's Suffrage Organizations, 1866–1914." *Social Forces* 80(2): 449–480.

McCarthy, John D., and Mark Wolfson. 1996. "Resource Mobilization by Local Social Movement Organizations: Agency, Strategy, and Organization in the Movement Against Drinking and Driving." *American Sociological Review* 61(6): 1070–1088.

Polletta, Francesca. 2002. *Freedom Is an Endless Meeting: Democracy in American Social Movements*. Chicago: University of Chicago Press.

Reger, Jo, and Suzanne Staggenborg. 2006. "Patterns of Mobilization in Local Movement Organizations: Leadership and Strategy in Four National Organization for Women Chapters." *Sociological Perspectives* 49(3): 297–323.

Sangaramoorthy, Thurka, Amelia M. Jamison, Meleah D. Boyle, Devon C. Payne-Sturges, Amir Sapkota, Donald K. Milton, and Sacoby M. Wilson. 2016. "Place-Based Perceptions of the Impacts of Fracking Along the Marcellus Shale." *Social Science and Medicine* 151: 27–37.

Sarge, Melanie A., Matthew S. VanDyke, Andy J. King, and Shawna R. White. 2015. "Selective Perceptions of Hydraulic Fracturing." *Politics and the Life Sciences* 34(1): 57–72.

Simonelli, Jeanne. 2014. "Home Rule and Natural Gas Development in New York: Civil Fracking Rights." *Journal of Political Ecology* 21(1): 258–278.

Smydo, Joe. 2010a. "City OKs Ban on Gas Drilling." *Pittsburgh Post-Gazette*, November 17: B1-B2.

Smydo, Joe. 2010b. "City Moves Ahead with Controversial Drilling Ban." *Pittsburgh Post-Gazette*, November 10: A1, A14.

Staggenborg, Suzanne. 1989. "Stability and Innovation in the Women's Movement: A Comparison of Two Movement Organizations." *Social Problems* 36(1): 75–92.

Staggenborg, Suzanne. 1998. "Social Movement Communities and Cycles of Protest: The Emergence and Maintenance of a Local Women's Movement." *Social Problems* 45(2): 180–204.

Theodori, Gene L. 2013. "Perception of the Natural Gas Industry and Engagement in Individual Civic Actions." *Journal of Rural Social Sciences* 28(2): 122–134.

Vasi, Ion Bogdan, Edward T. Walker, John S. Johnson, and Hui Fen Tan. 2015. "'No Fracking Way!' Documentary Film, Discursive Opportunity, and Local Opposition against Hydraulic Fracturing in the United States, 2010 to 2013." *American Sociological Review* 80(5): 34–59.

Weigle, Jason L. 2011. "Resilience, Community, and Perceptions of Marcellus Shale Development in the Pennsylvania Wilds: Reframing the Discussion." *Sociological Viewpoints* 27: 3–14.

Wilber, Tom. 2015. *Under the Surface: Fracking, Fortunes, and the Fate of the Marcellus Shale.* Ithaca, NY: Cornell University Press.

Wright, Marita. 2013. "Making It Personal: How Anti-Fracking Organizations Frame Their Messages." *Journal of Politics and Society* 24(2): 107–125.

Zald, Mayer N., and John D. McCarthy. 1987. *Social Movements in an Organizational Society: Collected Essays.* New Brunswick, NJ: Transaction Books.

5

Engines, Sentinels, and Objects

Assessing the Impacts of Unconventional Energy Development on Animals in the Marcellus Shale Region

CAMERON THOMAS WHITLEY

Introduction

The U.S. unconventional energy boom has received widespread attention from sociologists and social scientists. Numerous studies have documented its differential social and environmental impacts on extractive communities and what citizens perceive to be the major benefits and costs of hydraulic fracturing (e.g., Boudet et al. 2014; Ladd 2013, 2014; Malin 2014). However, researchers have largely ignored the intersection between animals, humans, and energy development (for an exception, see Bamberger and Oswald 2012, 2014a, 2014b). Animals are an integral part of American society. It is estimated that there are approximately 312 million domestic animals in the United States, many of which live in areas experiencing unconventional oil and gas development. Each year, over 9.2 billion land animals are slaughtered for food, but when shellfish and fish are accounted for, the total approaches 59 billion (Humane

Society of the United States 2016a; U.S. Department of Agriculture 2010). Livestock also experience unparalleled exposure to contaminants from unconventional energy development, which may enter human food systems (see, e.g., Royte 2012). Wildlife is especially vulnerable to energy-related risks, including the 1,441 animals listed as threatened or endangered in the United States (U.S. Fish & Wildlife Service 2016b). Given that little attention has been paid to assessing the impacts of fracking on wild and domestic animals, this chapter explores the past and present history of animals in energy development, outlines what we know about the impacts of unconventional energy development on animals broadly, and specifically identifies what existing research shows about the impacts to animals living in the Marcellus Shale region. The conclusion discusses the need for regulatory frameworks to include animals and opportunities for future research.

Historical Context: Animals as Energy Producers and Sentinels

Before fossil fuels became a dominant energy source, societies used animals to perform various kinds of work. People created animal engines, or animal-powered machines, which harnessed energy by driving animals around a center post (see, e.g., Major 1978, 2008). Of the few books that thoroughly explore animals as engines, *Animal-Powered Engines* (Major 1978) and *Animal-Powered Machines* (Major 2008) are the most widely referenced. In these rather short accounts, Major suggests that while animals have been instrumental in the advancement of energy development, they have seldom been recognized as such. The documentary *The Ghosts in Our Machine* also describes this relationship, and today the term is increasingly being used to denote the many hidden contexts in which animals are used, going beyond energy development to include other social structures like product development (Marshall 2013). While Major's books do an excellent job of describing specific animal engines, they capture only one component of a larger narrative about the use of animals in our demand for energy.

A Brief History of Animals as Sentinels

Sentinels are animals that are sensitive to environmental disturbances and can signal danger through the display of physical or cognitive distress markers. The most iconic sentinel is the "canary in the coal mine." John Scott Haldane argued that small animals could be used to assess risks at a lower cost and with greater accuracy than technological advancements (Goodman 2007). He asserted that canaries were ideal because of their rapid breathing, small size, high metabolism, and ease of handling. As a result, these animals were placed in small cages, carried into coalmines, and used as toxic gas exposure indicators. The term can also be applied to the use of animals in cosmetics and medical testing. Each year in the United States, about 26 million animals are used for medical testing and another 100,000 to 200,000 animals die each year for cosmetics production (Hastings Center 2011; Humane Society International 2016).

Even with technological innovation, sentinels continued to be used in mining operations well into the 20th century. It was not until 1986, for example, that British mining pits replaced canaries with electronic gas detectors (Goodman 2007). Today, animals are seldom used as energy providers or as intentional sentinels in *developed* countries; however, animals have become unintentional sentinels in energy development, signaling environmental contamination and its potential impacts on human health.

Unintentional Sentinels: Exploring the Impacts of Hydraulic Fracturing on Animals

Domestic and wild animals are likely to face risks from unconventional energy development and specifically from high-volume hydraulic fracturing (HVHF) processes. To date, researchers in the natural sciences have identified many risks animal populations face: contact with toxic chemicals, air quality problems, landscape fracturing, water issues (contamination and depletion), noise, and light pollution. Despite the breadth of these problems, relevant research is limited. Of the studies that have been conducted, few address unconventional energy development's impacts on domestic animals and livestock—and none use a social science perspective (for an exception, see Whitley 2017). Nonetheless, what is clear from the research that does exist is that unconventional energy development impacts animals and these impacts offer insights into their potential consequences for humans. As a result, animals are once again being used as sentinels in energy development, albeit unintentionally.

Water and Exposure to Chemicals. How HVHF water use and contamination affects human and animal life has become a leading concern among scholars and activists due to the enormous amount of water required for fracking oil or gas wells (up to 20 million liters during the lifetime of a well (Entrekin et al. 2011; Howarth, Ingraffea, and Engelder 2011). This high rate of consumption leads most companies to draw on groundwater sources, as well as nearby lakes and streams, altering seasonal flows and causing some systems to run dry. When this happens, animals that rely on water or marshlands for survival may be harmed (Kiviat 2013). One of the most important ways that animals are affected is through contaminated water stores and toxic spills and leaks that occur during the transport, storage, and fracturing processes (Rozell and Reaven 2012). Chemicals used in HVHF contain a mix of organic and inorganic additives, but the exact composition is treated as "proprietary" by federal law and varies by region and company. However, Ferrer and Thurman (2015) provide an extensive review of the literature analyzing chemical compounds and the potential impacts these chemicals can have on humans and animals. In general, acids are used to dissolve minerals and enhance cracks, biocides kill bacteria and limit corrosion, gels influence water viscosity, and distillates reduce friction. In a recent study evaluating 240 substances, 157 were associated with developmental or reproductive toxicity. Among these, 67 are considered potential public health hazards and have health-based use guidelines. Some of the more recognizable chemicals used in HVHF include arsenic, benzene, cadmium, lead, formaldehyde, chlorine, and mercury (Elliott et al. 2016).

Direct contact with these chemicals often occurs when animals consume contaminated water resulting from spills, leaks, and wastewater ponds (Bamberger and Oswald 2012). This type of direct harm has been documented with exposure to both fracturing fluids and the chemicals used for fracking-related transportation and development (Colborn et al. 2011; Davis and Robinson 2012; Davis 2012; Entrekin et al. 2011; Gillen and Kiviat 2012; Kiviat 2013; Rozell and Reaven 2012). In most cases, accidents occurred because of inadequate protection policies or a failure to follow safety precautions. Even relatively minor spills or leaks can have dramatic impacts on ecosystems. In many instances, wastewater fluids have a chloride content well above acceptable levels (56,900 mg L^{-1}) and are fatal to plant and animal species. Further, for species that are particularly sensitive to chloride, high levels can render "breeding habitats unsuitable" (Kiviat 2013: 3). In addition, toxic wastewater ponds are used as open holding basins for water returned from well fracturing and wildlife may attempt to utilize pools as water sources. Toxin exposure poses greater risk to animals during migratory seasons and the effects are likely compounded during droughts (see, e.g., Ramirez Jr. 2009).

Beyond exposure in natural habitats, animals, like humans, are at risk from consuming contaminated drinking water. Drinking water can become contaminated through methane leaks, chemical spills, and injection wells. To get rid of wastewater, companies usually inject it into underground storage wells and it is estimated that over 30 trillion gallons of wastewater have been injected into roughly 680,000 underground wells nationwide (Lustgarten 2012). Inspections have revealed that leaking wells are common (one in six), causing 17,000 violations nationally. Importantly, these wells are not just located in rural areas and can pose a greater risk to animal and human health in highly populated areas. For example, in 2010, contaminants from an injection well bubbled up into a west Los Angeles dog park (Lustgarten 2012).

Though work is still in the early stages, scholars have begun to document animal exposure to HVHF contaminants. Studying 24 cases in six states (Colorado, Louisiana, New York, Ohio, Pennsylvania, and Texas), Bamberger and Oswald (2012) identified the leading source of chemical exposure as contaminated wells and springs. The animals involved in these cases, ranging from cows to salamanders, experienced reproductive, neurological, urological, gastrointestinal, dermatological, upper respiratory, and musculoskeletal impacts, and death. Furthermore, Bamberger and Oswald (2012) describe one case in which wastewater leaked into a cattle pasture, causing direct exposure to toxic chemicals. Within one hour of the leak, 17 cows died. In another case, a leak created a natural experiment whereby 60 cattle had access to a contaminated creek, while a second group of 36 cattle was at pasture and did not have access to the creek. Of the 60 cattle that had access to the contaminated water, 21 died and 16 failed to produce calves the following spring; the 36 that were not exposed showed no symptoms or abnormal health problems. The authors note that because they have restricted movement and continuous exposure, "animals, especially livestock, are sensitive to the contaminants released into the environment by drilling and by its cumulative impacts" (Bamberger and Oswald 2012: 72).

Air Quality. Beyond direct exposure, animals are likely to be exposed to rogue airborne particulate matter and emissions from HVHF. Diesel exhaust from compressors and trucks, volatile organic compounds (VOCs) from fracturing fluids, chemical interactions, and road particulate matter affect air quality (Colborn et al. 2014). Because it takes roughly 6,800 truck trips to fracture a well, truck traffic creates dust that is likely to become airborne (Garti 2012). Particulates may also enter the waterways, posing risks to aquatic invertebrates (Kiviat 2013). The extent to which animals are harmed by air quality issues remains a topic of concern, but one that has been under-researched.

Landscape Fragmentation. As with other forms of energy development, forest loss and landscape fragmentation can affect animals in key respects, particularly their "dispersal, pollination, herbivory, and predation" (Kiviat 2013: 3). Species respond differently to unconventional energy development. Some adapt by moving to safer locations or altering migratory patterns, while others cannot. In these instances, forest fragmentation can contribute to range restriction and species extinction (Davis and Robinson 2012; Drohan et al. 2012). Bayne and Dale (2011) found that many forest songbirds avoid newly built connective pathways like roadways and pipelines. In a longitudinal study assessing habitat selection during and after unconventional development, Sawyer and colleagues (Lendrum et al. 2012; Sawyer et al. 2006; Sawyer, Kauffman, and Nielson 2009) found that mule deer relocate to areas farther from well pads once development begins. The authors suggest this is a clear indication of habitat loss and subsequent studies have documented increased population mortality (Sawyer and Nielson 2010). Habitat diminution also affects multiple species, not just mule deer, as demonstrated by Doherty and colleagues (2008), who found that sage grouse also avoid developed areas. Importantly however, not all species experience these consequences. Looking at lizard populations in Texas, Smolensky and Fitzgerald (2011) found no relationship between HVHF and population size. Though the study was limited by site selection, it suggests that numerous factors contribute to species abundance and distribution and that, as with conventional forms of energy development, the effects are likely to vary by species.

Noise and Light Pollution. Noise and light pollution accompany the initial stages of HVHF. Noise created by HVHF installation comes from compressors, truck traffic, and machinery and can be heard from great distances, especially in areas with few competing noises (Mellott 2011). As Kiviat (2013: 5) reminds us, "continuous loud noise from, for example, transportation networks, motorized recreation, and urban development can interfere with acoustic communication of frogs, birds, and mammals, and cause hearing loss, elevated stress hormone levels, and hypertension in various animals" (see also Barber, Crooks, and Fristrup 2010).

Given previous studies, it is unsurprising that Habib, Bayne, and Boutin (2007) found noise pollution from HVHF drilling in Alberta, Canada reduced bird-pairing success. Another study identified nuance in noise's effects, showing that breeding birds' survival rates were higher near well pads with compressors, likely due to low predation, while species richness was greater near well pads without compressors (Francis, Ortega, and Cruz 2009). Offshore, noise created from drilling platforms and transportation disrupts underwater animal communication channels (Schlossberg 2016). Additional research shows that the influence of noise varies by species (Bayne, Habib, and Boutin 2008).

In addition to noise, pad construction and well fracturing requires bright lights. In analyzing the impacts HVHF has on biodiversity, Kiviat (2013: 5) notes that light pollution may impact "mortality, reproduction and foraging." However, the full impacts of HVHF light pollution have not yet been studied. In fact, there has been limited research on the impact of noise and light pollution on animal populations in general. Of those that have been published, most examine bird populations and it is difficult to identify variation in sensitivity to noise and light across species.

Case Study: Animals and Unconventional Energy Development in the Marcellus Shale

Over a dozen shale formations sit below North America's surface. Of these, one of the largest and most productive "plays" (the industry term for these shale locations) is the Marcellus Shale (King 2015). Rich in organic materials, black shale grew from marine environments during the Middle Devonian period over 390 million years ago (Harper 2008). It is a deep formation, covering a total of 95,000 square miles under the states of Pennsylvania, New York,

Ohio, West Virginia, Maryland, Virginia, New Jersey, Kentucky, and Tennessee (Curtis 2011; King 2015). Despite a 2002 U.S. Geological Survey estimate that the Marcellus Shale contained 1.9 trillion cubic feet of undiscovered gas (Milici et al. 2003), prior to the year 2000, the region had seen limited activity due to the perceived difficulty of accessing the gas deposits. In 2003, a conventional well was drilled in Washington County, Pennsylvania that produced an expansive flow of natural gas. The fact that so much natural gas could be gathered without using HVHF techniques challenged the idea that the Marcellus Shale's abundant resources were too difficult to access. The well is often credited with launching the Marcellus Shale gas play (Harper 2008).

Across the country, industry professionals began developing and using HVHF technology to extract shale reserves in challenging environments. As revelations about the magnitude of untapped natural gas emerged, companies took advantage of the Marcellus Shale's energy resources and profit possibilities. In 2005, industry experts erected the first gas production well in the Marcellus Shale by utilizing HVHF methods already established in the Barnett Shale in Texas (Harper 2008). By the end of 2007, more than 375 gas wells had been permitted in Pennsylvania (Harper 2008). In 2008, two professors at Pennsylvania State University estimated that the original projections by USGS grossly underestimated the production potential of the Marcellus Shale, which could contain more than 500 trillion cubic feet of natural gas (Messer 2008). They further estimated that by using HVHF techniques, the industry could only recover about 10% of these stores (50 trillion cubic feet) (Messer 2008). The U.S. Energy Information Agency reassessed the untapped gas reserves in 2011 and reported that the Marcellus Shale contained approximately 410 trillion cubic feet of *recoverable* natural gas, a number that was revised down to 141 trillion cubic feet in 2012 (USEIA 2012). As of 2014, the Marcellus Shale is the largest gas producing region in the United States, exceeding 15 billion cubic feet per day (Bcf/d) and over 40% of all U.S. shale gas production (USEIA 2014). Drilling has expanded rapidly and in Pennsylvania alone, the number of new wells increased from 50 in 2007 to 1,370 in 2014 (King 2015).

Impacts to Wildlife in the Marcellus Shale

The Marcellus Shale region is home to numerous wildlife species, many of which are endangered or threatened and rely on environmental stability for success. Pennsylvania itself has 69 endangered species and 16 that are threatened (Pennsylvania Fish & Boat Commission 2016). Further, the Marcellus Shale sits underneath both agricultural and forest lands, including 33 National Park systems that are on or adjacent to the Marcellus Shale (this does not include National Register of Historic Places or National Trails, among other designated areas [National Parks Service U.S. Department of the Interior 2008]). As previously discussed, many scholars argue that HVHF poses significant risks to

wildlife biodiversity (Gillen and Kiviat 2012; Kiviat 2013). Because of its proximity to numerous national parks, development in the Marcellus Shale impacts
a wide range of species (National Parks Conservation Association 2013). These
effects are likely to be more pronounced than in other regions because of the
national parks' biodiversity and historically protected habitats.

Habitat fragmentation, which occurs when contiguous blocks of habitat
are broken up by various land uses, could have the greatest impact on ecosystems (U.S. Fish & Wildlife Service 2016a). In the Marcellus Shale region, fragmentation occurs as roads, transmission lines, pipelines, and well pads are laid.
Some of the most significant land fragmentation is happening in the Allegheny Plateau region of Pennsylvania, the state's largest block of contiguous
forest. This type of fragmentation not only contributes to direct habitat loss,
but also indirectly changes the landscape (Brittingham 2016). For instance,
between 2004 and 2010, shale development disturbed and destroyed 280,000
acres of forest in Bradford and Washington counties, causing considerable
impacts on rare and endangered species (Slonecker et al. 2012; Slonecker and
Milheim 2015). More specifically, scholars have argued that habitat fragmentation will have cascading effects on wildlife in the Marcellus Shale region by
alternating dispersal, pollination, herbivory, and predation patterns (Davis
and Robinson 2012; Drohan et al. 2012). Although numerous species are likely
to be impacted by habitat loss in the region, few have been thoroughly investigated. Of the species that have been assessed, research shows that amphibians
in the Marcellus Shale are less diverse in habitats that have sustained unconventional development than those which have not (Popescu and Hunter Jr.
2011; Rothermel and Semlitsch 2006). Such a reduction in biodiversity, even
within one species, can have impacts on other species within the same habitat.

Further, water allocation, contamination, and consumption by HVHF
are critical concerns for wildlife located in the Marcellus Shale region. Water
withdrawal for development can exacerbate drought conditions and limit
water availability for wildlife. While these concerns are most dramatic in the
western United States, they could pose a problem for other highly developed
locations such as those in the Marcellus region. For the Marcellus Shale, one of
the more pressing concerns is that HVHF water withdrawal may unknowingly
transfer invasive aquatic species and change or damage ecosystems. Moreover,
there exists the possibility that wildlife may be exposed directly and indirectly
to wastewater in impoundment pools or open pits. Chemicals in frack fluids
have been shown to have adverse health effects. Berlekamp (2013) reported that
75% of the documented chemicals can lead to respiratory and gastrointestinal
distress; 40–50% can impair the kidneys, nervous, and cardiovascular systems;
37% can affect the hormone system; and at least 25% can be associated with
cancer or growth mutations in mammalian cells. Several scholars have begun to
assess wastewater's impact on wildlife both in and outside of the Marcellus Shale

region. In a study evaluating the toxicity and potential carcinogenic effects of hydraulic fracturing wastewater, for instance, Yao and colleagues (2015) found that mammalian cells exposed to flowback water from Marcellus Shale wells showed increased levels of barium and strontium at exposure levels that have been shown to produce tumors in mice. Similarly, Latta and colleagues (2015) found that riparian songbirds residing in watersheds near hydraulic fracturing sites in two shale plays tested positive for elevated levels of metals associated with HVHF. The species in question is considered a top predator, implying significant amounts of cross-species contamination (Latta et al. 2015).

Beyond direct exposure through impoundment pools and spills, humans can expose animals to wastewater through seemingly innocuous government channels. For example, some states like West Virginia, located in the Marcellus Shale, have historically allowed fracturing wastewater to be used as a de-icing agent (Adams 2011). In a study assessing the impact of this action, Adams (2011) found that wastewater application contributed to widespread and rapid mortality of vegetation (e.g., 56% of trees within 2 years). Though specific to plant life, the study suggested that the practice likely influences other living creatures, including wild animals, though specific impacts on the ecosystems are unknown. As another example, contaminants can be passed through treatment plants. Many wastewater treatment plants are not equipped to handle wastewater's toxicity or salinity, which can be lethal to aquatic life when it is passed through sewage plants into streams and rivers (U.S. Fish & Wildlife Service 2016a). The release of hydraulic fracturing fluid into streams, even from treatment facilities, has resulted in direct mortality and stress on aquatic invertebrates (Papoulias and Velasco 2013). In a study assessing the conservation and restoration of native brook trout in the Marcellus Shale, researchers found that HVHF processes posed a significant risk to aquatic environments and specifically the conservation of native brook trout.

Because studies of this type are limited, more research is needed examining how aquatic life is exposed to wastewater and the ways in which existing processes can mitigate exposures (see, e.g., Drohan et al. 2012). In turn, evidence from other shale plays (specifically west of the Mississippi) and laboratory experiments show that air, noise, and light pollution, and increased truck traffic pose a risk to wildlife. However, more research specific to the Marcellus Shale is needed (Blickley, Blackwood, and Patricelli 2012; Martin et al. 2011; Watkinson et al. 2001).

Impacts to Farm Animals in the Marcellus Shale

In addition to forests and wildlife habitats, rural and agricultural lands overlay much of the Marcellus Shale region and are home to numerous agricultural animals. For example, in 2016, the Marcellus Shale region accounted for approximately 11.3% of all cattle and calves produced in the United States (U.S.

Department of Agriculture 2016). Several scholars argue that hydraulic fracturing impacts livestock and livestock production in the Marcellus Shale; however, limited research has been done to assess when and how this impact occurs. Nonetheless, with only slight differences, livestock are subject to many of the same concerns facing wildlife. One of the main differences is that agricultural animals face greater potential risks because while wildlife can leave an environment that is being developed or contaminated, livestock are geographically limited. Bamberger and Oswald (2012) argue that water could be the leading source of contamination in livestock, as exposure may occur through drinking water from wells, springs, ponds, creeks, spills, and open pits. Cattle that die from exposure on site are not introduced into the human food system, yet those that ingest or are exposed to wastewater and survive frequently continue their original path into the food supply. It remains unclear how this contamination might influence human health and more research on this topic is needed (Bamberger and Oswald 2012; Bamberger and Oswald 2014a). Although most research assessing HVHF's impacts on livestock rely on case studies or subjective narratives, industries associated with agriculture have begun to respond to the potential risks. For example, Nationwide Mutual Insurance, an agricultural insurer, will no longer cover damages related to hydraulic fracturing and Rabobank, which specializes in banking services for famers, will no longer sell mortgages to farmers with gas leases on their property (Royte 2012).

Beyond livestock contamination, the Marcellus Shale region faces challenges to the strength and diversity of its surrounding agricultural system. For example, dairy farming is an important part of Pennsylvania's economy and in 2014 the state was the only one in the nation to increase its number of dairy farms. Despite this growth, between 2013 and 2014 the number of cows in the state fell by about 6,000 head (Center for Dairy Excellence 2016). Scholars suggest this decrease provides preliminary evidence that unconventional energy development has had an impact on agricultural environments in the state. More specifically, the decrease in the number of cows is consistent with a past study examining milk production and dairy farming between 1996 and 2011 in Pennsylvania counties that had extreme growth in gas development. In the study, researchers found that milk production and the number of cows decreased across most of the counties where unconventional energy development was taking place (Finkel et al. 2013). Although a causal association cannot be determined, this study gives insight into the ongoing and potential impacts that unconventional development may have on livestock, food and milk production, and associated agricultural economies.

Pets and Companion Animals

Like livestock and wild animals, domestic animals such as pets and companion animals may also be routinely exposed to the effects of unconventional energy development. Although there are no official estimates of the number of companion animals at risk from fracking operations nationwide, a 2015 American Pet Products Survey suggests that 65% of U.S. households have at least one pet (Humane Society of the United States 2016b). The Marcellus Shale region reaches into nine states and is home to approximately 30 million households (U.S. Census Bureau 2016). If 65% of all households have at least one pet then there are approximately 19.5 million households with pets in the region, then conceivably at least 19.5 million companion animals are at risk of experiencing impacts from unconventional energy development in the Marcellus Shale alone. Companion animals are more likely to be susceptible to contaminant exposure because they are closer to the ground, they may drink or bathe in pooled water near developments, they generally spend a significant amount of time outside, and compared to wildlife and agricultural animals, their skin and paws have fewer protective barriers to limit exposure (Bamberger and Oswald 2012; Slizovskiy et al. 2015). The proximity that pets share with humans and their increased potential for exposure means that these animals have been and will continue to be unintentional sentinels, showcasing the possible effects that exposure poses for humans (Bamberger and Oswald 2012, 2014a, 2014b; Slizovskiy et al. 2015).

Presenting a sequence of case studies, Bamberger and Oswald (2012, 2014a, 2014b) show how canines can be exposed to HVHF chemicals through well, spring, pond, and creek contamination, in storm water runoff from well pads, through wastewater spread on roads for de-icing, or through uncontained wastewater impoundments. In one case, Bamberger and Oswald (2014a) connect the spread of wastewater on a road to the almost instantaneous sickness, deterioration, and death of an otherwise healthy one and a half-year-old dog that appeared to have been poisoned by waste products after drinking from and playing in a contaminated puddle. Bamberger and Oswald's observations and findings were further validated in the results of a community health survey done by Slizovskiy and colleagues (2015) in Washington County, Pennsylvania. In the study, researchers collected health data on 2,452 companion and backyard animals residing in 157 households. Survey participants reported 127 health concerns, mostly affecting dogs. Isolating reports for canines, Slizovskiy and colleagues (2015) found a relationship between health conditions and proximity to the nearest gas well, with increased proximity leading to increases in reported health conditions. Skin conditions were the most common health issue reported, providing valuable information for understanding potential impacts to humans. They suggest that companion animals near unconventional development should be monitored frequently and that health concerns should be systematically recorded to help predict the likelihood of similar conditions appearing in human populations.

Unlike wildlife and livestock, companion animals may also be exposed to contaminated in-home tap water. Although oil and gas industry personnel have routinely denied that drinking water has been contaminated by fracking operations (see, e.g., Urbina 2011; Yergin and Ineson 2009), there have been numerous claims of contamination across the Marcellus Shale region, including over 160 documented cases in Pennsylvania alone (Beaver 2014). In a study assessing drinking water quality in the region, 141 wells were analyzed for concentrations of the isotopic signatures of methane and propane. High concentrations of methane were found in 82% of the wells and these concentrations were six times higher for wells located less than one kilometer from a drilling site compared to those located farther away. Similarly, concentrations of methane and propane were also high for wells within 1 km of a drilling site (Jackson 2013). In addition, air pollution is likely to have a significant impact on pets and companion animals, but has not been well studied in the Marcellus Shale region.

Conclusions, Policy, and Recommended Research

Historically, animals have been an important part of energy development. Before mechanized operations, animals were used as energy providers. As machine power increased, the focus and use of animals within energy development changed from providers to intentional sentinels. Today, animals remain an important tool for assessing the impacts that HVHF has on human health, well-being, and the natural environment, though their present role as sentinels is less intentional than in previous eras. As Royte (2012) notes, cattle and other farm animals are the new proverbial canaries in the coalmine, unintentional victims of environmental destruction and contamination caused by human hands.

Reports of wildlife and domestic animals impacted by HVHF have surfaced. Among other things, wildlife experience habitat fragmentation, water shortages, exposure to water and air contamination, increased collisions with truck traffic, and potential impacts due to noise and light pollution. Although livestock may not experience habitat fragmentation or increased collision risk, their confined quarters make exposure to water and air pollutants a problem. The exposure of livestock to unconventional energy development pollution raises questions about the impacts to human food systems (Royte 2012). Like wildlife and livestock, companion animals are also susceptible to environmental contaminants. Although there is no risk that companion animal exposure will enter human food systems, pollution exposure could contaminate home environments if it is transferred from the paws or skin of animals sharing homes with humans.

Stronger regulations are needed to minimize risks posed to animal life. In theory, federal oversight of HVHF could occur through a variety of existing acts and regulations such as the National Environmental Protection Act, the Clean Air Act, and the Endangered Species Act, among others. In practice, however, HVHF has been largely excluded from federal oversight due to the

Energy Policy Act of 2005, which was touted as legislation that would help address energy shortages and promote more energy independence. The legislation exempted unconventional development from the erosion control provisions of the Clean Water Act and from disclosing the contents of fracturing fluids. Even if the Energy Policy Act had not gutted federal regulation, few existing provisions directly apply to animal protection, as most are general environmental or human health protections. Nonetheless, there are possibilities for greater animal protection in a handful of federal regulations, and the remainder of this section discusses the basic principles underlying some of these regulatory acts.

The Safe Drinking Water Act of 1974 (SDWA; 42 U.S.C. § 300F) regulates drinking water quality, which includes ground and surface water sources. Ideally, this act limits underground contaminant injection but does not apply to fracturing fluids, which are considered well stimulation techniques instead of hazardous waste (EPA 2012). Excluding fracturing fluids from the SDWA was challenged in the late 1990s, which led to an EPA study that concluded that fracturing fluids posed no substantial risks to drinking water. This study provided justification for the Energy Policy Act of 2005 (42 USC § 13201 et seq.) that specifically exempted HVHF fluids from SDWA regulation. States and localities may still regulate drinking water sources and place higher regulations and restrictions on unconventional energy development, but there is no federal support within the SDWA for these types of restrictions. Developing legislation to counter the Energy Policy Act and include fracturing fluids in the SDWA would protect not only humans, but domestic animals as well. A further step would be to adjust acceptable contaminant levels to reflect those tolerated by domestic animals, as well as humans.

In addition, the Clean Water Act (CWA) of 1972 (33 U.S.C. §1251 et seq.; formally known as the Federal Water Pollution Control Act) is designed to regulate disposal and discharge of wastewater into surface water. It sets target limits for pollutants, but allows for additional limits to be set by states. Entities desiring to release pollutants must obtain a permit from the appropriate regulatory agency in each location. Because of the Energy Policy Act, HVHF sediment in runoff cannot be classified as wastewater discharge and thus is not regulated by the CWA; however, hydraulic fracturing wastewater or flowback *is* included in the regulation and requires a permit for disposal. Developing legislation to regulate sediment would help protect wild and agricultural animals.

The National Environmental Protection Act of 1969 (NEPA; 42 U.S.C. § 4321 et seq.) requires development on federal lands to be evaluated for environmental impacts. This is important for areas like national forests, which require environmental evaluations before development begins. Evaluations are then reviewed by the EPA or an alternative agency like the U.S. Fish and

Wildlife Service. Although providing some protection, the act does not apply to areas that are adjacent to federal lands, which is problematic since research demonstrates that wildlife on federal lands are impacted by adjacent development (National Parks Conservation Association 2013). Strengthening this act to include setbacks for adjacent land could go a long way toward protecting animals within and surrounding national parks.

One of the only acts that directly protects animals from unconventional energy development is the Endangered Species Act (ESA) (7 U.S.C. § 136, 16 U.S.C. § 1531 et seq.), which is designed to prohibit the death, injury, or even harassment of vulnerable species. Its primary limitation, however, is that it only applies to species that are already designated as threatened or endangered. The ESA has been essential in protecting listed species, but recent research suggests that unconventional development may be contributing to biodiversity decline, increasing the number of species added to the list each year in high development areas such as the Marcellus Shale region (Drye 2012). However, tension exists between enforcement agents of the ESA and industry professionals. For instance, in Pennsylvania, energy development companies are required to conduct a habitat review to assess endangered or threatened animal presence in a proposed site (Colaneri 2013). Although beneficial for ecosystem stability, criticisms of the process give rise to proposed changes to make it more difficult to put a species on the list. To date, proposed legislation to change classification systems has failed (Colaneri 2013). A further limitation of the ESA is that it only protects wild animals, as there are no federal regulations for livestock or companion animal exposure.

Although the above discussion focuses on federal actions, much of the regulatory framework is limited to states and localities, where differences among regulatory enforcement in adjacent states can create problems. For example, there is no regional authority that oversees regulation of the Ohio River watershed. Given that Ohio allows the spread of fracturing wastewater on roads as a de-icing agent, this runoff could lead to contamination in West Virginia, which has eliminated the disposal of wastewater on roads. In the Marcellus Shale region, states differ in their regulatory approaches. Pennsylvania has embraced natural gas development overall, while New York has a statewide moratorium against fracking. Most states have similar regulatory frameworks that govern well spacing from water sources, as well as setback distances from buildings, but enforcement varies across jurisdictions.

At all levels, animal rights/welfare organizations need to take a greater initiative to protect animals from the negative effects of unconventional energy development. Since animals and national forests do not have legal standing (see Wells 2007), activists must take up the fight to protect animals from HVHF. One possible route to do this is Americans Against Fracking (AAF), which is a national coalition dedicated to banning fracking throughout the United States.

Several large national environmental organizations have pledged support for AAF, including Food & Water Watch, Environmental Action, and Greenpeace. Glaringly absent from this list are animal welfare organizations such as the American Humane Association, American Society for the Prevention of Cruelty to Animals (ASPCA), and the American Veterinary Medical Association. None of these organizations have taken up HVHF as a technological issue that threatens animals. In their Welfare Position Statement, for instance, the American Humane Association makes several broad statements about what is environmentally necessary to raise livestock (such as clean water, air, the ability to move, etc.), yet nowhere identifies HVHF as a concern (American Humane Association 2012). In contrast, the American Veterinary Medical Association (AVMA) has recognized the energy industry's push to expand HVHF, but also encourages continued research on its impacts on animal health and wellbeing (American Veterinary Medical Association 2016). Additionally, only the World Wildlife Fund (WWF 2013) has explicitly come out against hydraulic fracturing in Europe, while the Nature Conservancy has been involved in compiling sources for best management practices (Bearer et al. 2012).

The lack of involvement in antifracking activism and campaigns among animal welfare and animal rights organizations can be explained by at least three factors. First, HVHF is a complex and highly politicized issue that many peripheral nonprofits may not want to address for fear of losing bipartisan support or funding. To take a critical antifracking stance, an organization might be required to publicly confront the oil and gas industry in some way, a strategy that could be financially risky. Second, most animal activist organizations try to address what they see as their most immediate issues, as well as deal with these challenges simultaneously. Issues like animal abuse, cruelty, animal husbandry, or animal fighting are all considered immediate concerns because the impacts are visible and recognizable. In contrast, HVHF, which seems geographically remote for most Americans, has impacts that are not always visible, and therefore may not be considered of immediate concern. Third, there is a lack of data regarding the impacts of energy development on animals at all levels of existence, both domestic and wild.

This chapter has highlighted the dearth of research on the effects of hydraulic fracking on animals and shown that the research that does exist is often complicated and inaccessible to laypersons and policy makers. Still, concern about the impact of HVHF on nonhuman species has not gone unnoticed. However, much of this attention has centered on assessing the consequences of energy extraction on animals because of their importance to human food systems. For instance, Christopher Portier, director of the National Center for Environmental Health at the Centers for Disease Control and Prevention, has called for studies to examine and "include all the ways people can be exposed [to fracking] such as through air, water, soil, plants and animals" (Royte 2012:

6–7). More research is needed to directly assess the impacts of unconventional energy development on specific wildlife species. Researchers need to consider how all impacts of oil and gas production, including chemical contamination, groundwater depletion, habitat fragmentation, noise, light, and increased truck traffic, among other issues, can influence animal behaviors and survival. In addition, evolving research needs to assess how development adjacent to national parks impacts ecosystems and tourism. Given that HVHF activity may reduce biodiversity, additional studies should be conducted to assess how various species respond to energy development and evaluate whether it makes some species more vulnerable to becoming threatened or endangered. Although we know little about the impacts to wildlife, our understanding of fracking's effects on livestock and associated sources of food is even more limited. Comprehensive long-term studies need to assess if and how proximity to unconventional energy facilities alters livestock production and whether rural water and soil contamination will pose new threats to farms and agricultural systems from which humans draw their daily sustenance.

In the final estimation, it seems clear that animals will continue to function as sentinels for humans, signaling us about the new environmental dangers of oil and gas fracking as we move into the Third Carbon Era (Ladd 2016). Indeed, in an age of extreme energy production, all species serve as "canaries in the coalmines" in alerting us to the technological risks of industrial society. Whether we choose to act on the warning signs or not, the reputed words of Chief Seattle over a century and a half ago still provide us with a cautionary tale worth remembering: whatever happens to the animals of the world, will also happen soon to human beings.

References

Adams, Mary Beth. 2011. "Land Application of Hydrofracturing Fluids Damages a Deciduous Forest Stand in West Virginia." *Journal of Environmental Quality* 40(4): 1340–44.

American Humane Association. 2012. *American Humane Association Animal Welfare Policy and Position Statements*. Washington, D.C.

American Veterinary Medical Association. 2016. "Extractive Industries." AVMA. Retrieved March 20, 2016 (https://www.avma.org/KB/Resources/Reference/Pages/ExtractiveIndustries.aspx).

Bamberger, Michelle, and Robert E. Oswald. 2012. "Impacts of Gas Drilling on Human and Animal Health." *New Solutions: A Journal of Environmental and Occupational Health Policy* 22(1): 51–77.

Bamberger, Michelle, and Robert E. Oswald. 2014a. "Unconventional Oil and Gas Extraction and Animal Health." *Environmental Science: Processes & Impacts* 16(8): 860–65.

Bamberger, Michelle, and Robert E. Oswald. 2014b. *The Real Cost of Fracking: How America's Shale Gas Boom Is Threatening Our Families, Pets, and Food*. Boston, MA: Beacon Press.

Barber, Jesse R., Kevin R. Crooks, and Kurt M. Fristrup. 2010. "The Costs of Chronic Noise Exposure for Terrestrial Organisms." *Trends in Ecology & Evolution* 25(3): 180–89.

Bayne, Erin M., and Brenda C. Dale. 2011. "Effects of Energy Development on Songbirds." Pp. 95–114 in *Energy Development and Wildlife Conservation in Western North America*, edited by David E. Naugle. Washington, D.C.: Island Press.

Bayne, Erin M., Lucas Habib, and Stan Boutin. 2008. "Impacts of Chronic Anthropogenic Noise from Energy-Sector Activity on Abundance of Songbirds in the Boreal Forest." *Conservation Biology* 22(5): 1186–93.

Bearer, Scott, Emily Nicholas, T. Gagnolet, M. DePhilip, Tara Moberg, and Nels Johnson. 2012. "Evaluating the Scientific Support of Conservation Best Management Practices for Shale Gas Extraction in the Appalachian Basin." *Environmental Practice* 14(4):308–19.

Beaver, William. 2014. "Environmental Concerns in the Marcellus Shale." *Business and Society Review* 119(1): 125–46.

Berlekamp, Lauren. 2013. "EPA to Allow Consumption of Toxic Fracking Wastewater by Wildlife and Livestock." *EcoWatch*. July 11, 2013. Retrieved March 20, 2016 (http://ecowatch.com/2013/07/11/epa-fracking-wastewater-agriculture/).

Blickley, Jessica L., Diane Blackwood, and Gail L. Patricelli. 2012. "Experimental Evidence for the Effects of Chronic Anthropogenic Noise on Abundance of Greater Sage-grouse at Leks." *Conservation Biology* 26(3): 461–71.

Boudet, Hilary, Christopher Clarke, Dylan Bugden, Edward Maibach, Connie Roser-Renouf, and Anthony Leiserowitz. 2014. "'Fracking' Controversy and Communication: Using National Survey Data to Understand Public Perceptions of Hydraulic Fracturing." *Energy Policy* 65: 57–67.

Brittingham, Margaret. 2016. "Habitat Fragmentation." *Penn State Marcellus Shale Electronic Field Guide*. Retrieved March 3, 2016 (http://www.marcellusfieldguide.org/index.php/guide/ecological_concepts/habitat_fragmentation/).

Center for Dairy Excellence. 2016. "Pennsylvania Dairy Industry Overview." *Center for Dairy Excellence*. Retrieved March 20, 2016 (http://centerfordairyexcellence.org/pennsylvania-dairy-industry-overview/).

Colaneri, Katie. 2013. "Marcellus Shale Industry Lobbies New Endangered Species Law." *State Impact: Pennsylvania Energy. Environment. Economy*. Retrieved March 23, 2016 (https://stateimpact.npr.org/pennsylvania/2013/09/10/marcellus-shale-industry-lobbies-for-new-endangered-species-laws/).

Colborn, Theo, Carol Kwiatkowski, Kim Schultz, and Mary Bachran. 2011. "Natural Gas Operations from a Public Health Perspective." *Human and Ecological Risk Assessment: An International Journal* 17(5): 1039–56.

Colborn, Theo, Kim Schultz, Lucille Herrick, and Carol Kwiatkowski. 2014. "An Exploratory Study of Air Quality near Natural Gas Operations." *Human and Ecological Risk Assessment: An International Journal* 20: 86–105.

Curtis, Rachel. 2011. "Where Is Marcellus Shale Located in Pennsylvania?" *Institute for Energy & Environmental Research for Northern Pennsylvania*. Retrieved May 23, 2016 (http://energy.wilkes.edu/pages/153.asp).

Davis, Charles. 2012. "The Politics of 'Fracking': Regulating Natural Gas Drilling Practices in Colorado and Texas." *Review of Policy Research* 29(2): 177–91.

Davis, John B., and George R. Robinson. 2012. "A Geographic Model to Assess and Limit Cumulative Ecological Degradation from Marcellus Shale Exploitation in New York, USA." *Ecology and Society* 17(2): 25.

Doherty, Kevin E., David E. Naugle, Brett L. Walker, and Jon M. Graham. 2008. "Greater Sage-Grouse Winter Habitat Selection and Energy Development." *Journal of Wildlife Management* 72(1): 187–95.

Drohan, Patrick. J., Margaret Brittingham, Joseph Bishop, and Kendra Yoder. 2012. "Early Trends in Landcover Change and Forest Fragmentation due to Shale-Gas Development in Pennsylvania: A Potential Outcome for the Northcentral Appalachians." *Environmental Management* 49(5): 1061–75.

Drye, Kelley. 2012. "Endangered Species Act Finding with Potential Fracking Implications." *Fracking Insider*. Retrieved March 20, 2016 (http://www.frackinginsider.com/regulatory /endangered-species-act-finding-with-potential-fracking-implications/).

Elliott, Elise G., Adrienne S. Ettinger, Brian P. Leaderer, Michael B. Bracken, and Nicole C. Deziel. 2016. "A Systematic Evaluation of Chemicals in Hydraulic-Fracturing Fluids and Wastewater for Reproductive and Developmental Toxicity." *Journal of Exposure Science and Environmental Epidemiology*: 1–10.

Entrekin, Sally, Michelle Evans-White, Brent Johnson, and Elisabeth Hagenbuch. 2011. "Rapid Expansion of Natural Gas Development Poses a Threat to Surface Waters." *Frontiers in Ecology and the Environment* 9(9): 503–11.

Environmental Protection Agency. 2012. *Hydraulic Fracturing Background Information*. Retrieved March 20, 2016 (http://water.epa.gov/type/groundwater/uic/class2 /hydraulicfracturing/wells_hydrowhat.cfm).

Ferrer, Imma, and E. Michael Thurman. 2015. "Chemical Constituents and Analytical Approaches for Hydraulic Fracturing Waters." *Trends in Environmental Analytical Chemistry* 5: 18–25.

Finkel, Madelon L., Jane Selegean, Jake Hays, and Nitin Kondamudi. 2013. "Marcellus Shale Drilling's Impact on the Dairy Industry in Pennsylvania: A Descriptive Report." *New Solutions: A Journal of Environmental and Occupational Health Policy* 23(1): 189–201.

Francis, Clinton D., Catherine P. Ortega, and Alexander Cruz. 2009. "Noise Pollution Changes Avian Communities and Species Interactions." *Current Biology* 19(16): 1415–19.

Garti, Anne Marie. 2012. "The Illusion of the Blue Flame: Water Law and Unconventional Gas Drilling in New York." *Environ Law New York* 23(1): 159–65.

Gillen, Jennifer L., and Erik Kiviat. 2012. "Environmental Reviews and Case Studies: Hydraulic Fracturing Threats to Species with Restricted Geographic Ranges in the Eastern United States." *Environmental Practice* 14(04): 320–31.

Goodman, Martin. 2007. *Suffer and Survive: Gas Attacks, Miners' Canaries, Spacesuits and the Bends: The Extreme Life of Dr. J.S. Haldane*. New York: Simon & Schuster.

Habib, Lucas, Erin M. Bayne, and Stan Boutin. 2007. "Chronic Industrial Noise Affects Pairing Success and Age Structure of Ovenbirds Seiurus Aurocapilla." *Journal of Applied Ecology* 44(1): 176–84.

Harper, John A. 2008. "The Marcellus Shale—An Old 'New' Gas Reservoir in Pennsylvania." *Pennsylvania Geology* 38(1): 2–13.

Hastings Center. 2011. "Animals Used in Research in the U.S." *Ethics of Medical Research with Animals*. Retrieved June 7, 2016 (http://animalresearch.thehastingscenter.org/facts -sheets/animals-used-in-research-in-the-united-states/).

Howarth, Robert W., Anthony Ingraffea, and Terry Engelder. 2011. "Natural Gas: Should Fracking Stop?" *Nature* 477(7364): 271–75.

Humane Society International. 2016. "About Cosmetic Animal Testing." *Humane Society International*. Retrieved June 15, 2016 (http://www.hsi.org/issues/becrueltyfree/facts /about_cosmetics_animal_testing.html).

Humane Society of the United States. 2016a. "Farm Animal Statistics: Slaughter Totals." *Humane Society of the United States*. Retrieved June 20, 2016 (http://www.humanesociety .org/news/resources/research/stats_slaughter_totals.html).

Humane Society of the United States. 2016b. "Pets by the Numbers: U.S. Pet Ownership, Community Cat and Shelter Population Estimates." *Pets by the Numbers*. Retrieved May 20, 2016 (http://www.humanesociety.org/issues/pet_overpopulation/facts/pet _ownership_statistics.html).

Jackson, Robert B., Avner Vengosh, Thomas H. Darrah, Nathaniel R. Warner, Adrian Down, Robert J. Poreda, Stephen G. Osborn, Kaiguang Zhao and Jonathan D. Karr. 2013. "Increased Stray Gas Abundance in a Subset of Drinking Water Wells near Marcellus Shale Gas Extraction." *Proceedings of the National Academy of Sciences* 110(28):11250–55.

King, Hobart. 2015. "Marcellus Shale–Appalachia Basin Natural Gas Play." *Geology.com*. Retrieved March 22, 2016 (http://geology.com/articles/marcellus-shale.shtml).

Kiviat, Erik. 2013. "Risks to Biodiversity from Hydraulic Fracturing for Natural Gas in the Marcellus and Utica Shales." *Annals of the New York Academy of Sciences* 1286(1): 1–14.

Ladd, Anthony E. 2013. "Stakeholder Perceptions of Socioenvironmental Impacts from Unconventional Natural Gas Development and Hydraulic Fracturing in the Haynesville Shale." *Journal of Rural Social Sciences* 28(2): 56–89.

Ladd, Anthony E. 2014. "Environmental Disputes and Opportunity-Threat Impacts Surrounding Natural Gas Fracking in Louisiana." *Social Currents* 1(3): 293–311.

Ladd, Anthony E. 2016. "Meet the New Boss, Same as the Old Boss: The Continuing Hegemony of Fossil Fuels and Hydraulic Fracking in the Third Carbon Era." *Humanity and Society* (DOI:10.1177/0160597616628908).

Latta, Steven C., Leesia C. Marshall, Mack W. Frantz, and Judith D. Toms. 2015. "Evidence from Two Shale Regions That a Riparian Songbird Accumulates Metals Associated with Hydraulic Fracturing." *Ecosphere* 6(9): 1–10.

Lendrum, Patrick E., Charles R. Anderson Jr., Ryan A. Long, John G. Kie, and R. Terry Bowyer. 2012. "Habitat Selection by Mule Deer during Migration: Effects of Landscape Structure and Natural Gas Development." *Ecosphere* 3(9): 82.

Lustgarten, Abrahm. 2012. "Injection Wells: The Poison Beneath Us." *ProPublica*. Retrieved May 20, 2016 (http://www.propublica.org/article/injection-wells-the-poison-beneath-us).

Major, J. Kenneth. 1978. *Animal-Powered Engines*. London: Batsford.

Major, J. Kenneth. 2008. *Animal-Powered Machines*. London: Shire.

Malin, Stephanie. 2014. "There's No Real Choice but to Sign: Neoliberalization and Normalization of Hydraulic Fracturing on Pennsylvania Farmland." *Journal of Environmental Studies and Sciences* 4(1): 17–27.

Marshall, Liz. 2013. *The Ghosts in Our Machine: Animals Are Hidden in the Shadows of Our Highly Mechanized World*. Canada: Ghost Media.

Martin, Randal, Kori Moore, Marc Mansfield, Scott Hill, Kiera Harper, and Howard Shorthill. 2011. *Final Report: Uinta Basin Winter Ozone and Air Quality Study December 2010– March 2011. Energy Dynamics Laboratory, Utah State University Research Foundation*. EDL/11–039. Vernal: Utah State University.

Mellott, Cody. 2011. "Natural Gas Extraction from the Marcellus Formation in Pennsylvania: Environmental Impacts and Possible Policy Responses for State Parks." Master's Thesis, Center for Environmental Policy, Bard College, Annadale, NY.

Messer, Andrea. 2008. "Unconventional Natural Gas Reservoir Could Boost U.S. Supply." *Penn State News*, January 17. Retrieved January 17, 2008 (http://news.psu.edu/story /191364/2008/01/17/unconventional-natural-gas-reservoir-could-boost-us-supply).

Milici, Robert C., Robert T. Ryder, Christopher S. Swezey, Ronald R. Charpentier, Troy A. Cook, Robert A. Crovelli, Timothy R. Klett, Richard M. Pollastro, and Christopher J. Schenk. 2003. *USGS Assessment of Undiscovered Oil and Gas Resources of the Appalachian*

Basin Province, 2002. United States Geological Survey (USGS) Fact Sheet FT-009003. Washington D.C.: Government Printing Office.

National Parks Conservation Association. 2013. *National Parks and Hydraulic Fracturing: Balancing Energy Needs, Nature, and America's National Heritage*. Washington, D.C.: NPCA Center for Park Research. Retrieved May 23, 2016 (http://www.npca.org/assets /pdf/Fracking_Report.pdf).

National Parks Service U.S. Department of the Interior. 2008. *Potential Development of the Natural Gas Resources in the Marcellus Shale*. Retrieved March 20, 2016 (https://www .nps.gov/frhi/learn/management/upload/GRD-M-Shale_12-11-2008_high_res.pdf).

Papoulias, Diana M. and Anthony L. Velasco. 2013. "Histopathological Analysis of Fish from Acorn Fork Creek, Kentucky, Exposed to Hydraulic Fracturing Fluid Releases." *Southeastern Naturalist* 12(4): 92–111.

Pennsylvania Fish & Boat Commission. 2016. "Chapter 75: Endangered Species." *Threatened and Endangered Species*. Retrieved March 23, 2016 (http://fishandboat.com/endang1 .htm).

Popescu, Viorel D., and Malcolm L. Hunter, Jr. 2011. "Clear-Cutting Affects Habitat Connectivity for a Forest Amphibian by Decreasing Permeability to Juvenile Movements." *Ecological Applications* 21(4): 1283–95.

Ramirez, Pedro, Jr. 2009. "Reserve Pit Management: Risks to Migratory Birds." *U.S. Department of Interior, Fish and Wildlife Service*. Cheyenne, WY: Government Printing Office.

Rothermel, Betsie B., and Raymond D. Semlitsch. 2006. "Consequences of Forest Fragmentation for Juvenile Survival in Spotted (Ambystoma Maculatum) and Marbled (Ambystoma Opacum) Salamanders." *Canadian Journal of Zoology* 84(6): 797–807.

Royte, Elizabeth. 2012. "Fracking Our Food Supply." *The Nation*. Retrieved May 23, 2016 (http://www.thenation.com/article/fracking-our-food-supply/).

Rozell, Daniel J., and Sheldon J. Reaven. 2012. "Water Pollution Risk Associated with Natural Gas Extraction from the Marcellus Shale." *Risk Analysis* 32(8): 1382–93.

Sawyer, Hall, Matthew J. Kauffman, and Ryan M. Nielson. 2009. "Influence of Well Pad Activity on Winter Habitat Selection Patterns of Mule Deer." *Journal of Wildlife Management* 73(7): 1052–61.

Sawyer, Hall, and Ryan M. Nielson. 2010. "Mule Deer Monitoring in the Pinedale Anticline Project Area: 2010 Annual Report." *Western Ecosystems Technology, Cheyenne, Wyoming, USA*.

Sawyer, Hall, Ryan M. Nielson, Fred Lindzey, and Lyman L. McDonald. 2006. "Winter Habitat Selection of Mule Deer before and during Development of a Natural Gas Field." *Journal of Wildlife Management* 70(2): 396–403.

Schlossberg, Tatiana. 2016. "A Plan to Give Whales and Other Ocean Life Some Peace and Quiet." *New York Times*, June 5, A4.

Slizovskiy, I. B., L. A. Conti, S. J. Trufan, J. S. Reif, V. T. Lamers, M. H. Stowe, J. Dziura, and P. M. Rabinowitz. 2015. "Reported Health Conditions in Animals Residing near Natural Gas Wells in Southwestern Pennsylvania." *Journal of Environmental Science and Health, Part A* 50(5): 473–81.

Slonecker, E. Terrence, and Lesley E. Milheim. 2015. "Landscape Disturbance from Unconventional and Conventional Oil and Gas Development in the Marcellus Shale Region of Pennsylvania, USA." *Environments* 2(2): 200–20.

Slonecker, E. T., L. E. Milheim, M. Roig-Silva, A. R. Malizia, and G. B. Fisher. 2012. *Landscape Consequences of Natural Gas Extraction in Bradford and Washington Counties, Pennsylvania, 2004–2010*. U.S. Geological Survey Open-File Report 2012–1154. Reston, Virginia: Government Printing Office.

Smolensky, Nicole L., and Lee A. Fitzgerald. 2011. "Population Variation in Dune-Dwelling Lizards in Response to Patch Size, Patch Quality, and Oil and Gas Development." *Southwestern Naturalist* 56(3): 315–324.

Urbina, Ian. 2011. "A Tainted Water Well, and Concern There May Be More." *New York Times*, August 3, A1.

U.S. Census Bureau. 2016. "United States Households 2009–2013 by State." *United States Household, 2009–2013 by State*. Retrieved (http://www.indexmundi.com/facts/united -states/quick-facts/all-states/households#map).

U.S. Department of Agriculture. 2010. Livestock Slaughter 2009 Summary. Washington, D.C.: USDA National Agriculture Statistics Service. Retrieved June 6, 2016 (http://usda .mannlib.cornell.edu/usda/current/LiveSlauSu/LiveSlauSu-04-29-2010.pdf).

U.S. Department of Agriculture. 2016. "January 1, 2016 Inventory vs. 2015 Inventory." *All Cattle & Calves*. Retrieved May 20, 2016 (http://www.cattlerange.com/cattle-graphs/all -cattle-numbers.html).

U.S. Energy Information Administration. 2012. *Annual Energy Outlook 2012*. United States Department of Energy. Washington, DC: Government Printing Office.

U.S. Energy Information Administration. 2014. *Marcellus Region Production Continues Growth*. Retrieved March 22, 2016 (http://www.eia.gov/todayinenergy/detail.cfm?id= 17411).

U.S. Fish & Wildlife Service. 2016a. "Marcellus Shale Drilling." *U.S. Fish & Wildlife Service New York Field Office*. Retrieved May 23, 2016 (http://www.fws.gov/northeast/nyfo/fwc /marcellus.htm).

U.S. Fish & Wildlife Service. 2016b. Fish & Wildlife Service ECOS Environmental Conservation Online System. Retrieved June 20, 2016 (http://ecos.fws.gov/tess_public/reports /ad-hoc-species-report?kingdom=V&kingdom=I&status=E&status=T&status=EmE& status=EmT&status=EXPE&status=EXPN&status=SAE&status=SAT&mapstatus=3 &fcrithab=on&fstatus=on&fspecrule=on&finvpop=on&fgroup=on&header=Listed +Animals).

Watkinson, William P., Matthew J. Campen, Julianne P. Nolan, and Daniel L. Costa. 2001. "Cardiovascular and Systemic Responses to Inhaled Pollutants in Rodents: Effects of Ozone and Particulate Matter." *Environmental Health Perspectives* 109 (Suppl 4): 539- 46.

Wells, Stephen. 2007. "Standing for Animals." Animal Legal Defense Fund. Retrieved June 7, 2016 (http://aldf.org/blog/standing-for-animals/).

Whitley, Cameron Thomas. 2017. "Altruism, Risk, Energy Development and the Human-Animal Relationship." Doctoral Dissertation, Michigan State University, East Lansing, MI.

World Wildlife Fund. 2013. "*WWF Position on Shale Gas in the EU: Keep Pandora's Box Firmly Shut*." Retrieved March 30, 2016 (http://awsassets.panda.org/downloads/wwf _shale_gas_position.pdf).

Yao, Yixin, Tingting Chen, Steven S. Shen, Yingmei Niu, Thomas L. DesMarais, Reka Linn, Eric Saunders, Zhihua Fan, Paul Lioy, Thomas Kluz, Lung-Chi Chen, Zhuangchun Wu, and Max Costa. 2015. "Malignant Human Cell Transformation of Marcellus Shale Gas Drilling Flow Back Water." *Toxicology and Applied Pharmacology* 288(1): 121–30.

Yergin, Daniel, and Robert Ineson. 2009. "America's Natural Gas Revolution." *Wall Street Journal*, November 3, A1.

6

Motivational Frame Disputes Surrounding Natural Gas Fracking in the Haynesville Shale

ANTHONY E. LADD

Introduction

Environmental disputes surrounding unconventional energy development utilizing horizontal drilling and hydraulic fracturing techniques—i.e., "fracking"—have been on the rise in U.S. shale regions since 2010 (Fox 2010; Hauter 2016; Ladd 2013, 2014; Vasi et al. 2015). As citizens have grown more apprehensive about the social and ecological consequences of fossil fuel consumption, increasing numbers of communities have found themselves embroiled in controversy over the perceived benefits and risks surrounding shale gas and oil production at the local level. Indeed, the term "fracking" has come to symbolize a wide range of contested industrial processes associated with the exploration, drilling, extracting, transporting, and disposal of hazardous wastewaters required for the development of unconventional hydrocarbons (Evensen et al. 2014; Humes 2012). In an age of protracted technological risk, the debate over the use of hydraulic fracturing to extract natural gas and oil deposits from deep underground shale formations represents one of the most contentious socioenvironmental issues of our time (Ladd 2016).

For proponents, the shale energy revolution and hydraulic fracking constitute an economic and energy "game changer." Trumpeted by the fossil fuel industry, allied petro-interests, many politicians, and some mainstream environmental organizations, shale gas has been carefully sold to the American public as a safe, abundant, patriotic energy source that can revitalize rural communities, produce hundreds of thousands of new jobs, reduce U.S. reliance on foreign oil, and generate badly needed tax revenues (America's Natural Gas Alliance 2013). While recent sociological research has critiqued some of the key discursive narratives of the oil and gas industry portraying fracking as a "bridge fuel" to economic wealth, energy independence, a renewable energy future, or climate stability (see Ladd and Perrow 2016; York 2015), supporters celebrate the growing role of fracking in U.S. energy policy and its alleged ability to provide a century's worth of energy to American households (Gold 2014; Ladd 2014; Wright 2012; Yergin 2011)

Conversely, the rapid development of unconventional shale gas (and oil) reserves through high-volume hydraulic fracturing (HVHF) methods, along with other forms of "extreme energy" production (e.g., tar sands, coalbed methane, mountaintop coal removal, Arctic drilling), has also helped launch an ascendant antifracking movement that has been gaining political clout and scientific credibility nationwide. In shale communities like the Marcellus, Barnett, Tuscaloosa, Utica, Woodford, Eagle Ford, Niobrara, and Monterey, citizens have mobilized to voice concerns or publicly oppose the perceived threats of fracking to surface and groundwater supplies, air quality, roads, property values, rural landscapes, climate change, public health and safety, farm animals, and sustainable economic development (Bamberger and Oswald 2014; Food and Water Watch 2015; Ladd 2013, 2014; Sangaramoorthy et al. 2016; Willow et al. 2014). In turn, the rising tide of protest against hydraulic fracking has resulted in statewide fracking moratoriums in New York, Vermont, and Maryland, as well as over 475 local fracking bans in two dozen states. On the international stage, nationwide moratoriums exist in France, Germany, Luxembourg, Scotland, Ireland, and Bulgaria, and debates over shale drilling are growing in the European Union, South America, South Africa, and Australia (Food and Water Watch 2015; Ladd 2013; Wright 2012).

Environmental sociologists and social movement scholars have long utilized frame analysis concepts and similar analytic tools to examine how competing groups socially construct discursive interpretations of the environmental hazards, issues, and conflicts in their community (e.g., Brulle and Benford 2012; Capek 1993; Gray 2003; Krogman 1996; Ladd 2011; Shriver and Peaden 2009; Vincent and Shriver 2009). Typically, environmental frame disputes entail contrasting diagnostic, prognostic, and motivational narratives regarding what citizens view as the problems at hand, what solutions they propose to address such problems, and how these beliefs provide a motivation or rationale for adherents to take action on the issues driving the controversy (Benford 1993; Benford and Snow 2000). While the diagnostic and prognostic components of the larger frame dispute over natural gas fracking have been recently analyzed (see Ladd 2014), the motivational frames and discursive narratives that provide opposing rationales for the conflict have not received similar attention. In this chapter, drawing on sociological literature, archival sources, discursive documents, participant observation, and in-depth interview data from stakeholder groups in the region, I provide a qualitative analysis of the motivational frame disputes surrounding natural gas development and hydraulic fracturing in the Haynesville Shale region of Louisiana, as well as their implications for future mobilization efforts.

Controversy over Unconventional Energy Development and Hydraulic Fracking

Unconventional shale gas extraction utilizing hydraulic drilling and fracking has helped drive an unprecedented energy boom over the last decade that is predicted to account for almost one-half of total natural gas production by 2035 (Lavelle 2012; Theodori 2013; Wright 2012). First developed by Haliburton in the late 1940s, fracking represents a controversial well stimulation/completion technique in which millions of gallons of water, sand, and chemicals are injected under high pressure into deep underground shale formations to fracture the sedimentary rock and allow the gas and oil to flow into wells at the surface (Cosgrove at al. 2015; Ladd 2014). Made profitable in recent years by the advent of multidirectional drilling techniques and platforms, increased pipelines and export facilities, favorable energy legislation, tax subsidies, and rising natural gas prices, "unconventional" natural gas production (compared to older, "conventional" development methods involving single-directional, vertical drilling into shallower gas deposits) has risen twelve-fold since 2005 and is the only fossil fuel likely to increase its share of energy demand in the years to come (Yergin 2011). Today, hydraulic fracturing accounts for over 90% of shale gas (and oil) production in the United States, almost a half-million shale gas wells have been drilled in over 30 states, and over 25,000 wells are fracked each year. More than 15 million Americans now live within a mile or so of a gas or oil well that has been drilled since 2000 (Food and Water Watch 2015; Wilber 2015).

While hydraulic fracturing has played a beneficial role in creating new energy supplies, lower CO_2 emissions, jobs, increased state tax revenues, and rural economic growth, local and regional conflict over its negative social, economic, and environmental impacts has grown steadily in many shale communities across the country (Hauter 2016). Early opposition, for example, ranged from concerns over surface disruptions posed by the construction of drilling rigs on rural and metropolitan landscapes, to the differential signing bonuses and royalties paid by gas companies to landowners for their subsurface mineral rights (Eisenberg 2010). Galvanized by the national attention generated by the 2010 Academy Award–nominated documentary *Gasland* (Fox 2010; Vasi et al. 2015), fracking increasingly came under fire in the popular media for its potential to contaminate local groundwater supplies and deplete aquifers, induce seismic shocks and earthquakes through the deep-well injection of hazardous fracking fluids and wastewater, create well blowouts and worker injuries, harm farmland and livestock, industrialize the tranquility of rural landscapes with increased truck traffic, road damage, noise, stress, and crime, as well as generate unsustainable boom-and-bust impacts for local economies (see, for example, Brown 2013; Brune 2013; Cable 2012; Chapman 2010; Doe 2013; Dobb 2013; Goodell 2012: Lavelle 2012; Light 2015; Lohan 2013; Thetford 2013; Upton 2013; Urbina 2011; Zaitchik 2012).

As the issues surrounding the natural gas boom and shale development gained visibility in the public arena, various scientific studies, environmental assessments, and journalistic investigations began to emerge as well, particularly around the negative impacts of fracking on public health, climate change, drinking water supplies, biodiversity, renewable energy policy, job creation, and political governance (e.g., Bamberger and Oswald 2014; Bernd 2015; Concerned Health Professionals of New York & Physicians for Social Responsibility 2015; Finkel 2015; Food & Water Watch 2011, 2015; Gold 2014; Gullion 2015; Hauter 2016; Heinberg 2013; Hightower 2012; Howarth and Ingraffea 2011; Klein 2014; USEPA 2015; Wilber 2015). As a result, the word "fracking" has taken on multiple meanings in public discourse, often with biased connotations and lewd associations (Evensen et al. 2014). National political opinion polls suggest that the public is either unaware of the substantive issues surrounding fracking, or evenly divided in their views between support and opposition (Boudet et al. 2016; Clarke et al. 2015; Pew Research Center 2014; Sarge et al. 2015).

Review of the Literature

Recent sociological research on public perceptions of unconventional shale gas development and fracking has consistently identified a wide range of positive and negative impacts across different shale communities. In the Marcellus Shale region, for example (the nation's largest, most developed, and most studied shale basin), Weigle (2011) identified over four hundred different citizen concerns related to the socioeconomic, environmental, governmental, and public health and safety impacts of natural gas development. In other studies conducted in the Marcellus, Utica, Niobrara, and Barnett Shale communities, attitudes toward unconventional gas extraction varied in terms of the previous extractive history and current intensity of drilling in the region (Brasier et al. 2011), proximity to drilling and levels of trust in state and industry officials (Mayer 2016), how residents weighed the relative opportunities and risks associated with development (Eaton and Kinchy 2016; Kriesky et al. 2013; Schafft, Borlu, and Glenna 2013), residents' sense of place connections to land, water resources, and health problems (Jalbert, Kinchy and Perry 2014; Kinchy and Perry 2012; Poole and Hudgins 2014; Sangaramoorthy et al. 2016; Willow et al. 2014), exposure to different kinds of media discourse (Ashmoore et al. 2015; Sarge et al. 2015), as well as levels of economic vulnerability, socioeconomic constraints, or adherence to neoliberal economic policies (Hudgins and Poole 2014; Malin 2014; Malin and DeMaster 2016; Malin et al. 2017; Willow 2014).

Sociologists have also examined residents' perceptions of unconventional gas development in the Barnett Shale region of eastern Texas (Anderson and Theodori 2009; Brasier et al. 2011; Gullion 2015; Theodori 2009, 2013; Wynveen 2011), the Eagle Ford Shale region of south Texas (Ellis et al. 2016), as well as the Haynesville Shale region of northwest Louisiana (Ladd 2013, 2014). On the whole, residents in both neighboring states believed that shale development had positively benefited the local economy and community in terms of creating jobs, new businesses, landowner wealth, tax revenues, improved police and fire protection, better schools and teacher salaries, and improved health care services. At the same time, many residents also felt that shale development had created a wide a range of negative impacts for the community, including its potential to contaminate groundwater supplies and deplete aquifers; harm the environmental quality and aesthetic value of the landscape; disrupt the serenity of rural life with truck traffic; cause road damage and accidents; generate air pollution, noise, lighting, and crime; create conflicts over mineral rights; increase the danger for gas well leaks, explosions, and accidents; as well as alter the local power structure. In weighing the benefits and costs of development in the Barnett Shale, local perceptions tended to differ widely by site maturity and direct experience with gas development. In the Eagle Ford, differential perceptions tended to vary by whether residents were seniors, low-income, or landowners. In the Haynesville, while a majority of residents believed that the economic advantages of development had outweighed the overall socioenvironmental costs to the region, a substantial minority was unsure about or disagreed with whether the benefits to date had been worth the risks.

In summary, the extant literature suggests that local-level unconventional gas development represents a significant paradox and contradiction for most host communities. While residents generally appreciate and support the economic and/or service-related benefits that typically accompany such development, they also are ambivalent about or oppose what they view as the greater socioenvironmental challenges connected to shale drilling and fracking (Kriesky et al. 2013; Theodori 2013). As a result of the myriad opportunity-threat impacts perceived by citizens to accompany unconventional natural gas development today, increasing numbers of shale communities across the United States are deeply divided over whether the purported economic benefits of intensive fracking outweigh the social and ecological risks (Kreuze, Schelly, and Norman 2016; Ladd 2013, 2014; Mayer 2016; Schafft, Borlu, and Glenna 2013).

Frame Disputes over Shale Gas Development and Fracking

Drawing on the seminal work of prominent social movement scholars examining the role of collective action frames in mobilizing constituents for social action (e.g., Benford 1993, 1997; Benford and Snow 2000; Noakes and Johnston 2005; Snow and Benford 1988, 1992; Snow et al. 1986), environmental sociologists have often used frame concepts as interpretive lenses for analyzing the discursive narratives of different stakeholders in local environmental disputes (see, for example, Capek 1993; Edberg and Tarasova 2016; Gray 2003; Gunter and Kroll-Smith 2007; Krogman 1996; Ladd 2011; Messer, Shriver, & Kennedy 2009; Mika 2006; Mooney and Hunt 2009; Robertson 2009; Shriver 2001; Shriver, Adams, and Cable, 2013; Shriver, Cable, and Kennedy 2008; Shriver and Kennedy 2005; Shriver and Peaden 2009; Shriver, White, and Kebede 1998; Vincent and Shriver 2009). Snow and Benford (1988) argue that collective action frames generally involve three components: conflicting definitions of the causes of the problems at hand and who is to blame for them (diagnostic frames); the specific solutions, strategies, or policies that should be initiated to solve them (prognostic frames); and the larger social or ethical values that provide a rationale for people to act on their beliefs (motivational frames). While diagnostic and prognostic frame disputes have been explored in a number of environmental conflicts (Forsyth, Luthra, and Bankston 2007; Krogman 1996; Ladd 2011; Shriver and Peaden 2009; Vincent and Shriver 2009), far less attention has been paid to the motivational claims and narratives of stakeholders in local controversies, how they align with wider discursive frames disseminated by institutional actors and organizations, or their relative impact on social movement protest (Benford and Snow 2000; Gifford and Comeau 2011; Vaast et al. 2014).

Thus, by exploring the motivational frames that provide an ideological rationale and justification for stakeholder involvement in a local environmental dispute, this research attempts to contribute some broader insights into why shale energy development has been associated with both mobilization and nonmobilization at different levels of the controversy. With Louisiana at the forefront today of gas (and oil) production utilizing hydraulic drilling and fracturing technology (both onshore and offshore), the frame disputes over the opportunity-threat dimensions of Haynesville Shale development represent a viable social laboratory for examining the clash of worldviews that characterizes this natural resource controversy and how it compares to other shale regions.

Research Methods and Data

The Haynesville Shale formation covers a 9,000-square-mile area of north-western Louisiana and is located 10,000 feet beneath the surface of Caddo, Bienville, Bossier, DeSoto, Red River, and Webster parishes (counties), as well as a small portion of southwestern Arkansas and eastern Texas. The 150-million-year-old shale basin is estimated to contain over 66 trillion cubic feet (tcf) of recoverable natural gas, making it the second largest gas play in the United States and equivalent to a decade's worth of North American consumption (Lavelle 2012; USEIA 2015). While more than 2,700 wells have been drilled in the Haynesville Shale since the gas boom erupted in 2008, state officials project that some 10,000 wells may eventually be required to extract all the gas in the formation over the next 20–30 years (Louisiana Department of Natural Resources 2016; Schleifstein 2011).

The data for this project were drawn from sociological literature, archival documents, participant observation, and 35 in-depth, semistructured interviews conducted with residents, landowners, activists, industry spokespeople and professionals, business owners, scientists, and others who lived in the Haynesville Shale region (or other areas of the state) and had direct knowledge of the issues. Those selected for interviews were identified by purposeful and snowball methods of sampling involving referrals provided by key informants and individuals in various stakeholder groups or occupational networks, as well as names reported in media accounts and other sources relevant to the issue of natural gas development (see Kvale 2007). The prospective subjects were contacted by phone or email; provided with information about the purpose, scope, and format of the study; and subsequently asked to sign an Informed Consent Form (IRB) before the interviews were audio-recorded and transcribed verbatim. All of the interviews were conducted between July and October 2012, either at the subject's home or office, and generally lasted from 45 to 90minutes each.

The distribution of the stakeholders interviewed included: ten (10) environmental scientists/geologists; eight (8) concerned citizens/gas lease holders; six (6) oil/gas industry professionals/operators; four (4) environmental activists; two (2) gas industry representatives; two (2) parish/state government officials; one (1) newspaper editor/reporter; one (1) environmental attorney; and one (1) top state regulatory official. Although the sample included only five (5) female and one (1) African American respondents, the citizens interviewed for this study were nevertheless representative of the overwhelmingly white-male demographic base associated with oil and gas production in north-west Louisiana.

Each interview transcript was thematically analyzed to highlight those narrative accounts that articulated the stakeholder's motivational framing of the issues and differential impacts surrounding gas development and fracking in the Haynesville Shale. While the entire interview schedule consisted of more than 20 structured questions, the collective action frames analyzed here were derived from the narrative responses to a specific set of interview questions that probed the respondent's beliefs about his or her motivation and rationale for involvement in the issues connected to Haynesville gas development, their perceptions of the positive and negative impacts associated with such development, whether they believed the benefits outweighed the costs or not, as well as other attitudes related to the importance and operation of the gas industry in the region.

The data derived from this cross-section of stakeholder interviews allowed for a deeper thematic analysis of some of the discursive beliefs that frame the fracking dispute than the perceptions of impacts explored in previous qualitative studies (see Wynveen 2011; Ladd 2013; Malin and DeMaster 2016; Poole and Hudgins 2014). An additional strength of these data are that they were collected in the period following the initial Haynesville gas boom (2008–2011) after the pace of exploration and fracking episodes in the area had peaked, declined, and stabilized. As a result, residents were able to distance themselves somewhat from the most turbulent period of the fracking bonanza and reflect on the various energy development impacts on their community from a more comprehensive and balanced perspective (Wellborn 2011).

Data Analysis

Motivational Frames Expressing Support for Fracking

The motivational frame of an environmental controversy serves to provide a rationale and justification for social action by linking the issues at hand to wider ethical principles and shared values (Benford 1993; Ladd 2011; Benford and Snow 2000; Vincent and Shriver 2009). Most fracking supporters, for example, justified their involvement in the issue out of a conviction that the oil and gas industry was committed to the safe and responsible development of natural gas, as well as adequately regulated by existing state and federal laws:

> I don't have any questions about fracking; it's an industry procedure and it's safe and it doesn't have any detrimental effects. I feel a responsibility to do something about the misinformation that's out there. . . . We already have a lot of laws and a lot of regulations. We are swamped with them and I'd like to see less. (oil and gas operator)

The technology has changed immensely in the last 30 years. It's amazing how clean and how automated everything is. It's not like the John Wayne movies you see with the huge towers and the guys wrestling with the pipe and there is oil and grease all over the place. It isn't that way anymore and there is a great deal of care taken. I don't find anything wrong with the current policies generally. (environmental scientist)

Many stakeholders also emphasized the gas industry's strong environmental track record and maintained that no major studies had found a direct link between fracking operations and groundwater contamination. More emphatically, many interviewees claimed that it was virtually impossible for the fracking process to contaminate underground aquifers or drinking water supplies:

The Louisiana Geological Survey [has done] groundwater sampling in southern Bossier, southern Caddo, and northern Desoto (parishes) and over a thousand domestic wells have been sampled and analyzed. The data looks very good, the water quality is excellent. (geologist)

I'm a big believer that hydraulic fracturing is a very safe procedure [and] there has been no proof that the fracking has hurt anything. (oil and gas spokesperson)

It is impossible to frack into an aquifer. You cannot frack up almost a mile worth of rock and get into a drinking source. . . . Natural gas drilling has an incredibly small footprint on the environment. Once the drilling rig leaves and once completions have been done, it is just a small amount of equipment that is left behind. (energy company spokesperson)

There's really almost physically no chance that a frack would affect groundwater directly. (geologist)

Some respondents justified their active support for Haynesville natural gas fracking by framing it as simply the most recent chapter in a century-long history of oil and gas development that northwest Louisiana residents had long been adapted to dealing with:

People here are used to the industry and a lot of people have been involved in the industry, so they understand the technology of what hydraulic fracturing is and how it works. (energy company spokesperson)

Another discursive theme in the motivational frame of some supporters was the belief that fracking opponents were not only ignorant of the "science" behind hydraulic fracturing, but were largely environmental zealots using irrational "scare tactics" to bring down the fossil fuel industry:

> I think the scare tactics used by the environmental groups are just ludicrous and are not based on any type of science. The way the Haynesville wells are designed, it's impossible [for groundwater to be contaminated]. So the people up there (in New York) harping about the Marcellus screwing up their fresh water just don't understand the science, or they do and they're just using scare tactics. (oil and gas operator)

> My problem is that we have some real diehard activists that are against [fracking] no matter what. The news media will cover actors like Matt Damon, Mark Ruffalo—and they just don't know what they're doing. They don't have the knowledge; they're just following along. It [fracking] must be bad [if it's associated with] the bad oil and gas companies. (environmental scientist)

Motivational Frames Expressing Opposition to Fracking

On the opposing side of the fracking dispute, concerns over the depletion of aquifers and the potential threat of drinking water contamination provided the primary rationale for the skepticism that many stakeholders felt about unconventional gas development in the Haynesville. Not only did these residents believe that the deleterious impacts of fracking on water resources were the most important questions facing their community, but that other Louisiana residents needed to be more concerned about water-related issues as well. The following narratives were typical of the beliefs that motivated many of these stakeholders to speak out:

> I think that hydraulic fracking is different from the drilling that used to be because of the chemicals, because of the pressure, because of all those things that are now involved with it. While it is at a depth way below our water table, there are still some unknowns about what the impact can be. And that's the question, you know? Can you afford one accident? Can you have any margin for error when it comes to the potential for polluting water? (oil and gas professional)

> I don't know of any other place in the state of Louisiana where the potential for negative impacts on groundwater resources is greater than it is here. (concerned citizen)

> I think when you feel earthquakes in our area that we've never felt before in our lives, then people should be concerned. If you're seeing water wells contaminated that have been in existence for 50+ years and you've never had a problem and you can't get water out of your well, then you need to be concerned. (parish official)

We need to put a lot more value on the water that we have in this state. I think it's severely undervalued and we should make sure that we're protecting it before we reach a crisis like a lot of other places throughout the country. (environmental activist)

Beyond its impact on water resources and increased seismicity in the region, stakeholders expressed fears about whether shale fracking could have other unknown consequences for the environment:

We're talking about [shale] formations that were laid down hundreds of millions of years ago in some cases and we don't know what the effects are going to be. Like most environmental issues, whatever we do is irreversible. Once you start fracturing layers of shale, we don't know what the effect is going to be, 20 years from now, 100 years from now, 500 years from now. So it's the unknown nature of what we're doing that bothers me the most. (environmental scientist)

Environmentally, there's a potential downside that I don't think we know what it could be. That's the concern, the unknown risk. That's where people should be concerned about potential impacts (environmental scientist)

Many other stakeholders justified their opposition to fracking out of their larger ethical and moral concerns about the huge profits generated by gas development in the Haynesville, environmental injustice, and the ability of the energy industry to use their economic power to derail the political regulatory process and influence local and state politicians:

I think it's [fracking] a corrupting influence on the politicians and bureaucrats. I firmly believe that Haynesville Shale gas development can be done responsibly, but it's not because anybody's requiring it. There's nobody asking questions. . . . This [issue of fracking] has made me more cynical than maybe any other environmental issue I've dealt with. (environmental activist)

I think the potential for ethical dilemmas are much higher in this industry, being that it is so profitable; it touches natural resources, it touches people, their culture, and the environment. (environmental scientist)

Most people trust that our governing body has taken care of us in matters such as this, but they haven't. [The gas industry] has taken advantage of people tremendously in this area and a lot of blacks have really been taken advantage of. (concerned citizen)

The moral and ethical issues are the absolute incapacity of politicians and businessmen in this area to address the environmental side of producing oil and gas. I think their statutory and constitutional mandates to conserve, protect, and replenish the natural resources of Louisiana are not even being considered. (environmental activist)

Finally, some residents framed their skepticism about fracking around wider concerns about the unknown health risks, as well as scientific and geological uncertainties, accompanying the technological processes required for shale drilling and fracturing:

> I think that gas fracking is not as safe as it's perceived because there are great health risks that are not yet exposed, but only time will expose those problems. I believe that there is a risk with everything we do, but I believe there is a greater risk with fracking. (concerned citizen)

> It's a huge unknown. The whole issue of what happens when you mess with a whole geological layer down there and then you take the gas out of it—there are issues of subsidence, movement, and earthquakes. We have no concept of a precautionary principle and the burden of proof is on the wrong foot. If you can't be sure it won't hurt people, then don't do it. (environmental scientist)

Taken together, these motivational frames and narratives illustrate a variety of discursive themes that help explain citizens' differential levels of involvement in the issues surrounding natural gas fracking, including their willingness to take part in this study. On the one hand, adherents of the process strongly believed that fracking represents a safe and adequately regulated technology that could increase U.S. wealth and energy independence if only its irrational opponents would cease their unfounded attacks on the industry. Skeptics or opponents, on the other hand, viewed fracking as a potential threat to local water resources, as well as contributing to increased political corruption, environmental injustice, and unknown health risks. On both sides of the fracking debate, stakeholders articulated a set of discursive claims that reflected not only their own concerns about fracking, but why other Louisiana residents should care about the issues as well.

Discussion

As these narratives reveal, there were distinct opposing accounts in the motivational frames constructed by local stakeholders regarding the differential impacts of shale development, as well as their rationale for involvement in the issues. On the one hand, fracking supporters believed that the natural gas industry was strongly committed to the safe and responsible development of shale energy and was more than adequately regulated by existing state and federal laws. They did not believe that fracking had harmed or could result in any harm to local water resources or the environment and saw it as just another chapter in Louisiana's long history of oil and gas extraction that most residents have been comfortably adapted to for decades. In this respect, Haynesville stakeholders framed their support for natural gas extraction with essentially the same rationale used by coastal Louisiana residents in their support for offshore oil development. As argued in past research, Louisianans tend to be reflexively enthusiastic about the benefits of energy extraction out of what they perceive as their lived historic experience—not ignorance—and believe that they understand the oil and gas industry better than people in other parts of the country (see Freudenburg and Gramling 1993; Gramling and Freudenburg 1992).

Conversely, many stakeholders who were skeptical or opposed to fracking framed their justification for involvement primarily out of concerns over water availability and potential contamination, issues they saw as far too important to entrust to state or industry officials. Indeed, given the profitability of the natural gas boom for the Haynesville region, some viewed its presence as yet another corrupting influence on Louisiana politicians and regulators in a state already infamous for its political manipulation by the oil and gas industry (Freudenburg and Gramling 2011; Zebrowski and Leach 2014). Moreover, many residents expressed alarm over the long-term geological risks posed by fracking, its potential to create earthquakes from drilling and deep-well waste injection techniques, as well as its unknown and irreversible public health risks for local citizens.

Typically, motivational frames serve to highlight two important discursive opportunities in an environmental dispute: they express "vocabularies of motive" that stress the urgency of the situation and severity of the threat, as well as a "call to arms" that provides a rationale for ameliorative collective action (Benford and Snow 2000: 617). On both sides of the Haynesville debate, however, neither supporters nor opponents articulated a strong sense of urgency, threat severity, or collective agency that called for any conscious efforts to expand or halt shale fracking in the region. None of the narratives utilized alarmist language that labeled the views of opposing stakeholders as "dangerous" or "disastrous," nor did they convey the respondent's motivation to engage opponents in public forums, demonstrations, court battles, legislative initiatives, economic boycotts, or other actions that could significantly expedite or lead to moratoriums on natural gas fracking at the community, state, or national level. Moreover, while more than four out of ten Haynesville stakeholders expressed ambivalence about the presence of fracking in their community, as well as whether the benefits of development had been worth the larger risks to the region, virtually none of the respondents expressed an interest in participating in acts of protest against gas well operators, nor belonged to any local organizations that had publicly opposed shale exploration or fracking in the region. Indeed, whether they viewed energy development as a relative opportunity or threat, the prevailing narrative among residents was that it was "probably a good thing," an "economic boost," or even a "godsend" in bringing jobs and tax dollars to a poor region of the state already hard pressed by the national economic recession. In a region of Louisiana and the Gulf South long familiar with the exploration and extraction of oil and gas reserves (Freudenburg and Gramling 1993; Gramling and Brabant 1986), Haynesville residents were generally inclined to view unconventional gas development through a wide and supportive lens that normalized fracking and rationalized it on the grounds of economic necessity and vulnerability (Malin 2014). In addition, the motivational frames supporting fracking paralleled the corporate official frame (see Messer, Adams, and Shriver 2012; Messer et al. 2009) of the U.S. oil and gas industry, a hegemonic narrative extolling the myriad economic and energy benefits of the "Natural Gas Revolution" (America's Natural Gas Alliance 2013; Ladd 2016).

By extension, these data also help explain why the perceived socioenvironmental threats from fracking in the Haynesville have not generated the level of environmental controversy or community conflict witnessed in other key shale regions like the Marcellus (Malin 2014; Malin and DeMaster 2016; Poole and Hudgins 2014; Willow et al. 2014). Given the longstanding fossil fuel–friendly culture of Louisiana (including the fact that the first natural gas well in the state was drilled in Caddo Parish over a century earlier), most of its citizens have a high level of comfort with the social and ecological impacts of the oil and gas industry compared to citizens of other parts of the United States. Due to the state's comparatively lower educational levels, extractive orientation toward the environment, favorable patterns of contact with industry personnel, and extensive prior adaption to energy development, it has been argued that to know the oil and gas industry in Louisiana is to love it (see Freudenburg and Gramling 1993: 345; Zebrowski and Leach 2014). Conversely, even when stakeholders were skeptical or opposed to the negative impacts of fracking on their water resources, landscape, or public health, they rarely articulated narratives that were consistent with those of the larger antifracking movement. Indeed, some researchers have pointed out that grassroots actors in environmental disputes like fracking do not always share the same ideological perspectives, nor are they necessarily aware of the national frame discourse and movement strategies that align with their own privately held convictions and solutions (Bamberger and Oswald 2014; Staggenborg 2011; Vasst et al. 2014). In any given frame dispute, citizens can express selective diagnostic or prognostic narratives that capture the severity and urgency of a perceived environmental threat, yet hold complacent motivational beliefs that indicate a diminished sense of efficacy and reluctance to challenge the status quo (Benford and Snow 2000; Noakes and Johnston 2005).

Finally, recent studies of opposition to energy projects and facilities have emphasized that most at-risk communities do not experience protest mobilization, or for long periods, nor does collective resistance occur unless community members are both motivated and capable of mobilizing (Eaton and Kinchy 2016; McAdam and Boudet 2012; Wright and Boudet 2012; Vasi et al. 2015). In communities where residents are facing economic hardships, yet are familiar with a particular industry, they are more likely to emphasize the economic benefits of energy development and underestimate or deemphasize its potential drawbacks. Moreover, when a community typically views the surrounding energy infrastructure as beneficial to the citizenry at large, mobilization is also unlikely to occur (Wright and Boudet 2012). In view of these factors, among others, citizens' response to shale development and fracking in the Haynesville to date has been characterized by quiescence—the absence of grievance perception in the face of risks—rather than the kind of collective protest, resistance, or political mobilization witnessed in other states. Emphasizing the "quiet voices" of residents that often underlie the fracking debate, Eaton and Kinchy (2016: 24) stress that: "(N)onmobilized communities are not necessarily sites of consent. People remain quiet about their grievances with the unconventional oil and gas industry when they generally support community economic growth and sense that collection action is unlikely and ineffective."

Despite the expressed ambivalence and skepticism toward fracking felt by many residents in this study (largely because of the anonymity and confidentiality provided by the interview process), there was a pronounced reluctance to openly challenge the Louisiana gas industry's exploration practices or impacts to any significant degree. Poole and Hudgins (2014), for example, found that while some residents in the Marcellus Shale were personally critical of fracking, they also believed that energy development had already degraded their land and water resources to the point where expressions of protest were pointless. To the extent that quiescence is facilitated by informal systems of social control that are both anticipatory and invisible, state and corporate elites in Louisiana appear to have been successful in muting public criticism of the oil and gas industry by fostering acceptance of its normative claims (see Cable, Shriver, and Hastings 1999). As a case in point, one oil and gas professional who was skeptical about fracking commented "off the record" that he would be hesitant to say anything negative about the Haynesville fracking boom in public, out of fear that such remarks could negatively affect his wife's real estate business in a town where the oil and gas industry generates a lot of jobs and home sales for employees.

On the whole, the environmental frame dispute over natural gas fracking in the Haynesville points to how a community can be socially divided over issues of energy development without residents being motivated to engage in ameliorative social action or organized resistance. Consequently, future researchers need to question the often unchallenged assumption that ideological cleavages and competing narratives in an environmental controversy tend to lead to the emergence of overt schisms, political protest, movement mobilization, civil discord, or rancorous policy debate. For these Louisiana residents at least, the dispute over shale fracking in the Haynesville constitutes a veritable "double-edged sword" of overlapping benefits and risks to the local community (Ladd 2013, 2014).

Conclusion

Today, stakeholders in natural resource–dependent communities like the Haynesville Shale face a growing number of social, economic, and environmental uncertainties tied to the boom and bust cycles of energy production (Malin and DeMaster 2016). Feelings of isolation, powerlessness, and economic vulnerability in such settings can work to deter citizens from voicing their complaints about the impacts of shale fracking, much less demand greater regulatory oversight of an industry that is already laying off thousands of Louisiana workers due to low energy prices and drilling activity (Adler 2016). Yet, since April 2014, a surprising new antifracking movement has been on the rise in St. Tammany Parish, some 200 miles southeast of the Haynesville in the Tuscaloosa Marine Shale formation of central Louisiana. Described by the Louisiana Oil and Gas Association as "the next big shale oil play" in the south, the Tuscaloosa Shale is estimated to contain 7 billion barrels of oil and at least 40 active wells have been drilled in the region (Louisiana Oil & Gas Association 2016). For over two years, residents, local environmental groups, and the St. Tammany Parish government legally battled the state of Louisiana to deny Helis Oil & Gas Company the necessary permits to drill a 13,000 ft. exploratory oil well on undeveloped, residentially zoned land near the town of Mandeville (Rhoden 2016a). In 2014, the parish government filed suit against both the state Department of Resources and the Office of Conservation in an effort to block Helis's drilling proposal, but were unsuccessful at every stage of deliberative process.

On June 18, 2016, the Louisiana State Supreme Court, in a 4–3 decision, upheld the lower appellate court ruling that St. Tammany Parish could not use its zoning regulations to block the state's regulation of oil and gas activity. In a strongly worded two-page dissent, one of the justices wrote that the opposing writs "present important, difficult, and challenging issues that this Court should address" (Rhoden 2016b: B4). The decision cleared the way for Helis to begin drilling the exploratory well at the end of June and potentially frack the 960-acre well site should the well prove to be commercially profitable. Many parish officials and citizens have held meetings, staged protest rallies,

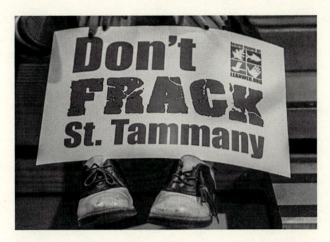

FIGURE 6.1 A popular antifracking poster viewed around Abita
Springs, LA. Source: Louisiana Environmental Action Network:
www.leanweb.org

signed petitions, supported local political candidates, and initiated lawsuits to
oppose the project on environmental and economic grounds. To date, mobilization efforts have been largely driven by citizen fears that fracking could
damage the aquifer that supplies the parish's drinking water, create air and
water pollution, open the door to the industrialization of the parish, and harm
property values (Rhoden 2016b). At various locations throughout the parish,
billboards and signs can be seen proclaiming, "Don't FRACK St. Tammany"
and "Keep the FRACK Out of My Water" (Ladd 2015). While Helis ended
up withdrawing their plans to frack the well site in September 2016, based on
cost-benefit calculations that did not appear profitable for the company, many
St. Tammany residents are worried about future oil development in the parish and what risks lie ahead for their communities and environment. Taking a
proactive stand against future fracking threats, the mayor and town council of
Abita Springs recently announced the city's goal to transition to 100% renewable energy by 2035, joining at least 23 other U.S. cities that have made similar
commitments (Becktold 2017: 68).

Paralleling the national divide over fracking, Louisiana is marked by varying degrees of quiescence and resistance in its two dominant shale regions,
the Haynesville and the Tuscaloosa. While the disputes there are more muted
and less contentious than many of those playing out in the Marcellus, Barnett, Eagle Ford, Niobrara, or Monterey communities, thousands of Louisiana citizens are nevertheless finding themselves caught between the dictates
of global energy markets on the one hand and the growing recognition that
their communities are disproportionately paying the hidden costs of national
energy production on the other. Clearly, more extensive sociological research

is needed that examines how citizens are framing the impacts of shale development and fracking on their communities if we are to fully grasp the manner in which our society is being increasingly fractured by a new generation of unconventional fossil fuels and extractive technologies.

References

Adler, Ben. 2016. "Wanna See What Happens When You Rely on the Fossil Fuel Sector and Slash Taxes? Check Out Louisiana." *Grist Magazine*, March 7. Retrieved March 8, 2016. (http://grist.org/politics/wanna-see-what-happens-when-you-rely-on-the...a/?utmmedium=email&utm source=newsletter&utm campaign=daily=horizon).

America's Natural Gas Alliance. 2013. "Think About It." Retrieved August 26, 2013 (http://thinkaboutit.org).

Anderson, Brooklynn J., and Gene L. Theodori. 2009. "Local Leaders' Perceptions of Energy Development in the Barnett Shale." *Southern Rural Sociology* 24(1): 113–129.

Ashmoore, Olivia, Darrick Evensen, Chris Clarke, Jennifer Krakower, and Jeremy Simon. 2015. "Regional Newspaper Coverage of Shale Gas Development across Ohio, New York, and Pennsylvania: Similarities, Differences, and Lessons." *Energy Research & Social Science* 11: 119–132.

Bamberger, Michelle, and Robert Oswald. 2014. *The Real Cost of Fracking: How America's Shale Gas Boom Is Threatening Our Families, Pets, and Food*. Boston: Beacon Press.

Becktold, Wendy. 2017. "Abita Springs Takes the Lead." *Sierra* 102(3): 68.

Benford, Robert D. 1993. "Frame Disputes within the Nuclear Disarmament Movement." *Social Forces* 71(3): 677–701.

Benford, Robert D. 1997. "An Insider's Critique of the Social Movement Framing Perspective." *Sociological Inquiry* 67: 409–430.

Benford, Robert D., and David A. Snow. 2000. "Framing Processes and Social Movements: An Overview and Assessment." *Annual Review of Sociology* 26: 611–639.

Bernd, Candice. 2015. "Republicans Aim to Preempt Local Democracy, Target Fracking Bans." *Truthout*, May 8. Accessed May 8, 2015. (http://truth-out.org/new/item/30670-republicans-aim-to-preempt-local-democracy-targeting-frackingbans#).

Boudet, Hilary, Dylan Bugden, Chad Zanocco, and Edward Maibach. 2016. "The Effect of Industry Activities on Public Support for 'Fracking.'" *Environmental Politics* (http://dx.doi.org/10.1080/09644016.2016.11537710).

Brasier, Kathryn J, Matthew R Filteau, Diane K. McLaughlin, Jeffrey Jacquet, Richard C. Stedman, Timothy W. Kelsey, and Stephan W. Kelsey. 2011. "Residents' Perceptions of Community and Environmental Impacts from Development of Natural Gas in the Marcellus Shale: A Comparison of Pennsylvania and New York Cases." *Journal of Rural Social Sciences* 26(1): 32–61.

Brown, Chip. 2013. "North Dakota Went Boom." *New York Times Magazine*. February 3: 22–31.

Brulle, Robert J., and Robert D. Benford. 2012. "From Game Protection to Wildlife Management: Frame Shifts, Organizational Development, and Field Practices." *Rural Sociology* 77(1): 2–88.

Brune, Michael. 2013. "The Opposition's Opening Remarks." *The Economist*, February 5. Retrieved June 2, 2014 (http://www.economist.com/debate/days/view/934/print).

Cable, Sherry. 2012. *Sustainable Failures: Environmental Policy and Democracy in a Petro-Dependent World*. Philadelphia, PA: Temple University Press.

Cable, Sherry, Thomas Shriver, and Donald Hastings. 1999. "The Silenced Majority: Quiescence and Government Social Control on the Oak Ridge Nuclear Reservation." *Research in Social Problems and Public Policy* 7: 59–81.

Capek, Stella M. 1993. "The Environmental Justice Frame: A Conceptual Discussion and Application." *Social Forces* 40: 5–24.

Chapman, Karen. 2010. "Trashing the Planet for Natural Gas; Shale Gas Development Threatens Freshwater Sources, Likely Escalates Climate Destabilization." *Capitalism Nature Socialism* 21(4): 72–82.

Clarke, Christopher E., Philip S. Hart, Jonathan P. Schuldt, Darrick T. N. Evensen, Hilary S. Boudet, Jeffrey B. Jacquet, and Richard C. Stedman. 2015. "Public Opinion on Energy Development: The Interplay of Issue Framing, Top-of-Mind Associations, and Political Ideology." *Energy Policy* 81: 131–140.

Concerned Health Professionals of New York & Physicians for Social Responsibility. 2015. Compendium of Scientific, Medical, and Media Findings Demonstrating Risks and Harms of Fracking (unconventional gas and oil extraction) (3rd ed) (http://concernedhealthny.org/compendium/).

Cosgrove, Brendan M., Daniel R. LaFave, Sahan T. M. Dissanayake, and Michael R. Donihue. 2015. "The Economic Impact of Shale Gas Development: A Natural Experiment along the New York/Pennsylvania Border." *Agricultural and Resource Economics Review* 44(2): 20–39.

Dobb, Edwin. 2013. "The New Oil Landscape." *National Geographic* 223(3): 28–58.

Doe, Phillip. 2013. "Tsunami of Public Outrage Builds in Colorado as Fracking Invades Cities." *Alternet*, July 17. Retrieved July 17, 2013. (http://www.alternet.org/print/fracking/tsunami-public-outrage-builds-colorado-fracking-invades-cities).

Eaton, Emily, and Abby Kinchy. 2016. "Quiet Voices in the Fracking Debate: Ambivalence, Nonmobilization, and Individual Action in Two Extractive Communities." *Energy Research and Social Science* 20: 22–30.

Edberg, Karin, and Ekaterina Tarasova. 2016. "Phasing Out or Phasing In: Framing the Role of Nuclear Power in the Swedish Energy Transition." *Energy Research & Social Science* 13: 170–179.

Eisenberg, Nora. 2010. "A New Generation of Natural Gas Drilling Is Endangering Communities from the Rockies to New York." *The Nation*, June 22. Accessed July 3, 2010. (http://www.alternet.org/module/printversion/147298).

Ellis, Colter, Gene L. Theodori, Peggy Petrzelka, Douglas Jackson-Smith, and A. E. Luloff. 2016. "Unconventional Risks; The Experience of Acute Energy Development in the Eagle Ford Shale." *Energy Research & Social Science* (http://dx.doi.org/10.1016/j.erss.2016.05.006).

Evensen, Darrick, Jeffrey B. Jacquet, Christopher E. Clarke, and Richard C. Stedman. 2014. "What's the 'Fracking' Problem? One Word Can't Say It All." *Extractive Industries and Society* 1: 130–136.

Finkel, Madelon L. (ed.). 2015. *The Human and Environmental Impact of Fracking: How Fracturing Shale for Gas Affects Us and Our World*. Santa Barbara, CA: Praeger.

Food & Water Watch. 2011. *The Case for a Ban on Gas Fracking*. Washington, DC: Food & Water Watch.

Food & Water Watch. 2015. *The Urgent Case for a Ban on Fracking*. Retrieved December 6, 2015. (www.foodandwaterwatch.org/sites/default/files/urgent_case_for_ban_on_fracking.pdf.)

Forsyth, Craig J., Asha D. Luthra, and William B. Bankston. 2007. "Framing Perceptions of Oil Development and Social Disruption." *Social Science Journal* 44: 287–299.

Fox, Josh. 2010. *Gasland*. (Film Documentary). Wow Productions: New York.

Freudenburg, William R., and Robert Gramling. 1993. "Socio-environmental Factors and Development Policy: Understanding Opposition and Support for Offshore Oil." *Sociological Forum* 8(3): 341–365.

Freudenburg, William R., and Robert Gramling. 2011. *Blowout in the Gulf: The BP Oil Spill Disaster and the Future of Energy in America*. Cambridge, MA. MIT Press.

Gifford, Robert, and Louise A. Comeau. 2011. "Message Framing Influences on Perceived Climate Change Competence, Engagement, and Behavioral Intentions." *Global Environmental Change* 21(4): 1301–1307.

Gold, Russell. 2014. *The Boom: How Fracking Ignited the American Energy Revolution and Changed the World*. New York: Simon & Schuster.

Goodell, Jeff. 2012. "The Big Fracking Bubble: The Scam Behind the Gas Boom." *Rolling Stone*, March 15. Accessed March 25, 2012. http://www.rollingstone.com/politics/news/the-big-fracking-bubble-the-scam-behind-the-gas-boom-20120301?print=true).

Gramling, Robert, and Sarah Brabant. 1986. "Boom-towns and Offshore Energy Impact Assessment: The Development of a Comprehensive Model." *Sociological Perspectives* 2(9): 177–201.

Gramling, Robert, and William R. Freudenburg. 1992. "Opportunity-Threat, Development, and Adaption: Toward a Comprehensive Framework for Social Impact Assessment." *Rural Sociology* 57(2): 216–234.

Gray, Barbara. 2003. "Framing of Environmental Disputes." Pp. 11–34 in Roy J. Lewicki, Barbara Grey, and Michael Elliot (eds.), *Making Sense of Intractable Environmental Conflicts: Concepts and Cases*. Washington, D.C.: Island Press.

Gullion, Jessica Smartt. 2015. *Fracking the Neighborhood: Reluctant Activists and Natural Gas Drilling*. Cambridge, MA: MIT Press.

Gunter, Valerie, and Steve Kroll-Smith. 2007. *Volatile Places: A Sociology of Communities and Environmental Controversies*. Thousand Oaks, CA: Pine Forge Press.

Hauter, Wenonah. 2016. *Frackopoly: The Battle for the Future of Energy and the Environment*. New York: New Press.

Heinberg, Richard. 2013. *Snake Oil: How Fracking's False Promise of Plenty Imperils Our Future*. Santa Rosa, CA: Post-Carbon Institute.

Hightower, Jim. 2012. "Oil and Gas Marauders Are Destroying Our Land, Water, and Communities All Over America." *The Hightower Lowdown* 14(7): 1–8.

Howarth, Robert W., and Anthony Ingraffea. 2011. "Should Fracking Stop?" *Nature* 477: 271–3.

Hudgins, Anastasia, and Amada Poole. 2014. "Framing Fracking: Private Property, Common Resources, and Regimes of Governance." *Journal of Political Ecology* 21: 303–319.

Humes, Edward. 2012. "Fractured Lives: Detritus of Pennsylvania's Shale Gas Boom." *Mother Jones* 97(4): 52–59.

Jalbert, Kirk, Abby J. Kinchy, and Simona L. Perry. 2014. "Civil Society Research and Marcellus Shale Natural Gas Development: Results of a Survey of Volunteer Water Monitoring Organizations." *Journal of Environmental Studies and Science* 4(1): 78–86.

Johnson, Erik W., and Scott Frickel. 2011. "Ecological Threat and the Founding of U.S. National Environmental Movement Organizations, 1962–1998." *Social Problems* 58(3): 305–329.

Kinchy, Abby J. and Simona L. Perry. 2012. "Can Volunteers Pick Up the Slack? Efforts to Remedy Knowledge Gaps about the Watershed Impacts of Marcellus Shale Gas Development." *Duke Environmental Law & Policy Forum* 22(2): 303–339.

Klein, Naomi. 2014. *This Changes Everything: Capitalism vs. The Climate*. New York: Simon & Schuster.

Kreuze, Amanda, Chelsea Schelly, and Emma Norman. 2016. "To Frack or Not to Frack: Perceptions of the Risks and Opportunities of High-Volume Hydraulic Fracturing in the United States." *Energy Research & Social Science* (http://dx.doi.org/10.1016/j.erss.2016 .05.010).

Kriesky, J. B., D. Goldstein, K. Zell, and S. Beach. 2013. "Differing Opinions about Natural Gas Drilling in Two Adjacent Counties with Different Levels of Drilling Activity." *Energy Policy* 58: 228–236.

Krogman, Naomi T. 1996. "Frame Disputes in Environmental Controversies: The Case of Wetlands Regulations in Louisiana." *Sociological Spectrum* 16: 371–400.

Kvale, Steiner. 2007. *Doing Interviews*. London: Sage Publications.

Ladd, Anthony E. 2011. "Feedlots of the Sea: Movement Frames and Activist Claims in the Protest over Salmon Farming in the Pacific Northwest." *Humanity and Society* 35(4): 343–375.

Ladd, Anthony E. 2013. "Stakeholder Perceptions of Socioenvironmental Impacts from Unconventional Natural Gas Production and Hydraulic Fracturing in the Haynesville Shale." *Journal of Rural Social Sciences* 28(2): 56–89.

Ladd, Anthony E. 2014. "Environmental Disputes and Opportunity-Threat Impacts Surrounding Natural Gas Fracking in Louisiana." *Social Currents* 1(3): 293–312.

Ladd, Anthony E. 2015. "A Tale of Two Louisiana Shale Communities: Differential Responses to Unconventional Energy Development and Fracking in the Haynesville and Tuscaloosa Shale Regions." Paper represented at the American Sociological Association meetings, August 21–25, 2015, Chicago, IL.

Ladd, Anthony E. 2016. "Meet the New Boss, Same as the Old Boss: The Continuing Hegemony of Fossil Fuels and Hydraulic Fracking in the Third Carbon Era." *Humanity and Society* (http://dx.doi.org/10.1177/0160597616628908).

Ladd, Anthony E., and Charles Perrow. 2016. "Institutional Dilemmas of Hydraulic Fracking: Economic Bonanza, Renewable Energy Bridge, or Gangplank to Disaster?" Paper presented at the American Sociological Association meetings, August 21–24, 2016, Seattle, WA.

Lavelle, Marianne. 2012. "Bad Gas, Good Gas." *National Geographic* 222 (December): 90–109.

Light, John. 2015. "Fracking Is Definitely Causing Earthquakes, Another Study Confirms." *Grist Magazine*, January 7. Retrieved January 7, 2015. (http://grist.org/news/fracking -is-definitely-causing-earthquakes-anot...tter&utm_medium=email&utm_term=Daily %2520lan%25207&utm).

Lohan, Tara. 2013. "5 Reasons Natural Gas Won't Be an Environmental and Economic Savior." *Alternet*, January 10. Retrieved January 10, 2013 (http://www.alternet.org/print /fracking/5-reasons-natural-gas-wont-be-environmental-and-economic-savior).

Louisiana Department of Natural Resources. 2016. "Haynesville Shale." Accessed March 14, 2016. (http://www.dnr.louisiana.gov/index.cfm?md=pagebuilder&tmp=home&pid= 442&pnid=0&nid=170).

Louisiana Oil & Gas Association. 2016. "Tuscaloosa Marine Shale." Accessed March 14, 2016. (http://www.loga.la/louisiana-shale-plays/tuscaloosa-marine-shale).

Malin, Stephanie A. 2014. "There's No Real Choice but to Sign: Neoliberalization and Normalization of Hydraulic Fracturing on Pennsylvania Farmland." *Journal of Environmental Studies and Science* 4: 17–27.

Malin, Stephanie A., and Kathryn Teigen DeMaster. 2016. "A Devil's Bargain: Rural Environmental Injustices and Hydraulic Fracturing on Pennsylvania's Farms." *Journal of Rural Studies* 47: 278–290.

Malin, Stephanie A., Adam Mayer, Kelly Shreeve, Shawn K. Olson-Hazboun, and John Adgate. 2017. "Free Market Ideology and Deregulation in Colorado's Oil Fields: Evidence for Triple Movement Activism?" *Environmental Politics* (http://dx.doi.org/10.1080 .09644016.2017.1287627).

Mayer, Adam. 2016. "Risk and Benefits in a Fracking Boom: Evidence from Colorado." *Extractive Industries and Society* (http: //dx.doi.org/10.1016/j/exis.2016.04.006).

McAdam, Doug, and Hilary Boudet. 2012. *Putting Social Movements in Their Place: Explaining Opposition to Energy Projects in the United States, 2000–2005.* New York: Cambridge University Press.

Messer, Chris M., Alison E. Adams, and Thomas E. Shriver. 2012. "Corporate Frame Failure and the Erosion of Elite Legitimacy." *Sociological Quarterly* 53(3): 475–499.

Messer, Chris M., Thomas E. Shriver, and Dennis Kennedy. 2009. "Official Frames and Corporate Environmental Pollution." *Humanity and Society* 33(4): 273–291.

Mika, Marie. 2006. "Framing the Issue: Religion, Secular Ethics and the Case of Animal Rights Mobilization." *Social Forces* 85(2): 915–941.

Mooney, Patrick H., and Scott A. Hunt. 2009. "Food Security: The Elaboration of Contested Claims to a Consensus Frame." *Rural Sociology* 74(4): 469–497.

Noakes, John A., and Hank Johnston. 2005. "Frames of Protest: A Road Map to a Perspective." Pp. 1–29 in Hank Johnston and John A. Noakes (eds.), *Frames of Protest: Social Movements and the Framing Perspective.* Lanham, MD: Rowman & Littlefield.

Pew Research Center. 2014. "Views on Increased Use of Fracking Tilt Negative." November 12. Accessed February 13, 2016. (http://www.people-press.org/2014/11/12/little -enthusiasm-familiar-divisions-after-the-gops-big-midterm-victory/).

Poole, Amanda, and Anastasia Hudgins. 2014. "'I Care More about This Place, Because I Fought for It': Exploring the Political Ecology of Fracking in an Ethnographic Field School." *Journal of Environmental Studies and Science* 4(1): 37–46.

Rhoden, Robert. 2016a. "Fracking Opponents Lose Round in Court." *Times-Picayune,* March 11: B1–B2.

Rhoden, Robert. 2016b. "Tammy Fracking Opponents Lose Appeal." *Times-Picayune,* June 19: B1, B4.

Robertson, Erin. E. 2009. "Competing Frames of Environmental Contamination: Influences on Grassroots Community Mobilization." *Sociological Spectrum* 29: 3–27.

Sangaramoorthy, Thurka, Amelia M. Jamison, Meleah D. Boyle, Devon C. Payne-Sturges, Amir Sapkota, Donald M. Milton, and Sacoby M. Wilson. 2016. "Place-Based Perceptions of the Impacts of Fracking Along the Marcellus Shale." *Social Science & Medicine* 151: 27–37.

Sarge, Melanie A., Matthew S. VanDyke, Andy J. King, and Shawna R. White. 2015. "Selective Perceptions of Hydraulic Fracturing: The Role of Issue Support in the Evaluation of Visual Frames." *Politics and the Life Sciences* 34(1): 57–71.

Schafft, Kai A., Yetkin Borlu, and Leland Glenna. 2013. "The Relationship between Marcellus Shale Gas Development in Pennsylvania and Local Perceptions of Risk and Opportunity." *Rural Sociology* 78(2): 143–66.

Schleifstein, Mark. 2011. "Boomville, Louisiana." *Times-Picayune,* March 27: A1–A10.

Shriver, Thomas E. 2001. "Environmental Hazards and Veterans' Framing of Gulf War Illness." *Sociological Inquiry* 71(4): 403–420.

Shriver, Thomas E., Alison E. Adams, and Sherry Cable. 2013. "Discursive Obstruction and Elite Opposition to Environmental Activism in the Czech Republic." *Social Forces* 91(3): 873–893.

Shriver, Thomas E., Sherry Cable, and Dennis Kennedy. 2008. "Mining for Conflict and Staking Claims: Contested Illness at the Tar Creek Superfund Site." *Sociological Inquiry* 78(4): 558–579.

Shriver, Thomas E., and Dennis Kennedy. 2005. "Contested Environmental Hazards and Community Conflict over Relocation." *Rural Sociology* 70(4): 491–513.

Shriver, Thomas E., and Charles Peaden. 2009. "Frame Disputes in a Natural Resource Controversy: The Case of the Arbuckle Simpson Aquifer in South-Central Oklahoma." *Society and Natural Resources* 22: 143–157.

Shriver, Thomas E., Debbie White, and Alem Seghed Kebede. 1998. "Power, Politics, and the Framing of Environmental Illness." *Sociological Inquiry* 68(4): 458–475.

Snow, David A., and Robert D. Benford. 1988. "Ideology, Frame Resonance, and Participant Mobilization." *International Social Movement Research* 1: 197–218.

Snow, David A., and Robert D. Benford. 1992. "Master Frames and Cycles of Protest." Pp. 133–155 in Aldon Morris and Carol McClung Mueller (eds.), *Frontiers of Social Movement Theory*. New Haven, CT: Yale University Press.

Snow, David A., E. Burke Rochford, Jr., Stephen K. Worden, and Robert D. Benford. 1986. "Frame Alignment Processes, Micromobilization, and Movement Participation." *American Sociological Review* 51: 464–481.

Staggenborg, Suzanne. 2011. *Social Movements*. New York: Oxford University Press.

Theodori, Gene L. 2009. "Paradoxical Perceptions of Problems Associated with Unconventional Natural Gas Development." *Southern Rural Sociology* 24(3): 97–117.

Theodori, Gene. 2013. "Perception of the Natural Gas Industry and Engagement in Individual Civic Actions." *Journal of Rural Social Sciences* 28(2): 122–134.

Thetford, Kyle. 2013. "The Natural Gas Boom: Processes, Production, and Problems." *The Atlantic*, Retrieved August 26, 2013. (http://www.theatlantic.com/technology/print/2013/08/the-natural-boom-processes-production-and-problems/278913).

United States Environmental Protection Agency. 2015. *Assessment of the Potential Impacts of Hydraulic Fracturing for Oil and Gas on Drinking Water Resources* (External Review Draft). U.S. Environmental Protection Agency, Washington, DC, EPA/600/R-15/047, 2015.

Upton, John. 2013. "Ohio Fracking Boom Has Not Brought Jobs." *Grist Magazine*, March 22. Retrieved June 16, 2013. (http://grist.org/news/ohio-fracking-boom-has-not-brought-jobs/?utm_campaign=daily&utm_medium=email&utm_source=newsletter).

Urbina, Ian. 2011. "Regulation Lax as Gas Wells' Tainted Water Hits Rivers." *New York Times*, February 26, 2011. (http://www.nytimes.com/2011/02/27/gas.html?pagewanted=print).

Vaast, Emanuelle, Hani Safadi, Bogdan Negoita, and Liette LaPointe. 2014. "Grassroots Versus Established Actors' Framing of a Crisis: Tweeting the Oil Spill." *Academy of Management Annual Meeting Proceedings*, 2014, pp. 904–909.

Vasi, Ion Bogdan, Edward T. Walker, John S. Johnson, and Hui Fen Tan. 2015. "No Fracking Way! Documentary Film, Discursive Opportunity, and Local Opposition against Hydraulic Fracturing in the United States, 2010–2013." *American Sociological Review* 80(5): 934–959.

Vincent, Shirley G., and Thomas E. Shriver. 2009. "Framing Contests in Environmental Decision-making. A Case of the Tar Creek (Oklahoma) Superfund Site." American *Journal of Environmental Sciences* 5(2): 164–178.

Weigle, Jason L. 2011. "Resilience, Community, and Perceptions of Marcellus Shale Development in the Pennsylvania Wilds: Reframing the Discussion." *Sociological Viewpoints* 27: 3–14.

Wellborn, Vickie. 2011. "Haynesville Shale Reaches Milestone While Fracking Debate Continues." *Shreveport Times*, March 20, 2011. (http://pqasb.pqarchiver.com /shreveporttimes/access/2300584901.htm).

Wilber, Tom. 2015. *Under the Surface: Fracking, Fortunes, and the Fate of the Marcellus Shale.* Ithaca, NY: Cornell University Press.

Willow, Anna J. 2014. "The New Politics of Environmental Degradation: Un/Expected Landscapes of Disempowerment and Vulnerability." *Journal of Political Ecology* 21(1): 237–257.

Willow, Anna J., Rebecca Zak, Danielle Vilaplana, and David Sheely. 2014. "The Contested Landscape of Unconventional Energy Development: A Report from Ohio's Shale Gas Country." *Journal of Environmental Studies and Science* 4(1): 56–64.

Wright, Rachel A., and Hilary Schaffer Boudet. 2012. "To Act or Not to Act: Context, Capability, and Community Response to Environmental Risk." *American Journal of Sociology* 118(3): 728–777.

Wright, Simon. 2012. "An Unconventional Bonanza." *The Economist*, July 12: 3–15.

Wynveen, Brooklynn J. 2011. "A Thematic Analysis of Local Respondents' Perceptions of Barnett Shale Energy Development." *Journal of Rural Social Sciences* 26(1): 8–31.

Yergin, Daniel. 2011. *The Quest: Energy, Security, and the Remaking of the Modern World.* New York: Penguin Press.

York, Richard. 2015. "How Much Can We Expect the Rise in U.S. Domestic Energy Production to Suppress Net Energy Imports?" *Social Currents* 2(3): 222–230.

Zaitchik, Alexander. 2012. "The Fight over Fracking." *Rolling Stone*, March 17, 2012. (http:// www.rollingstone.com/politics/news/the-fight-over-fracking-josh-fox-vs-big-gas -20110517?print=true).

Zebrowski, Ernest, and Mariah Zebrowski Leach. 2014. *Hydrocarbon Hucksters: Lessons from Louisiana on Oil, Politics, and Environmental Justice.* Jackson: University of Mississippi Press.

7

Denial, Disinformation, and Delay

Recreancy and Induced
Seismicity in Oklahoma's
Shale Plays

TAMARA L. MIX AND

DAKOTA K. T. RAYNES

Introduction

On a November 2011 night, the rural communities of Prague, Sparks, and Jones, Oklahoma heard what sounded like the crash of a commercial jet-liner, followed by dizzying shaking. A 5.7 magnitude earthquake at 10:53 p.m. startled many from sleep and shook the towns to their foundations. Felt by local residents across nine states, the earthquake damaged homes, structures, and roadways. Now known as the Prague earthquakes in the Jones swarm (a number of earthquakes clustered in space and time), it was one of the largest human-caused swarms of earthquakes associated with wastewater injection in the United States to date. The seismic events were an early touchstone in the growing controversy about the hazards of injection-induced seismicity linked to hydraulic fracturing and wastewater disposal practices.

The oil and gas (O&G) industry is central to Oklahoma communities. An increase in extraction activity has occurred across the state as unconventional resource plays have become "unlocked" to development. Six major plays (areas in which hydrocarbon accumulations occur) account for approximately 90% of O&G extraction in Oklahoma. All have been subject to horizontal drilling and hydraulic fracturing techniques. The most active play is the Woodford Shale, located in the southeastern and central western areas of the state. The Mississippian Lime is located in the north central part of the state. The Granite Wash, Tonkawa, Cleveland, and Marmaton are located in the western part of the state.

Hydraulic fracturing (HF or fracking), a well stimulation method to enhance the recovery of oil and gas reserves, is the process of injecting large volumes of chemically laced fluids into O&G reservoirs at high pressure, causing fractures in the geologic formation to increase output. HF is not new to Oklahoma; an early version of the technology arrived in 1949 when one of the first commercial HF operations in the nation was undertaken in Stephens County (Montgomery and Smith 2010). What is new to Oklahoma is the expanding development and close proximity of multiwell sites using horizontal drilling and HF technologies. Unconventional extraction technologies are poorly regulated in current local and regional statutes, and require the use of millions of gallons of scarce water resources each time a well is fractured.

Central Oklahoma is experiencing a boom in HF, with permits being granted almost daily. Water-rich shale conditions in the region require reliance on wastewater disposal methods. While HF in some parts of the state results in approximately 10 barrels of wastewater per barrel of extracted oil, plays like the Mississippian Lime yield 40 or more barrels of wastewater per barrel of oil. Wastewater disposal wells (also called saltwater disposal wells, recognized by the Environmental Protection Agency [EPA] as Class II Underground Injection Wells) inject fluids at a rapid rate deep into the earth. Scientific uncertainty exists regarding the implications of injecting large volumes of contaminated wastewater underground at high pressure. Little research addresses where the water goes, impacts on fault lines, and the potential to perpetuate an upsurge in seismic activity, referred to as induced seismicity, injection-induced seismicity, or triggered seismicity.

Historically, earthquakes in the central United States were rare. Prior to 2009, the year marking enhanced seismic activity in the Jones region, Oklahoma experienced an average of 50 earthquakes a year, with 5 or fewer magnitude 2.5 or greater earthquakes felt by residents in most years. The United States Geological Survey (USGS) and Oklahoma Geological Survey (OGS) attempt to verify the location and intensity of each recorded quake, currently a difficult feat given the number of earthquakes, the rate of occurrence, and understaffing at the OGS. Several earthquake scales exist and the placement and proximity of seismographs impacts measurement. The unexpected exponential increase in seismic activity has been attributed to the quantity and rate of fluid injection into disposal wells associated with hydrocarbon extraction processes. Shaking occurs primarily during or after wastewater disposal, but also during HF operations (Ellsworth 2013; Holland 2011, 2013; Keranen et al. 2013; Keranen et al. 2014; Zoback et al. 2012).

Hough and Page (2015) suggest that in the past century, larger earthquakes in Oklahoma were likely induced by O&G production activities. As rates of induced seismicity increase and Oklahoma homes shake, residents have directed frustration toward local emergency planners, city councils, county commissioners, state representatives, the O&G industry, the OGS, the Oklahoma Corporation Commission (OCC), and the EPA. Fearing for the structural integrity of their homes, as well as risks to safety and well-being, residents who have been historically unwilling to challenge powerful O&G interests are now taking action.

We use the concept of *recreancy* to examine the erosion of institutional trust among Oklahoma residents in the wake of the rapid increase in earthquakes and induced seismicity from wastewater injection disposal methods. Recreancy refers to situations of technological risk or disaster whereby "the behaviors of persons and/or institutions that hold positions of trust, agency, responsibility, fiduciary or other forms of broadly expected obligations to the collectivity . . . behave in a manner that fails to fulfill the obligations or merit the trust" (Freudenburg 1993: 917). We argue that preventable, anthropogenically induced earthquakes and the lack of state response constitute recreancy—or perpetuate what has been termed a "culture of normative recreancy" (Edelstein 2013)—that serves as a catalyst for critical discourse on risks and hazards associated with HF and related processes, community impacts, and the potential for protest mobilization. We make use of archival news materials from a statewide newspaper outlet, contextualized with participant observation and semistructured stakeholder interviews, to ask: In what ways do state actors respond to increasing induced seismicity? Do residents experience the increase in induced seismicity and resulting state response as recreancy? How does the anthropogenic nature of the risk and subsequent state response influence mobilization? While research on fracking and antifracking movements focus on shale plays in densely populated areas, we highlight less populated communities in Oklahoma where O&G development is historically embedded in the culture, institutional structures, and landscape of the state. Additionally, we address a newly emerging risk: wastewater injection-induced earthquakes associated with unconventional O&G development utilizing HF technology.

Risk, Recreancy, and Hydraulic Fracturing

Technological disasters such as the *Exxon Valdez* and BP *Deepwater Horizon* oil spills are different from natural disasters in that they are the product of human actions. Technological hazards not only cause immediate physical damage, but frequently involve contaminating toxins, negative long-term temporal and spatial impacts, and enhanced uncertainty. Experts debate the existence of the risk, its magnitude, mitigation and response, and the possibility and duration of recovery. Studies demonstrate the ubiquity of several key risk dynamics, including difficulty in assessing risk (Gamero et al. 2011); ability of news media to distribute information (Hornig 1993); the impact of culture and institutional structure on risk dynamics (Cable, Shriver, and Mix 2008); and how public input is ignored in decision-making processes (Pilisuk, Parks, and Hawkes 1987).

Effective risk communication poses challenges to policy makers and corporations reliant on the maintenance of public trust (Cvetkovich 2013; Renn 2008; Slovic 1987). Unconventional O&G extraction and production technologies pose unique concerns for the public, including unfamiliarity with technology and terminology, external and often unjust imposition of harm, delimited individual or community-level control (Gupta, Fischer, and Frewer 2012), and disruption of community and place-based identities (Jacquet and Stedman 2014). Researchers assert that HF is positioned to become a defining challenge of modern society in terms of risk assessment, perception, and communication, resulting in sociopolitical impacts and environmental threats linked to the emergence of the "Third Carbon Era" of extreme energy production (Jaspal and Nerlich 2014; Ladd 2016; Wheeler et al. 2015). Perceived benefits of unconventional hydrocarbon extraction and production center on economic gains such as increased jobs, greater tax revenues, expanded services, and new business development. Perceptions of negative impacts include degraded water resources, infrastructure damage, noise, and traffic accidents, as well as threats to public health, livestock, wildlife, and destruction of rural landscapes (Anderson and Theodori 2009; Ladd 2013, 2014). Furthermore, in agricultural-based communities dependent on natural resources, residents often report corporate bullying, an absence of procedural equity, exposure to environmental risks, and lack of transparency on the part of corporations, their representatives, state officials, and regulatory bodies (Malin and DeMaster 2016).

Empirical studies of hazards linked to unconventional resource extraction technologies increased significantly beginning in 2014. Researchers have addressed a wide range of human health concerns, including site proximity, water and air quality, and asthma and cancer risk (McKenzie et al. 2012; Meng and Ashby 2014; Rabinowitz et al. 2015; Wheeler et al. 2015). Studies of community-based social movement organization (SMO) responses to technological risk have emerged and include insights into construction and communication of risk (Bostrom et al. 2015; Nelkin 1989), public engagement processes (Wheeler et al. 2015), use of social media (Starbird et al. 2015), and creation of citizen science teams (Penningroth et al. 2013).

As our reliance on technology grows, we encounter the likelihood of technological risk at an ever-increasing rate. Perrow (1999) refers to the inevitability of technological failure as "normal accidents" or "system accidents." Given the inescapability of risks linked to emerging technologies, the public has expectations of behavior and action from state and local institutions. Anticipated responses include strategies to mitigate harm, trust in institutions to resolve concerns, and implementation of policies and practices to limit future hazards. Often, to protect their own interests, institutions engage in actions that create recreancy, reflected in "a retrogression or failure to follow through on a duty or trust" (Freudenburg 1996: 47). Sociological research has examined recreant institutional responses at nuclear sites (Cable, Shriver, and Hastings 1999; Freudenburg and Youn 1993), with respect to food access and security (Geisler and Currens 2014), in environmental justice cases (Çapek 1999), in terrorist-related disasters (Maret 2013), and with respect to natural resources (Edelstein 2013; Ritchie, Gill, and Farnum 2012). In recent research on Alberta's tar sands, Edelstein offers useful insights into how addictive economies (Freudenburg 1992) like resource extraction encourage normative recreancy. He notes that the "growth and hard technological drivers of our prosperity and longevity depend on recreant regulation for permitting and operation outside of the boundaries of what one would consider appropriate risk . . . as a result, recreancy is hardly the exception but rather the rule" (Edelstein 2013: 121).

Research Design

Our analysis draws from archival newspaper material to gain insight into the public discourse—involving information provided to the public, as well as responses from the public—surrounding induced seismicity in Oklahoma. The *Tulsa World*, a daily newspaper, was selected for its high circulation rate and extensive use of investigative journalism. Content was drawn from November 5, 2011 (the date of the Prague quakes) to June 30, 2016 (one month after the end of the most recent Oklahoma legislative session). A search for "earthquakes" or "quakes" anywhere in the *Tulsa World* during this time period revealed 918 items. To specify our selection for this chapter, we identified articles and editorials in which "earthquakes" or "quakes" appeared in the title and only examined those focused on Oklahoma, resulting in an analysis of 257 articles and editorials. We contextualized archival material with data from a larger project, most notably two years of participant observation at public meetings, SMO events, and semistructured interviews representing a broad spectrum of stakeholders. A diverse range of archival materials addressing earthquakes linked to O&G production were examined and the data were organized and analyzed according to procedures broadly outlined by Charmaz (2006). Interpretive line-by-line thematic coding was systematically used to identify prominent themes and narratives.

In the following sections, we illustrate how the rapid increase, frequency, and severity of earthquakes in Oklahoma prompted community response, the emergence of antifracking SMO actions, and recreant rejoinders from state and local actors. We further explore how the lack of adequate institutional response initially facilitated a suppressive impact on recruitment, mobilization, and collective action efforts, but later prompted even greater mobilization. Finally, we consider the impact of anthropogenic hazards on perception and communication of risk, and we offer insights into citizen mobilization around contentious issues in a hostile context within a culture of normative recreancy whereby state and local actors routinely employ denial, disinformation, and delay.

Oklahoma in Context: The Power of Oil and Gas

Oklahoma's economic cogs turn on the exploration, production, and extraction of hydrocarbons and agriculture. While the state was still a territory, a crew in Mayes County discovered oil accidentally in 1859. Soon Oklahoma was producing more oil than any other U.S. territory or state. In 2010, Oklahoma's second largest city, Tulsa, designated its central downtown area the "Oil Capital Historic District" in the National Register of Historic Places. Mining, including O&G production, contributed 3% of the nation's gross domestic product (GDP) but 15% of Oklahoma's GDP in 2014, signifying the industry's importance to Oklahoma's economic base.

Oklahoma has practiced innovative resource extraction strategies for the last 60 years, making it difficult to assess the impacts of today's more invasive forms of high-volume, high-pressure HF. Many states, including Oklahoma, do not adequately collect and/or release data on well stimulation activities (e.g., the number of times each well is fractured, how much freshwater is used, how much wastewater is created, how/where disposal occurs, etc.). The EPA has criticized the OCC for ongoing problems with collection, accessibility, and reliability of O&G extraction and production process data (Wertz 2015). Recent estimates suggest that more than 100,000 wells have been hydraulically fractured in Oklahoma. Approximately 2,660 active operators, 137,800 active wells (43,600 gas, 83,700 oil, and 10,500 injection/disposal), thousands of miles of gathering and transmission pipelines, and approximately 320,000 plugged and/or abandoned wells were in place in 2011 (State Review of Oil & Gas Environmental Regulations [STRONGER] 2011). According to an interview with Secretary of Energy and Environment Michael Teague, Oklahoma not only produces significant amounts of wastewater, but also takes in regional wastewater, injecting 1.5 billion barrels in 2015, with at least 545,000 barrels coming from Texas, Kansas, New Mexico, Colorado, and Arkansas (Summars 2016).

Companies, holding companies with subsidiaries, or companies that provide resources for O&G production—such as sand for the HF process or drilling pads—support Oklahoma politicians with millions of dollars each year, making the industry among the top political donors for the last decade (Open Secrets 2016). The industry lobbied for lower gross production (severance) tax rates (1% rather than the state standard of 7%) in the 1990s when innovative practices like horizontal and deep well drilling were new and considered risky. In 2015, the Oklahoma legislature renewed a 1% tax rate on horizontally drilled wells, regardless of the fact that the exemption costs to the state ballooned from $2 million in 2004 to $282 million in 2014 (Blatt 2015).

The Oklahoma Energy Education and Marketing Act (Oklahoma State Statute 52–288), facilitated by the Oklahoma Energy Resources Board (OERB), a state agency comprised of independent O&G producers and major oil company representatives, promotes programs such as "Little Bits" and "Petro Active" to educate K–8th graders about well site safety, products made from petroleum, and the formation and recovery of oil and natural gas through hands-on energy experiments (OERB 2016). The curriculum offers useful information, yet allows the energy industry an opportunity to shape beliefs, attitudes, values, and practices concerning energy extraction, production, and consumption. Despite how deeply embedded the O&G industry is in Oklahoma's history, economy, politics, and culture, its stature is being shaken today—quite literally—not because of concerns over human or environmental health, but because as the problem of induced seismicity grows, people's houses have begun to crack and crumble.

Seismic Swarms, Manufactured Uncertainty, and Mobilization

From the first official records in 1977 through 2008, Oklahoma experienced an average of 50 earthquakes annually, occurring primarily along the Central Oklahoma Fault Zone. Seismic activity began to increase exponentially both at and away from the zone in 2009, as did seismic monitoring activity funded by outside entities, with 10 of the 12 largest earthquakes in Oklahoma history having since occurred. Scholars attribute the increase in earthquakes to hydraulic fracturing and the use of wastewater disposal wells (Ellsworth 2013; Holland 2011, 2013; Keranen et al. 2013; Keranen et al. 2014; Zoback et al. 2012). Increased seismicity has primarily impacted smaller, rural communities in 23 counties in central Oklahoma, an area stretching from Kansas to Texas.

Table 7.1
Earthquakes in Oklahoma by Year

Year	1978–2008	2009	2010	2011	2012	2013	2014	2015	2016
Total									
1.0 or >	~2/YR	~50	~1,059	~1,542	~1,028	~2,850	~5,417	~5,691	~4,793
>3.0	~2/YR	~20	~42	~67	~40	~109	~585	~907	~622

Initial attention spiked when specific areas experienced repeated earthquakes, referred to as seismic swarms (e.g., the Jones swarm), and when especially large earthquakes occurred (e.g., the Prague quakes). Table 7.1 indicates the year, approximate total number of earthquakes, and the number of earthquakes measuring magnitude 3.0 or greater.

On Shaky Ground: The Mystery of Increased Seismicity

The Prague earthquakes startled residents, caused damage, and sparked state and local news coverage. Early news reporting in the *Tulsa World* (40 articles published between November 2011 and December 2013) revolved around three primary themes: debates over the cause of the earthquakes (50%); descriptions of residents' reactions, damage reports, and costs of repairs (38%); and property insurance concerns (10%). One editorial also appeared (2%).

In response to the sizeable quakes, nearly 64,000 people across nine states reported their experience via the USGS "Did You Feel It?" website. The Tulsa Police Department was inundated with calls from alarmed residents, many who reported feeling scared and vulnerable. Significant damage to homes and structures prompted current governor Mary Fallin (R) to seek low-interest disaster loan assistance from the U.S. Small Business Administration to aid recovery in Lincoln and adjacent counties. She simultaneously avoided addressing the cause of the quakes. Descriptive articles detailed individual and family experiences, including reactions, photos of damaged property, and plans for recovery. Later pieces highlighted the timing, location, and magnitude of earthquakes, residents' use of social media to share experiences, and reports of damage extent.

Insurance Commissioner Doak urged homeowners to consider earthquake insurance. Many families' first investment and single largest asset is often their home. Each year the *Tulsa World* ran at least one article featuring an extensive interview with Doak, who revealed that only 1% of Oklahoma property owners have earthquake insurance. Few companies offer it in Oklahoma, and while premiums are low, typically $100–$150 per year, deductibles are high, amounting to 10% or more of the property's replacement value (Overall 2011). Most importantly, many insurers have a 30–90-day waiting period for creation or modification of policies after a significant earthquake. Some whose homes or businesses were damaged in the Prague quakes were unable to afford repairs and simply packed up and left the rubble behind. Commissioner Doak urged homeowners to review available policies, educate themselves about earthquake insurance, and plan for adequate coverage.

Discourse on technological hazards, particularly debates concerning the extent of the risk and what could or should be done regarding mitigation, response, and recovery, were central. Days after the Prague quakes, the OGS discussed the major fault lines in Oklahoma. Other local geologists pointed to the relationship between local geologic formations and broader North American tectonic activity. Many of the narratives articulated the following theme:

> It doesn't make any sense to think hydraulic fracturing by petroleum exploration has anything to do with the weekend's excitement—no matter what you may have read on the Internet. Holland [an OGS research seismologist] said that hypothetically, fracking can cause small, localized earthquakes, but there's a much simpler explanation for these events—a known fault line and a continent's worth of geopressure. Keller [OGS Geologist] agreed, "It's just too deep . . . if somebody wants to believe something, they will, but it just isn't logical to believe anything going on 10,000 feet above you is coming down into the solid basement rock and causing things to move." (Greene 2011)

The OGS is housed at the University of Oklahoma's Mewbourne College, named after oilman and benefactor Curtis Mewbourne, and the recipient of significant donations from the O&G industry. At a Natural Gas and Energy Association of Oklahoma event just after the Prague quakes, State Representative John Sullivan (R-Tulsa) expressed skepticism about environmental theories linking induced seismicity to increased HF and disposal wells, stating, "It's ridiculous to think that . . . how could just a little oil in the ground have anything to do with that?" (Krehbiel 2011). Refutation of risk was common in spite of well-respected research establishing a correlation between deep well injection and seismicity conducted in Rangely, Colorado during the 1960s and 1970s. The study found that "earthquakes can be controlled wherever we can control the fluid pressure in a fault zone" (Raleigh, Healy, and Bredehoeft 1976: 1230). Regardless of existing scientific data, public officials planted seeds of doubt that grew into a seemingly organized campaign of denial, disinformation, and delay.

In April 2012, a USGS report on mid-continent seismicity asserted, "a naturally occurring rate change of this magnitude is unprecedented outside of volcanic settings or in the absence of a main shock, of which there were neither in this region" (Greene 2012a). In response to media coverage of the report, the OGS released a "triggered seismicity" counter-statement casting aspersions on the report's validity. Amplifying the state's position,

OGS geologist Randy Keller said, "We are taking a measured and scientific approach . . . so that any conclusions that earthquakes are linked to oil and gas activities can be scientifically defensible." He highlighted the powerful position of the industry within the state, noting, "We consider a rush to judgment about earthquakes being triggered to be harmful to state, public, and industry interests" (Greene 2012b).

· Nearly a year later in March 2013, the OGS issued a position statement regarding the 2011 earthquakes, claiming that all activity was the result of "natural" causes. A week later, Keranen and colleagues' (2013) examination of the same seismic events showed that the earthquakes began within hundreds of meters of disposal wells in the same geologic units into which the wells were injecting wastes, leading the researchers to conclude, "spatially we don't have much doubt, there is a direct spatial link" (Greene 2013). As the scientific evidence for injection-induced seismicity grew (Ellsworth 2013; Holland 2013), the OGS, OCC, state officials, and O&G industry could no longer outright deny the risk, yet continued to cast doubt by challenging researchers' findings and claiming more research was needed prior to any action. Minority House Democrats, in late 2013, called for a hearing, arguing that "the USGS and OGS have indicated that 'the injection of wastewater generated from O&G activities, such as fracking, may be a potential contributing factor to the over tenfold increase in earthquake frequency in Oklahoma since 2009'" (Casteel 2013).

During the first few years of enhanced induced seismicity two primary narratives were present. One was descriptive, providing individuals and communities information about the frequency and severity of earthquakes. The other presented peer-reviewed research findings alongside counterclaims made by scientists and public officials, as though each perspective were equally valid. The result perpetuated the perception of more conflict than consensus within the scientific community about injection-induced seismicity in general and case studies in Oklahoma specifically. Initial local assessments of causality painted seismic activity and risk as natural rather than technological hazards. The relative isolation of the Jones swarm and Prague quakes, combined with conflicting expert assessments of causation, led residents to question both the source of the quakes, as well as why state and local officials had not acted.

Injection-Induced Seismicity: Constant Shaking
Turns Fear to Anger and Action

Throughout 2014, the frequency and magnitude of earthquakes increased. People living near seismic swarms in rural Oklahoma, stretching from Medford near the Kansas border, to Stillwater and Guthrie in the state's center, to Marietta near the Texas border, grew frustrated with the constant shaking and daily damage inspections, but were unsure about what actions to take. Peer-reviewed studies linking the increase in seismicity with disposal wells became more prominent. Residents soon realized that neighboring states were facing similar issues, but handling them quite differently.

The 44 articles from 2014 were less easily categorized as descriptive (34%) or debate-focused (11%), appearing as a combination of the two (32%). The content of descriptive articles expanded to include details about HF processes, the OCC's regulatory responses, considerations of increased seismic risk, and information about recent lawsuits against several energy firms due to Prague quake damages. Few articles solely addressed concerns related to property insurance (4%), but advice to obtain earthquake insurance continued. Several more editorials (18%) appeared. In this section, we focus on warnings of increased seismic risk, the OCC's regulatory response, and the emergence of antifracking SMO action.

In May 2014, the USGS and OGS released a joint statement warning that "as a result of the increased number of small and moderate shocks, the likelihood of future, damaging earthquakes has increased for central and north-central Oklahoma" (Stancavage 2014). Residents, schools, and businesses were urged to develop earthquake preparedness plans. Building owners and government officials were advised to "have special concern for older, unreinforced brick structures, which are vulnerable to serious damage during sufficient shaking" (Stancavage 2014).

When asked if wastewater disposal wells could be contributing to increasing seismicity around the state, OGS seismologist Austin Holland stated, "We're still looking at that as a possibility, but we're waiting for data . . . those determinations can take a very long time to complete." OCC Commissioner Dana Murphy similarly stated, "You have to collect the data and do the research to study if there are connections . . . we have to have the data to make sure we're taking appropriate action" (Wilmoth 2014). Meanwhile, Ohio, Arkansas, and Colorado responded to injection-induced seismicity by examining available data and banning disposal well use, enacting moratoriums on new disposal wells, or drastically reducing daily volume, rate, and pressure limits. In Oklahoma, despite elevated seismic risk, permits for HF and disposal wells continued to be approved. The OCC instituted a "traffic-light" system to identify problematic wells in seismically active areas, a mitigation tactic supported by some (Ellsworth 2013; McGarr et al. 2015; Zoback et al. 2012) but considered utterly ineffective by others (Bommer, Crowley, and Pinho 2015). A 19-county "area of interest" was identified and disposal well operators in those counties were given 30 days to prove their wells were not injecting into the crystalline basement. The OCC, in contradistinction to similar agencies in other states, asserted that contact with basement rock created a greater seismic risk than high-volume, high-pressure injection. The OCC's actions created little to no change in permitting policies and seismic eruptions continued to shake communities across the state. Some residents took to the Internet to learn all they could about HF, disposal wells, earthquakes, and Oklahoma geology. Many grew tired of waiting for state action. To them, "the science was in." The daily quakes were human-made—rather than natural events—and therefore were seen as preventable.

The transformation in public attitudes was seen in *Tulsa World* editorials, the creation of local grassroots antifracking or anti-induced seismicity SMOs, comments made at town hall meetings, the circulation of a petition for a moratorium on deep well injection, and marches, demonstrations, and press conferences demanding action. Shortly after the USGS and OGS joint statement alerting the public that Oklahoma faced an elevated seismic risk, editorials expressed public frustration. One exclaimed, "now Oklahoma is under an earthquake warning . . . rather, over an earthquake warning . . . that's not even ironic . . . it's absurd . . . it's like somebody owes us money for this extra worry . . . nobody wants to live in Frackerville . . ." (Cronley 2014). Another editorial discussed key points of the Colorado earthquake experiment and lambasted the energy industry and local/federal politicians, saying they "had to be well aware of this incident . . . [but] it is less expensive to dispose of it in this way and, as our Supreme Court made clear, money talks in Washington. . . . they knew that they would not have to pay a significant fine for any serious consequences that might occur due to their irresponsible activities" (Bolton 2014). Later in the year, as the gubernatorial election loomed, another resident connected Governor Fallin's inaction to her ties with the O&G industry: "She is either one of the last to conclude that disposal wells are the cause, or she just doesn't want to risk offending her wealthy supporters. . . . we need a governor who will put the industry to work solving the problem and protecting the citizens, not just gathering more data" (Chambers 2014).

The realization that the increase in seismic activity represented a technological, rather than natural hazard, did not absolve residents of fear, but led many to reframe their understanding of each new quake. Residents became angry and outspoken. The ever-increasing number of earthquakes and burgeoning scientific evidence, coupled with what appeared to be an orchestrated program of denial, disinformation, and delay on the part of the OGS, OCC, state officials, and industry, coalesced to create a situation wherein institutional responses to induced seismicity were seen as recreant.

While some residents expressed outrage through the power of the pen, others engaged in the work of building grassroots antifracking or anti-induced seismicity SMOs. The group Stop Fracking Payne County emerged in early 2014 and organized consciousness-raising town hall meetings in small rural communities across the state and demonstrations at the state capitol. Town halls often featured a panel of speakers affording residents an opportunity to ask questions and make statements. Other events included screenings of documentary films such as *Gasland*, *Triple Divide*, and *Groundswell Rising*. Some in attendance expressed difficulty with technical jargon, requesting assistance in understanding the issue and help in determining what actions to take. Others were clearly knowledgeable, demanding that state agencies justify inaction. Occasionally, an O&G industry representative refuted induced seismicity as a bogus scientific claim, reminded residents of the energy industry's crucial economic function, said that patriotic Americans should support energy independence to avoid war for oil overseas, or heckled attendees for driving fuel-operated vehicles. Several events were staged at the state capitol coinciding with the presentation of signatures for a moratorium to the Office of the Governor, as well as hearings for an interim study on earthquakes, an unusual bipartisan effort by State Representatives Jason Murphey (R–Guthrie) and Cory Williams (D–Stillwater). During hearing testimony, Angela Spotts, co-founder of Stop Fracking Payne County, made an emotional plea that "Action needs to be taken sooner rather than later," adding that those affected were dealing with "an enormous level of stress" (Hoberock 2014).

Contacted in advance, Governor Fallin and her staff refused to meet with activists, leaders of local grassroots SMOs like Stop Fracking Payne County, the Oklahoma Coalition Against Induced Seismicity, Clean Energy Future Oklahoma, or directors of national coalition partners such as the National Association for the Advancement of Colored People, the Sierra Club, and Food and Water Watch. Chants and signs at events demanded a "Moratorium Now," asserting that "The fracksters have to go," and proclaimed "We Will NOT Be Collateral Damage." Near the end of 2014, an application permit request was submitted to site three HF wells near one of Oklahoma City's largest public recreational green spaces, Lake Hefner. More than 500 people protested the application, resulting in its denial.

Little to no mass organization occurred before large numbers of residents began to realize that the increase in seismicity was anthropogenically and technologically induced. Once residents perceived the harm as preventable, the fact that it was *not* being prevented generated charges of institutional recreancy. Inaction from institutional actors catalyzed mobilization, encouraging efforts for accountability and creating opportunities for networking between antifracking, anti-induced seismicity, and other more generally pro-environmental SMOs. As demonstrated by Raynes and colleagues (2016: 416), a confluence of factors spawned the emergence of fervent place-based activism, namely "increasing industrialization of a beautiful agricultural landscape, continuous encroachment of O&G activity within residential neighborhoods in small towns and cities dotting the Oklahoma countryside, the near-constant assault of injection-induced seismicity, or the potential for long-lasting environmental and public health impacts, such as pollution/contamination of the air, water, and soil." Nonetheless, by the end of 2014, Oklahoma surpassed California as the most seismically active area in the nation.

Failure to Protect People's Lives and Property: Debilitation and Revelation in a "Sacrifice Zone"

SMO mobilization and slow revelations from recreant state actors continued. Several cities sought to update O&G zoning regulations to more effectively control encroaching drilling and wastewater injection. In response, the state legislature passed a "ban the bans" bill (SB 809) limiting municipal authority to regulate or reject O&G activity. Interim studies on seismic activity and disposal wells continued. Other key developments included OGS and state officials' affirmation of the link between HF, disposal wells, and induced seismicity, and greater media coverage of induced seismicity broadly and antifracking SMO activities specifically. Earthquakes impacting urban areas catalyzed new movement adherents and greater national attention.

Of the 173 articles examined for 2015–2016, the majority were descriptive (26%), debate-focused (12%), or a combination of both (30%). As the debate over Oklahoma's ever-increasing seismicity began to dwindle, new disputes emerged. They included: differences in earthquake measurements; funding of earthquake research and O&G regulation enforcement including well closure and volume reduction; authority to control industry and provide public relief; continued requests for a moratorium; what agency should handle lawsuits; and designation of homeowners with earthquake damage in class action lawsuits. The remaining articles were editorials (18%), highlighted social movement activities (8%), or foregrounded property insurance issues (6%).

The practice of state and corporate recreancy through denial, disinformation, and delay was detailed in the *Tulsa World*'s special three-part "Quake Debate" investigative series, a review of thousands of documents received via the Freedom of Information Act (FOIA). Governor Fallin, key state officials, and energy industry groups continued to publicly cast doubt on the scientific consensus. Further demonstrating the power of the O&G industry, the investigation revealed that 5 of the 12 appointees to the Governor's Coordinating Council on Seismic Activity had direct energy industry connections, information not acknowledged publicly.

In March 2015, it was revealed that Oklahoma's chief seismologist, Austin Holland, experienced pressure from University of Oklahoma officials and O&G industry representatives to suppress information linking wastewater injection and induced seismicity (Branstetter 2015a). Simultaneously, Kansas declared injection-induced earthquakes an "immediate threat" and passed new, extensive restrictions on permits for disposal wells in high-risk areas (Branstetter 2015b). By April, the OGS and Governor Fallin, in chorus, stopped denying that Oklahoma's massive increase in seismicity was indeed induced. Local efforts to update O&G zoning regulations continued with grassroots anti-fracking organizations playing a key role in mobilizing community input and providing decision makers with peer-reviewed research. Many realized that more stringent regulations for setbacks, lighting, and fencing could set precedents for other municipalities. O&G industry representatives and lobbyists presented counter-narratives at city council meetings emphasizing economic aspects, while downplaying environmental concerns. They also critiqued activists and residents who opposed fracking as "illogical" and "overly emotional."

In response to local zoning efforts and increasing antifracking and anti-induced-seismicity SMO activity at the state capitol (Hoberock 2015), state legislators proposed eight separate bills seeking revocation of municipalities' constitutionally guaranteed right of "home rule" and ensuring that regulatory power over the O&G industry rested entirely with the OCC. City council meetings on pending zoning proposals were lengthy and decision-making processes lasted months. In the end, all city councils approved watered-down versions of initial ordinances due to implementation of SB 809. Under new legislation, all O&G regulations had to be either commensurate with those developed by the OCC or otherwise "reasonable," although the term reasonable has yet to be defined. The challenges in the process and less than desirable outcomes left many residents and activists feeling as if their work had been for nothing, with the land they loved declared a "sacrifice zone" (Raynes et al. 2016).

The state continued to engage in disinformation and delay in the coming months, with some minor wins for activists and major victories for the O&G industry. In July 2015, an Oklahoma Supreme Court ruling, in a case stemming from the 2011 Prague earthquakes, found that "Damages from high-pressure disposal wells—and, by extension, other oil-and gas-related activity—must be decided by the court system, not the Oklahoma Corporation Commission. Letting the Corporation Commission rule on such cases, some believe, would essentially give oil and gas companies a home-field advantage in a process that already figures to be an uphill climb for the plaintiffs" (Krehbiel 2015). Soon after the ruling, the public discovered that Governor Fallin's 2011 verbal response to the Prague tremblors had been scripted verbatim by Devon Energy Corporation (Soraghan 2015). Later in July, the OCC directed 211 wastewater injection wells to reduce their depths in response to earthquake concerns (Jones 2015), and OGS Chief Seismologist Holland left his position. Efforts toward limiting the volume of wastewater injection continued throughout the summer, meeting industry opposition in October when a Tulsa company refused to abide by the OCC's earthquake-related injection well regulations (Monies 2015a). The refusal prompted controversy about the OCC's ability to effectively enforce regulation of industry.

Uncertainty surrounding regulatory efficiency encouraged Washington-based Public Justice and the Oklahoma Sierra Club to bring a lawsuit under the federal Resource Conservation and Recovery Act (RCRA), a 1976 law allowing citizen lawsuits against hazardous waste facilities. The goal was to reduce the volume of wastewater injected into disposal wells and establish an independent body to "investigate, analyze and predict the cumulative effect of injecting production wastewater." The chairwoman of the Oklahoma Sierra Club stated, "I am angry and offended that the oil and gas industry has been so slow to protect Oklahoma and its citizens in the face of this earthquake crisis" (Monies 2015b). While outside advocates began to speak out about induced seismicity in Oklahoma, and the issue began to garner national media attention (including pieces on all major television networks, the *New York Times*, *Huffington Post*, and the *Daily Show*, among others), minority voices in the state leveled clear charges of recreancy as well. Left without a chief seismologist since July, a position that remained vacant until well into 2016, at a point of increasing seismic intensity and magnitude, State Representative Williams, an advocate for intervention regarding induced seismicity, noted in a December op-ed piece that he was:

> Weary of the overly supportive attitude state regulators—primarily Michael Teague, Oklahoma's Secretary of Energy and Environment—exhibit toward the oil and gas industry and their continual need to refer to the industry as "cooperative." If the industry, in collusion with state leadership, hadn't pursued a two or threeyear campaign of misinformation and continually claimed they were "waiting on the science," we could have put together a measured response long ago that might actually have had an impact. . . . The damage has already been done. . . . Now we're reaping the consequences. . . . We are continually reacting after the seismic activity happens. This is a terrible way to protect the lives and property of Oklahomans. (Williams 2015)

December 2015 closed with a 4.3 magnitude quake shaking Edmond, an urban bedroom community of Oklahoma City. By the end of 2015, Oklahoma had become the most seismically active area in the world.

The year 2016 began with an economic downturn related to reduced O&G prices, disposal well shutdowns and volume cutbacks, large-scale public meetings, greater calls for a moratorium, and challenges to the insurance industry, as residents of urban communities began to awaken and seek greater state accountability. Denial, disinformation, and delay continued, even in the face of challenges from actors inside and outside the state. Edmond residents filed a lawsuit against a dozen energy companies, claiming that saltwater disposal wells were in part to blame for the earthquakes that had hit central Oklahoma and that the companies acted negligently with the use of disposal wells constituting an "ultrahazardous activity" (Monies 2016). State Representative Richard Morrissette (D–Oklahoma City) held a public hearing with a standing room-only audience and a steady stream of residents' comments. Members of the public, antifracking groups, and anti-induced seismicity SMOs made their presence known. As one news story noted, "They came mad, frustrated, seeking answers and action," articulating their impression that "Fallin and lawmakers have turned their heads on the issue." Even Matt Skinner, spokesperson for the OCC, stated, "If your house is shaking, it [the state's response] isn't fast enough" (Hoberock 2016a). By mid-January, the *Tulsa World* reported an 81% increase in the volume of wastewater from O&G activities across six years, coinciding with the state's enhanced earthquake rate. "Even more notably, the wastewater injected into the state's deepest geologic formation, the Arbuckle, ballooned 141 percent during the same period" (Jones 2016a).

A stream of editorials, including one penned by Governor Fallin, addressed earthquakes. Topics ranged from the best way to protect the O&G industry, to assessments of who should pay for damages, and the role science should have in the state's response. On February 13, Fairview was rattled by a 5.1 earthquake, the state's third largest to date. Recreant state responses resulted in yet more lawsuits. Attorney Robin Greenwald with Weitz & Luxenberg, the law firm representing the Sierra Club's newest suit, stated, "The seismic activity of this past weekend is quickly becoming the new normal in Oklahoma. If the fracking industry doesn't change its ways, the next earthquake could be catastrophic. This lawsuit seeks to beat back immediately the amount of production waste that fracking creates, to reduce the deep well injection of that waste and, most important, to limit the amount of damage this process is causing across the Sooner State" (Staff 2016). Increasing litigation and requests from concerned citizens brought well-known environmentalist and consumer advocate Erin Brockovich to speak. In comments at the state capitol, she reflected the concerns of Oklahoma residents: "They want action from their politicians. . . . They have a lot of fears. They have a lot of frustration. They are not getting answers. They do worry that politics or something in the system prevents them from being heard" (Hoberock 2016b).

In spite of national attention, a pattern of denial, disinformation, and delay continued. In March, Governor Fallin's office released a statement arguing that current research supported state actions. "Recent declines in produced wastewater disposal in Oklahoma are not reflected in the USGS map. . . . This gives us even a stronger base in going forward and gives state regulators further justification for what they are doing." State Representative Morrissette took issue with Governor Fallin's statements, noting, "What the governor fails to mention is that her administration was more than a year late in responding to all of the seismic activity in Oklahoma. Foundations, walls and ceilings were cracking for at least two years before this governor and the Corporation Commission took this issue seriously" (Jones 2016b). Rallies at the state capitol continued with residents, antifracking groups, and anti-induced seismicity SMOs engaging and executing calls to action. Angela Spotts, co-founder of Stop Fracking Payne County, stated, "We have the right to vote. . . . We best get active and do it. Injustice is being done to many residents across the state. . . . Elected officials should help voters who put them in office, not the industries that paid to put them there. We have a right to clean air, clean water, and a safe environment" (Hoberock 2016c).

Today, Oklahoma remains the most seismically active region in the world. Approximately seven years after the initial uptick in seismicity and five years after the Prague earthquakes, very little action has been taken to protect the people of Oklahoma from the risks of induced seismicity. Notably, the Sierra Club and Public Justice RCRA–based lawsuit was dismissed in April 2017 after Judge Stephen Friot declared that the OCC had "energetically" responded to the issue of induced seismicity and the court was "ill-equipped to outperform" the agency in ensuring adequate and appropriate regulatory response (Wertz 2017). An economic downturn linked to reduced O&G prices negatively impacted the state's economy. Yet, industry grinds on and leaders continue to engage in denial and disinformation, delaying action and proactive planning as earthquakes remain a daily occurrence and constant threat. On September 3, 2016, a 5.8 magnitude earthquake, the largest recorded Oklahoma earthquake to date, rocked the town of Pawnee and was felt across six states. The potential for future greater magnitude earthquake events, national attention to induced seismicity including several *60 Minutes* and other media pieces, and recreant state and corporate responses continue to facilitate SMO mobilization. Several new class action lawsuits seeking reparations for earthquake damages have been filed by the Pawnee Nation and others, but none have been heard in court yet. As earthquakes increase in magnitude and frequency and impact wealthier communities near urban centers and surrounding states, cycles of consciousness-raising, town halls, and community mobilization seem likely to continue, with each wave of shaking acting as a new catalyzing event.

Conclusion

Following Edelstein (2013), we argue that recreancy can manifest as a systemic normative dynamic permeating the economic, political, and cultural arenas of a place or a particular historic period. We illuminate some of the structures and dynamics involved in the social construction of risk, the conditions and mechanisms by which recreancy occurs or is induced, and the cultural and institutional structures that facilitate or constrain early SMO emergence and mobilization. Our analysis of *Tulsa World* archival data illustrates a pattern of denial, disinformation, and delay, implemented on the part of the OGS, OCC, state officials, and the energy industry, resulting in a normative culture of recreancy that has inhibited actions to protect state residents. Oklahomans are expected to live with the risks and impacts associated with O&G activity, legitimated by the energy industry and state narratives about its centrality to local history and the economy. In spite of mounting scientific data about the health risks of unconventional O&G development, it was the spike in earthquakes that encouraged SMO participation in Oklahoma. Perrow (1999: 12) aptly describes the situation experienced by Oklahoma residents when he notes that people exposed to "normal accidents" are frequently "excluded from participating in decisions about the risks that the few have decided the many cannot do without. The issue is not risk, but power."

Initially, seismic swarms occurred in rural, less populated areas. The location of the swarms, combined with active risk refutation, resulted in some Oklahomans being shaken daily, others not experiencing any seismic activity, but nearly everyone assuming the quakes were natural, rather than technologically induced. As seismic swarms spread, earthquake magnitude increased, and peer-reviewed studies connected HF and disposal wells to the phenomenon of induced seismicity, residents began to realize the exponential increase in seismicity was preventable, facilitating transformation of their reactions from curiosity and fear to concern, anger, and action. The uncertainty of risk prompted many who would not have otherwise engaged in SMO action to do so, thereby accelerating mobilization against fracking and wastewater injection. Still, the push toward SMO activity was not seamless. It was heavily impacted by cultural settings, institutional structures, and state legislative measures retracting communities' rights to engage in meaningfully democratic local governance processes. Even though Oklahoma municipalities have been stripped of their power to limit O&G extraction and waste disposal, the sense of outrage has been reawakened through recreant institutional responses, admissions of industry influence over state scientists and officials, as well as legal successes, and reinvigorated when seismic swarms increase in intensity and impact larger, more populated urban areas.

Based on our findings, we argue that Oklahoma's institutional actors failed to exercise their duties in a manner that maintained the public trust, perpetuating a culture of normative recreancy, and that the recreant responses to induced seismicity sparked SMO activity. What will it take for the state to take protective action? What is required to encourage a groundswell of sustained mobilization in Oklahoma? Future studies should explore the dynamics necessary to maintain sustained SMO activity in the face of the state's entrenched O&G industry, while continuing to track the ebb and flow of protest mobilization resulting from the expansion of hydraulic fracking, wastewater injection, and related induced seismicity.

References

Anderson, Brooklynn J., and Gene L. Theodori. 2009. "Local Leaders' Perceptions of Energy Development in the Barnett Shale." *Southern Rural Sociology* 24(1): 113–129.

Blatt, David. 2015. "Understanding Oklahoma's New Tax Rates on Oil and Gas Production." *Oklahoma Policy Institute*, June 30. Retrieved April 25, 2016. (http://okpolicy.org /understanding-the-new-tax-rates-on-oil-and-gas/).

Bolton, Chris. 2014. "Letter to the Editor: Earthquakes and Disposal Wells." *Tulsa World*, July 1. Retrieved August 15, 2014. (http://www.tulsaworld.com/opinion/letters/letter -to-the-editor-earthquakes-and-disposal-wells/article_40bf5c13-ece0-5bb5-9e1b -a7681c8ff276.html).

Bommer, Julian J., Helen Crowley, and Rui Pinho. 2015. "A Risk Mitigation Approach to the Management of Induced Seismicity." *Journal of Seismology* 19: 623–646.

Bostrom, Ann, Susan Joslyn, Robert Pavia, Ann Hayward Walker, Kate Starbird, and Thomas M. Leschine. 2015. "Methods for Communicating the Complexity and Uncertainty of Oil Spill Response Actions and Tradeoffs." *Human and Ecological Risk Assessment* 21: 631–645.

Branstetter, Ziva. 2015a. "Under Pressure? Do Emails Tell of Earthquake Information Sharing or State, Industry Interference?" *Tulsa World*, March 4. Retrieved May 15, 2015. (http://www.tulsaworld.com/earthquakes/under-pressure-do-emails-tell-of-earthquake -information-sharing-or/article_a7df8a5c-47f7-540c-af46-faa04765bc5c.html).

Branstetter, Ziva. 2015b. "Kansas Declares Earthquakes 'An Immediate Threat' In Passing New Energy Restrictions." *Tulsa World*, March 30. Retrieved May 15, 2015. (http://www .tulsaworld.com/earthquakes/kansas-declares-earthquakes-an-immediate-threat-in -passing-new-energy/article_fa556dea-56d6-54d7-856c-4f18be10ecdc.html).

Cable, Sherry, Thomas E. Shriver, and Donald W. Hastings. 1999. "The Silenced Majority: Quiescence and Social Control on the Oak Ridge Nuclear Reservation." *Research in Social Problems and Public Policy* 6: 59–81.

Cable, Sherry, Thomas E. Shriver, and Tamara L. Mix. 2008. "Risk Society and Contested Illness: The Case of Nuclear Weapons Workers." *American Sociological Review* 73(3): 380–401.

Çapek, Stella. 1999. "Institutional Failure and the Demise of Carver Terrace." *Research in Social Problems and Public Policy* 7: 139–162.

Casteel, Chris. 2013. "House Democrats Call for Hearing on Fracking and Oklahoma Quakes." *Tulsa World*, December 19. Retrieved January 15, 2015. (http://www.tulsaworld .com/news/government/house-democrats-call-for-hearing-on-fracking-and-oklahoma -quakes/article_7f5be920-7302-57e0-b320-4ea544497407.html).

Chambers, Frank. 2014. "Letter to the Editor: Governor Needed to Address Earthquakes." *Tulsa World*, October 15. Retrieved November 15, 2014. (http://www.tulsaworld.com /opinion/letters/letter-to-the-editor-governor-needed-to-address-earthquakes/article _d5d1e758-6d3c-500a-b656-cafb87bd8266.html).

Charmaz, Kathy. 2006. *Constructing Grounded Theory*. Thousand Oaks, CA: Sage Publications.

Cronley, Jay. 2014. "Earthquakes in Oklahoma? State's Disaster List Growing a Lot Shakier." *Tulsa World*, May 9. Retrieved January 15, 2015. (http://www.tulsaworld.com/news /jaycronley/jay-cronley-earthquakes-in-oklahoma-state-s-disaster-list-growing/article _f7d262f0-d555-5df4-aea3-f8cb41bef815.html).

Cvetkovich, George. 2013. *Social Trust and the Management of Risk*. London: Routledge.

Edelstein, Michael R. 2013. "When Recreancy Becomes the Norm: Emergency Response Planning and the Case of Tar Sands Upgrading in the Alberta Industrial Heartland." Pp. 119–175 in *William R. Freudenburg, A Life in Social Research (Research in Social Problems and Public Policy, Volume 21)*, edited by S. Maret and T. I. K. Youn. Bradford, UK: Emerald Group.

Ellsworth, William. 2013. "Injection-Induced Earthquakes." *Science* 341(6142): 1225–1242.

Freudenburg, William R. 1992. "Addictive Economies: Extractive Industries and Vulnerable Locales." *Rural Sociology* 57(3): 305–332.

Freudenburg, William R. 1993. "Risk and Recreancy: Weber, the Division of Labor, and the Rationality of Risk Perceptions." *Social Forces* 71: 909–32.

Freudenburg, William R. 1996. "Risky Thinking: Irrational Fears about Risk and Society." *Annals of the American Academy of Political and Social Science* 545: 44–53.

Freudenburg, William R., and Ted I. K. Youn. 1993. "A New Perspective on Problems and Policy." *Research in Social Problems and Public Policy* 5: 1–20.

Gamero, Nuria, Josep Espluga, Ana Prades, Christian Oltra, Rosario Solá, and Jordi Farré. 2011. "Institutional Dimensions Underlying Public Trust in Information on Technological Risk." *Journal of Risk Research* 14(6): 685–702.

Geisler, Charles, and Ben Currens. 2014. "'Peak Farmland': Revealed Truth or Recreancy." Pp. 177–199 in *William R. Freudenburg, A Life in Social Research (Research in Social Problems and Public Policy, Volume 21)*, edited by S. Maret and T. I. K. Youn. Bradford, UK: Emerald Group.

Greene, Wayne. 2011. "Geologists Have No Easy Answers to Earthquake Questions." *Tulsa World*, November 7. Retrieved January 15, 2014. (http://www.tulsaworld.com/archives /geologists-have-no-easy-answers-to-earthquake-questions/article_da0200e0-98fc-5610 -b2d9-92563bf1f286.html).

Greene, Wayne. 2012a. "Report Shows Oklahoma Earthquakes May Be Man-Made." *Tulsa World*, April 7. Retrieved January 15, 2014. (http://www.tulsaworld.com/archives/report -shows-oklahoma-earthquakes-may-be-man-made/article_3075d83e-50d8-5199-9837 -b2f5070290c7.html).

Greene, Wayne. 2012b. "Oklahoma Geologist Disagrees with U.S. Report, Says It's Unlikely That Area's Earthquakes Man-Made." *Tulsa World*, April 10. Retrieved January 15, 2014. (http://www.tulsaworld.com/news/government/oklahoma-geologist-disagrees-with-u-s -report-says-it-s/article_e3f01d68-6f58-5b15-bfa4-99a02185caf7.html).

Greene, Wayne. 2013. "Oklahoma Earthquake in 2011 'Induced' By Injection Wells, OU Professor Says." *Tulsa World*, March 27. Retrieved January 15, 2014. (http://www.tulsaworld .com/news/government/oklahoma-earthquake-in-induced-by-injection-wells-ou -professor-says/article_99206475-9896-503e-80a0-7e4341ee40bd.html).

Gupta, Nidhi, Arnout R. H. Fischer, and Lynn J. Frewer. 2012. "Socio-psychological Determinants of Public Acceptance of Technologies: A Review." *Public Understanding of Science* 21: 782–795.

Hoberock, Barbara. 2014. "Payne County Woman Asks Lawmakers to Protect Residents from Earthquakes." *Tulsa World*, October 29. Retrieved November 15, 2014. (http://www .tulsaworld.com/news/capitol_report/payne-county-woman-asks-lawmakers-to-protect -residents-from-earthquakes/article_0e947bf2-b598-507e-81d0-0190fc4903fc.html).

Hoberock, Barbara. 2015. "Groups Call on State Leaders to Take Action to Prevent More Earthquakes." *Tulsa World*, May 12. Retrieved June 15, 2015. (http://www.tulsaworld.com /news/capitol_report/groups-call-on-state-leaders-to-take-action-to-prevent/article _a5e57ce0-d79a-56cf-ae17-a68db3eb7e93.html).

Hoberock, Barbara. 2016a. "Scores of Angry Oklahomans Call for Moratorium on Earthquake-causing Injection Wells." *Tulsa World*, January 16. Retrieved June 20, 2016. (http://www .tulsaworld.com/news/capitol_report/scores-of-angry-oklahomans-call-for-moratorium -on-earthquake-causing/article_1a912502-8009-5587-81a5-50eaa0116e5e.html).

Hoberock, Barbara. 2016b. "Erin Brockovich Talks to Oklahomans About Earthquakes." *Tulsa World*, February 25. Retrieved June 20, 2016. (http://www.tulsaworld.com

/news/government/erin-brockovich-talks-to-oklahomans-about-earthquakes/article
_9263fc44-ab78-5e6b-8d98-761ca6a7c6c2.html).

Hoberock, Barbara. 2016c. "Victims of Property Damage from Oklahoma Earthquakes Rally
in OKC Against Big Oil." *Tulsa World*, April 13. Retrieved June 20, 2016. (http://www
.tulsaworld.com/news/state/victims-of-property-damage-from-oklahoma-earthquakes
-rally-in-okc/article_ed9c0218-848d-5b06-b73c-c09ca1d88ec9.html).

Holland, Austin A. 2011. "Examination of Possibly Induced Seismicity from Hydraulic
Fracturing in the Eola Field, Garvin County, Oklahoma." *Oklahoma Geological Survey*
OF1–2011.

Holland, Austin A. 2013. *Bulletin of the Seismological Society of America* 103(3): 1784–1792.

Hornig, Susanna. 1993. "Reading Risk: Public Response to Print Media Accounts of Technological Risk." *Public Understanding of Science* 2(2): 95–109.

Hough, Susan E., and Morgan Page. 2015. "A Century of Induced Earthquakes in Oklahoma?" *Bulletin of the Seismological Society of America* 105(6): 2863–2870.

Jacquet, Jeffrey B., and Richard C. Stedman. 2014. "The Risk of Social-Psychological Disruption as an Impact of Energy Development and Environmental Change." *Journal of Environmental Planning and Management* 57(9): 1–20.

Jaspal, Rusi, and Brigitte Nerlich. 2014. "Fracking in the UK Press: Threat Dynamics in an
Unfolding Debate." *Public Understanding of Science* 23(3): 348–363.

Jones, Corey. 2015. "Citing Quake Concerns, Corporation Commission Directs 211 More
Wastewater Injection Wells to Reduce Depths." *Tulsa World*, July 18. Retrieved August
15, 2015. (http://www.tulsaworld.com/news/local/citing-quake-concerns-corporation
-commission-directs-more-wastewater-injection-wells/article_1d5cf0e8-8a8c-5db8-b974
-b52e487abd50.html).

Jones, Corey. 2016a. "Wastewater Disposal Volumes Rose 81 Percent in Six Years as Earthquakes Rumbled More Frequently." *Tulsa World*, January 17. Retrieved June 20, 2016.
(http://www.tulsaworld.com/earthquakes/wastewater-disposal-volumes-rose-percent-in
-six-years-as-earthquakes/article_bfd90706-b53a-5c0f-9a50-2a9c36efd019.html).

Jones, Corey. 2016b. "USGS Report: Man-Made Quakes Give Oklahoma Highest Risk,
Making State's Hazards Comparable to California." *Tulsa World*, March 29. Retrieved
June 20, 2016. (http://www.tulsaworld.com/earthquakes/usgs-report-man-made-quakes
-give-oklahoma-highest-risk-making/article_92b43bff-d570-53bf-a5a4-4c3d8dc15d89
.html).

Keranen, Kathleen M., Heather M. Savage, Geoffrey A. Abers, and Elizabeth S. Cochran.
2013. "Potentially Induced Earthquakes in Oklahoma, USA: Links between Wastewater
Injection and the 2011 Mw 5.7 Earthquake Sequence." *Geology,* March 26. (http://dx.doi
.org/10.1130/G34045.1).

Keranen, Kathleen M., Matthew Weingarten, Geoffrey A. Abers, Barbara A. Bekins, and
Shemin Ge. 2014. "Sharp Increase in Central Oklahoma Seismicity Since 2008 Induced
By Massive Wastewater Injection." *Science* 345(6195): 448–451.

Krehbiel, Randy. 2011. "Sullivan Scoffs at Suggested Fracking, Quake Connection." *Tulsa
World*, November 11, p. A11.

Krehbiel, Randy. 2015. "State Supreme Court Decision Seen as a Victory for Earthquake
Victims." *Tulsa World*, July 6. Retrieved August 15, 2015. (http://www.tulsaworld.com
/business/energy/state-supreme-court-decision-seen-as-a-victory-for-earthquake/article
_be91c7f1-5f7d-5286-b265-531f00956bf4.html).

Ladd, Anthony E. 2013. "Stakeholder Perceptions of Socioenvironmental Impacts from
Unconventional Natural Gas Production and Hydraulic Fracturing in the Haynesville
Shale." *Journal of Rural Social Sciences* 28(2): 56–89.

Ladd, Anthony E. 2014. "Environmental Disputes and Opportunity-Threat Impacts Surrounding Natural Gas Fracking in Louisiana." *Social Currents* 1(3): 293–312.

Ladd, Anthony E. 2016. "Meet the New Boss, Same as the Old Boss: The Continuing Hegemony of Fossil Fuels and Hydraulic Fracking in the Third Carbon Era." *Humanity and Society* (http://dx.doi.org/10.1170/0160597616628908).

Malin, Stephanie A., and Kathryn Teigen DeMaster. 2016. "A Devil's Bargain: Rural Environmental Injustices and Hydraulic Fracturing on Pennsylvania's Farms." *Journal of Rural Studies* 47: 278–290.

Maret, Susan. 2013. "Freudenburg Beyond Borders: Recreancy, Atrophy of Vigilance, Bureaucratic Slippage, and the Tragedy of 9/11." Pp. 201–223 in *William R. Freudenburg, A Life in Social Research (Research in Social Problems and Public Policy, Volume 21)*, edited by S. Maret and T. I. K. Youn. Bradford, UK: Emerald Group.

McGarr, A., B. Bekins, N. Burkhardt, J. Dewey, P. Earle, W. Ellsworth, S. Ge, S. Hickman, A. Holland, E. Majer, J. Rubenstein, and A. Sheehan. 2015. "Coping with Earthquakes Induced by Fluid Injection: Hazard May be Reduced by Managing Injection Activities." *Science* 347(6224): 830–831.

McKenzie, Lisa M., Roxana Z. Witter, Lee S. Newman, and John L. Adgate. 2012. "Human Health Risk Assessment of Air Emissions from Development of Unconventional Natural Gas Resources." *Science of the Total Environment* 424: 79–87.

Meng, Qingmin, and Steve Ashby. 2014. "Distance: A Critical Aspect for Environmental Impact Assessment of Hydraulic Fracking." *Extractive Industries and Society* 1(2): 124–126.

Monies, Paul. 2015a. "Tulsa Oil Company Challenges Corporation Commission's Earthquake-Related Injection Well Regulation." *Tulsa World*, October 13. Retrieved November 15, 2015. (http://www.tulsaworld.com/news/government/tulsa-oil-company-challenges-corporation-commission-s-earthquake-related-injection/article_427b4acb-c869-5ade-94ff-c406bc9bd662.html).

Monies, Paul. 2015b. "Environmental Groups Vow to File Lawsuit against Oklahoma Energy Firms for Quakes." *Tulsa World*, November 3. Retrieved December 15, 2015. (http://www.tulsaworld.com/business/energy/environmental-groups-vow-to-file-lawsuit-against-oklahoma-energy-firms/article_90f67a81-876a-59d7-9101-1e63166cf506.html).

Monies, Paul. 2016. "Edmond Residents File Earthquake Lawsuit Against 12 Oil Companies." *Tulsa World*, January 12. Retrieved June 20, 2016. http://www.tulsaworld.com/earthquakes/edmond-residents-file-earthquake-lawsuit-against-oil-companies/article_45f352b1-c8d8-5612-8192-73851c5dc714.html

Montgomery, Carl T., and Michael B. Smith. 2010. "Hydraulic Fracturing: History of an Enduring Technology." *Journal of Petroleum Technology,* December: 26–31.

Nelkin, Dorothy. 1989. "Communicating Technological Risk: The Social Construction of Risk Perception." *Annual Review of Public Health* 10(1): 95–113.

Oklahoma Energy Resources Board (OERB). 2016. "Home Room." Retrieved January 15, 2016. (http://oerbhomeroom.com/?ga=1.128968140.158420906.1458780115).

Oklahoma Oil and Gas Association. 2016. "Shale Plays and Unconventional Resources." Retrieved April 15, 2016 (http://okoga.com/sources/).

Oklahoma State Statute 52–288.3. 1992. "Oklahoma Energy Education and Marketing Act." Retrieved January 15, 2016. (https://www.oerb.com/uploads/oerb-statute.pdf).

Open Secrets: Center for Responsive Politics. 2016. "Oklahoma Congressional Delegation." Retrieved January 15, 2016. (https://www.opensecrets.org/states/delegatn.php?cycle=2016&state=OK).

Overall, Michael. 2011. "Demand for Quake Insurance Up." *Tulsa World*, November, 2011, p. A4.

Penningroth, Stephen M., Matthew M. Yarrow, Abner X. Figueroa, Rebecca J. Bowen, and Soraya Delgado. 2013. "Community-based Risk Assessment of Water Contamination from High-Volume Horizontal Hydraulic Fracturing." *New Solutions* 23(1): 137–166.

Perrow, Charles. 1999 (2nd ed.). *Normal Accidents: Living with High Risk Technologies*. Princeton, NJ: Princeton University Press.

Pilisuk, Marc, Susan Hillier Parks, and Glenn Hawkes. 1987. "Public Perception of Technological Risk." *Social Science Journal* 24(4): 403–413.

Rabinowitz, Peter M., Ilya B. Slizovskiy, Vanessa Lamers, Sally J. Trufan, Theodore R. Holford, James D. Dziura, and Peter N. Peduzzi. 2015. "Proximity to Natural Gas Wells and Reported Health Status: Results of a Household Survey in Washington County, Pennsylvania." *Environmental Health Perspectives (Online)* 123(1): 21.

Raleigh, C. B., J. H. Healy, and J. D. Bredehoeft. 1976. "An Experiment in Earthquake Control at Rangely, Colorado." *Science* 191(4233): 1230–1237.

Raynes, Dakota K. T., Tamara L. Mix, Angela Spotts, and Ariel Ross. 2016. "An Emotional Landscape of Place-based Activism: Exploring the Dynamics of Place and Emotion in Antifracking Actions." *Humanity and Society* 40(4): 401–423.

Renn, Ortwin. 2008. *Risk Governance: Coping with Uncertainty in a Complex World*. London: Earthscan.

Ritchie, Liesel Ashley, Duane A. Gill, and Courtney N. Farnham. 2012. "Recreancy Revisited: Beliefs About Institutional Failure Following the Exxon Valdez Oil Spill." *Society & Natural Resources: An International Journal* 26(6): 655–671.

Slovic, Paul. 1987. "Perception of Risk." *Science* 236(4799): 280–285.

Soraghan, Mike. 2015. "In Oil-Friendly Okla., Gov. Fallin Moved Slowly on 'Awkward' Issue of Quakes." *EnergyWire*, July 8. Retrieved January 1, 2016. (http://www.eenews.net/stories/1060021388).

Staff. 2016. "Sierra Club Sues Three Oklahoma Energy Firms over Wastewater Disposal, Quakes." *Tulsa World*, February 17. Retrieved June 20, 2016. (http://www.tulsaworld.com/earthquakes/sierra-club-sues-three-oklahoma-energy-firms-over-wastewater-disposal/article_e474ff16-f6f0-5e51-9629-be0be27ad49a.html).

Stancavage, John. 2014. "Agencies: Number of Oklahoma Earthquakes Raises Threat." *Tulsa World*, May 5. Retrieved January 15, 2015. (http://www.tulsaworld.com/earthquakes/agencies-number-of-oklahoma-earthquakes-raises-threat/article_ofc5a13a-d490-11e3-9ce2-0017a43b2370.html).

Starbird, Kate, Dharma Dailey, Ann Hayward Walker, Thomas M. Leschine, Robert Pavia, and Ann Bostrom. 2015. "Social Media, Public Participation, and the 2010 BP Deepwater Horizon Oil Spill." *Human and Ecological Risk Assessment* 21: 605–630.

State Review of Oil & Gas Environmental Regulations (STRONGER). 2011. "Oklahoma Hydraulic Fracturing State Review." Retrieved April 15, 2016. (http://www.occeweb.com/STRONGER%20REVIEW-OK-201-19-2011.pdf)

Summars, Emily. 2016. "Oklahoma Becomes 'An Intake State' for Disposal Wells." *Enid News*, January 24. Retrieved April 15, 2016, (http://www.enidnews.com/news/local_news/oklahoma-becomes-an-intake-state-for-disposal-wells/article_3bdf295e-d5c4-580b-8a80-a0a8ed7b3951.html?mode=jqm).

Wertz, Joe. 2015. "EPA Wants Oklahoma Oil and Gas Officials to Issue More Earthquake Restrictions" *State Impact*, October 14. Retrieved April 15, 2016. (https://stateimpact.npr.org/oklahoma/2015/10/14/epa-wants-oklahoma-oil-and-gas-officials-to-issue-more-earthquake-restrictions/).

Wertz, Joe. 2017. "Judge Dismisses Sierra Club Lawsuit Against Oil Companies over Oklahoma Quakes." *State Impact,* April 5, 2017. Retrieved June 23, 2017. (https://stateimpact

.npr.org/oklahoma/2017/04/05/judge-dismisses-sierra-club-lawsuit-against-oil
-companies-over-oklahoma-quakes/).

Wheeler, David, Margo MacGregor, Frank Atherton, Kevin Christmas, Shawn Dalton, Mau-
rice Dusseault, Graham Gagnon, Brad Hayes, Constance McIntosh, Ian Mauro, and Ray
Ritcey. 2015. "Hydraulic Fracturing—Integrating Public Participation with an Independent Review of the Risks and Benefits." *Energy Policy* 85: 299–308.

Williams, Cory. 2015. "State Officials Are Reactive, Not Proactive, to the Spate of Earthquakes That Are Growing in Intensity, Legislator Asserts." *Tulsa World*, December 3.
Retrieved April 15, 2016. (http://www.tulsaworld.com/communities/skiatook/news
/opinion/state-officials-are-reactive-not-proactive-to-the-spate-of/article_20d947a0
-cf24-5c9f-b05d-50e7325a3b6f.html).

Wilmoth, Adam. 2014. "Reason for Oklahoma Earthquakes Remains a Mystery." *Tulsa
World*, February 19. Retrieved January 15, 2015. (http://www.tulsaworld.com/business
/energy/reason-for-oklahoma-earthquakes-remains-a-mystery/article_70d3a9b3-e2c1
-52a2-8986-4c8c61c158ad.html).

Zoback, Mark D., Arjun Kohli, Indrajit Das, and Mark William Mcclure. 2012. "The Impor-
tance of Slow Slip on Faults During Hydraulic Fracturing Stimulation of Shale Gas
Reservoirs." In *SPE America's Unconventional Resources Conference*. Society of Petroleum
Engineers. (http://dx.doi.org/10.2118/155476-MS).

8

Contested Colorado

─────────────────────────●

Shifting Regulations and
Public Responses to
Unconventional Oil Production
in the Niobrara Shale Region

STEPHANIE A. MALIN, STACIA S.
RYDER, AND PETER M. HALL

Introduction

The recent boom in unconventional oil and natural gas extraction in U.S. shale regions has reinvigorated energy production in many states, but few as dramatically as Colorado. The industry has enjoyed rapid growth, boldly expanding production in hundreds of U.S. communities. In addition to technological advancements, favorable state legislation, and lucrative tax subsidies, federal deregulation has been a particularly important factor in facilitating unconventional oil and gas (hereafter, UOG) production nationwide. The 2005 Energy Policy Act, for instance, exempted unconventional processes such as hydraulic fracturing from assorted federal environmental regulations, including the Safe Drinking Water Act (Malin 2014; Shapiro and Warner 2013). States stepped in with their own zoning rules, regulatory approaches, and enforcement plans, many heavily influenced by the oil and gas industry (Toan 2015). But unlike previous waves of extraction that largely took place in rural areas, UOG production rapidly expanded to more heavily populated urban/suburban areas that lack the capacity to effectively oversee it. These communities face uncertainties and community tensions regarding how to best regulate the environmental risks, social disruptions, and uneven socioeconomic impacts, as well as the pads, compressor stations, and other infrastructural components that operate near neighborhoods (Brasier et al. 2011; Gullion 2015).

In Colorado, where energy extraction and mining have historically left large ecological footprints on the state's land, air, and water resources (Ubbelohde, Smith, and Benson 2006), UOG producers' freedom to set the pace of development has been particularly ubiquitous. The politically powerful oil and gas (O&G) industry has worked closely with Colorado lawmakers to minimize the opportunities for citizens, community groups, municipal leaders, and other stakeholders to participate equally in decision making about UOG production. Yet, Colorado citizens living within the Denver-Julesburg Basin,[1] atop the Niobrara Shale formation (USEIA 2016), have seen their daily lives change amid increased oil production.[2] When the UOG production boom began in Colorado in 2013, environmental, public health, and regulatory conflicts quickly emerged, leaving Coloradans to ask fundamental questions about related democratic processes and procedures (Jaffe 2013).

Why do UOG zoning and regulation matter? Procedural justice regarding land use decisions is one key process at stake—the meaningful, authentic involvement of multiple stakeholders in environmental decision making and the capabilities that Colorado communities and citizens have, or not, to realize environmental justice goals surrounding UOG production. At stake are the rights of Home Rule municipalities to participate equitably in these decisions; should the state of Colorado alone or municipalities within Colorado (which host UOG activities) regulate the industry and control UOG zoning? Lake (1996) and Schlosberg (2007: 74) assert that while the inequitable distribution of environmental risks and hazards remains an important issue in environmental justice struggles, increased attention must be paid to *procedural* injustices that create the "maldistribution of risks and resources" (see Mohai, Pellow, and Roberts 2009). In Home Rule states like Colorado, where many communities theoretically retain the power to locally control development decisions, jurisdictional uncertainties about UOG production have pitted the rights of municipalities against the rights of mineral owners, particularly as the state has claimed preemption in developing mineral rights. Northern Colorado communities have been fractured in response to these significant tensions over UOG development, but procedural justice issues and implications have yet to be systematically analyzed.

In this chapter, we show how fractures formed in the struggle between state power and local control, exploring understudied institutional contexts of decision and rule making (Rinfret, Cook, and Pautz 2014). We examine the state's attempt to facilitate mineral production and the uncertainty it created for community leaders and residents adapting to rapid industrial growth. Next, we explore how the state's support, in combination with the industry's historical influence in Colorado, created conditions where the industry set the pace of production and reaped a majority of benefits, even as communities struggled to adapt quickly, justly, and resiliently. We further document conflicts that grew between municipal leaders as they responded to UOG production in their communities, as well as analyze important tensions among activists.

Citizens in other communities face similar procedural marginalization. Middle-class activists fight environmental risks like air and water pollution as drilling rigs operate 100 feet from their homes in Texas (Gullion 2015). Daily life has been disrupted as people experience "collective trauma" (Perry 2012) and community tensions over land use and leases in rural Pennsylvania (Wilber 2015). North Dakota's Bakken Shale communities reel from social disruption (Weber et al. 2014). In these regions and others, UOG producers' economic and political power has allowed them to create a hegemonic discourse surrounding continued fossil fuel production, while communities and citizens struggle to have a place at the negotiating table (Ladd 2016).

In Colorado, the O&G industry's enhanced operational and regulatory flexibility has increased their status and power relative to local governments and citizens (Toan 2015). Strong protection of mineral owners' rights, and the presumption that they want to develop their minerals, have erected serious barriers to procedural justice in making decisions about land use, specifically how and where industrial UOG operations have been located in Colorado. Codified spatial inequalities, particularly for landowners who own their surface but not their mineral wealth due to systems of split estate, have created entrenched barriers to meaningful municipal or grassroots participation. These outcomes have appeared systemic: state preemption of municipal and local rights to zone and regulate repeatedly excluded various stakeholders from meaningful participation in rulemaking (Aguilar 2012, 2016; Cook 2015). Then, in a widely contested May 2016 decision, Colorado's state Supreme Court ruled that municipalities do *not* have the right to zone or regulate UOG production within their jurisdictions, with that responsibility falling to the state. The Court ruled in favor of state preemption over local bans and moratoriums and overturned local ordinances on the grounds that they were "invalid and unenforceable" (Finley 2016). This decision reinforced power disparities and systemic distributive injustices that have shaped struggles for communities and residents dealing with prodigious UOG activity near their homes, even as the industry has suffered a drilling downturn in Colorado (Svaldi 2016a, 2016b). Finally, the Colorado secretary of state's office rejected multiple attempts made by various activist organizations in fall 2016 to introduce ballot initiatives to assert local control over UOG production, blocking one of the last avenues citizens had to gain local control.

To realize procedural environmental justice in Colorado, citizens must be granted the power to play more meaningful roles in governing UOG activity and redress power imbalances in decision-making; thus far, they have been systematically denied this power. Yet, these public conversations represent important steps toward creating wider democratic discussions about energy policy, economic development, and the role that powerful O&G corporations play in making decisions that have long-term impacts on people, communities, and ecosystems (Faber 2008).

Research on the Impacts of Unconventional Oil and Gas Production

The literature to date has explored various issues related to UOG production, including pollution of water resources through drilling and wastewater disposal (Gregory et al. 2011; Vidic et al. 2013); air pollution involving methane leakage and diesel pollution (Howarth et al. 2011); general public health concerns (Adgate et al. 2014), including higher birth defect rates around UOG wells (McKenzie et al. 2015); and disruptive social effects on shale communities (Brasier et al. 2011; Perry 2012; Willow 2015). A number of studies have examined national perceptions of UOG activity and show that U.S. citizens have relatively little familiarity with and are uncertain about UOG development (Boudet et al. 2014; Crowe, Ceresola, and Silva 2015). Even those living near UOG operations hold divergent perceptions about socio-environmental and community impacts (Theodori 2009; Ladd 2013, 2014).

Land use, governance, and regulation issues have remained central to public debates over UOG production (Malin 2014; Minor 2014; Nolon and Gavin 2013; Negro 2012). Wiseman (2009) argued that the absence of uniform federal regulations surrounding UOG production led to a patchwork of regulatory polices across states, and recent legal reviews have compared how these states operate. Brady and Crannell (2012) compared significant distinctions in state-level regulations in Colorado, New York, Pennsylvania, Texas, Louisiana, and Wyoming. Davis (2012) compared Texas and Colorado regulatory frameworks and found that Colorado's approach was more rigorous. Since then, legal scholars have analyzed controversies between local and state policy makers over UOG regulatory authority in Ohio, Pennsylvania, New York, Texas, West Virginia, New Mexico, and Colorado (Armstrong 2013; Davis 2014; Walsh et al. 2015). Toan (2015) focused on Colorado court battles between the state, particularly the Colorado Oil and Gas Conservation Commission (COGCC), and local governments. In short, states remain uncertain and deeply divided about how to regulate and zone UOG production.

While Colorado may claim superiority in environmental regulations (e.g., Davis 2012), those claims have yet to be tested on a variety of fronts, including the space for authentic citizen participation in decision making. Testing such claims is indeed necessary. Citizen concerns have emerged across U.S. shale communities over the impacts associated with UOG development, especially about effects on air and water resources, rural livelihoods, human health, and long-term economic development (Malin and DeMaster 2016). Current research indicates that many citizens believe that they are not participating in just, equitable, and democratic processes, but in environmental disputes shaped by clear power imbalances and contentious environmental conflict (Gullion 2015; Perry 2011, 2012). While few analyses explore procedural (in)justice and power inequalities among states, communities, and citizens, researchers have documented the powerful influence of UOG producers in comparison to other Colorado stakeholders (Cook 2015; Opsal and Shelley 2014).

Still, limited research highlights places where procedural equity has thrived. Carre (2012), for instance, studied successful methods used by residents of Delta, Colorado, to exert local regulatory control over UOG activity. Heikkila et al. (2014) explored public responses to various strategies used by interest groups and policy makers during regulatory changes in Colorado. Wylie and Albright (2014) showed how Colorado residents used WellWatch to network across communities by generating digital databases and citizen narratives about living near UOG development. Willow and Wylie (2014) reviewed how in-depth, ethnographic research helped systematically analyze environmental health concerns across communities with UOG operations. Yet, these studies highlight exceptions. Citizens have few institutional mechanisms to realize procedural justice and address their concerns about setback rules that control drilling's proximity to residential areas; health risks like respiratory problems and birth defects; air pollution from methane leaks; marginalization from land use decision making; and other prominent socio-environmental risks.

Communities hosting significant UOG production face assorted environmental costs, yet usually lack the extensive wealth, political influence, cultural legitimacy, and technological infrastructure enjoyed by the O&G industry (Wylie and Albright 2014). Instead, in Colorado, the distribution of industry-related costs and benefits remains highly unequal. Communities have inherited numerous responsibilities, such as environmental remediation costs, social disruption, and economic instability associated with falling O&G prices (Finley and Murphy 2015; Jacquet 2014). Opsal and Shelley (2014) observed procedural inequities and inadequate regulatory responsiveness to citizen complaints by Colorado's COGCC. Environmental risks and toxins related to UOG production have not stopped at city, county, or state lines. Rather, they have often been felt first and foremost on the surface of the land—where comparatively few benefits accrue—or in exposure to air pollution in communities with no drilling. For instance, people living up against the foothills in Loveland and Fort Collins are exposed to concentrated air pollution from Weld County's 17,126 operating wells, a number that topped 22,000 wells when the oil economy was booming in 2014 (COGCC 2016a). Many citizens in these communities report feeling marginalized, disempowered, and excluded from related decision-making.

Apple (2014: 224) observes that "the outcomes of interactions sparked by fracking development will depend on the local-regional dynamics produced by local and regional actors' unique configuration of preferences and their imperfect navigation of uncertain circumstances." By rigorously examining the roles communities and citizens played in controlling the pace, scale, and siting of UOG production, we illuminate important localized attempts to realize procedural justice along the northern Colorado Front Range.

The Colorado Context: An Overview of State Oil and Gas History

O&G production has a significant history in Colorado, where drilling began in the mid-1800s. As national efforts were first underway to systematize mineral development, split estate situations were created. In split estates, legal ownership of minerals beneath the surface is severed from ownership of the surface land (Toan 2015). Mineral owners or lessees have "the right to enter, dig and carry them away, and all other such incidents thereto are necessary to be used for getting and enjoying them" (*Cowan*, 26 Tex. at 217). In natural resource–rich states like Colorado, where oil and railroad interests dominated mineral ownership, land use conflicts resulted almost immediately from split estate situations. Initially, courts ruled that the mineral rights holder "has the right of possession even as against the owner of the soil, so far as it is necessary to carry on mining operations" (*Turner v. Reynolds* 1854), and successive court cases have reinforced this ruling (i.e., *Cowan v. Hardeman* 1862; *Marvin v. Brewster Iron Mtn. Co.* 1874; *Clevenger v. Continental Oil Company* 1962). Federal policies such as the General Mining Act of 1872 quickly codified the legal dominance of split estates and privileged the rights of mineral holders over surface owners or dwellers in states like Colorado.

Oil production in Colorado exploded at the turn of the 20th century, increasing from 200 barrels per day in 1885 to over 2 billion barrels per day by the late 1920s (Colorado Energy Office 2013). The rush to drill also resulted in rampant waste and inefficiency, even after the Gas Conservation Commission (GCC) was established in Colorado in 1927 (Toan 2015). In 1951, the Colorado General Assembly created the Colorado Oil and Gas Conservation Commission (COGCC) and gave them police powers (§§ 34–60–101–129, C.R.S. [the "Act"]). The commission was established to protect mineral rights owners, develop minerals in an orderly and responsible manner, and prohibit the waste of oil or gas in the state of Colorado (Colorado Oil and Gas Conservation Act 1951; Randall 2012). The commission executed these functions until 1985, when the COGCC was further tasked with developing regulations to promote safety, welfare, and general public health related to O&G production.

Not until 1994 was the COGCC charged with environmental protection. Though the COGCC would not codify it until 2007, in 1997 the Colorado courts did address the power disparities between surface and mineral owners by creating the "accommodation doctrine," stipulating that parties with mineral interests must fairly accommodate surface owners (*Gerrity Oil and Gas Corp. v. Magness* 1997; Johnson 1998). Yet even now, most Colorado landowners live on split estate properties, wherein O&G operators or railroads control much of subsurface mineral wealth, limiting the power of citizens living on the surface.

Methodology

Data Collection and Analysis

This chapter presents integrated findings from distinct research projects conducted by each co-author. Qualitative and historical methods were employed across each of the projects, including in-depth interviews, ethnographic/participant observation, and document and archival analyses of regulatory frameworks, court cases, congressional minutes, and industry-related historical documents. Attendance at Colorado State Oil and Gas Task Force meetings and extensive travel to communities and households experiencing UOG activity also informed our historical and regulatory analyses.

The Colorado State University Water Center and the Rural Sociological Society's Early Career Award supported one strand of this research. Community-based fieldwork included 45 in-depth interviews, participant observation, and analyses of an extensive sample of archival documents covering Colorado's water law and split estate regime. Interviews included questions about activism, impacts of UOG production in daily life, and perceptions of industry regulations. Participant observation included attending community meetings, farm visits, and water and energy infrastructure tours conducted from 2013 to 2015. Colorado State University's (CSU) Center for Collaborative Conservation's Fellowship Program supported another strand of this research, related to multiyear (2012–2015) comparative fieldwork in two Colorado cities. Participatory methods were utilized, including attending city council meetings; observing events and meetings held by community organizations such as Protect Our Loveland; and viewing, coding, and analyzing audiovisual recordings and minutes of legislative meetings, such as city council and COGCC meetings. Network samples of city employees and council members participated in semistructured interviews that focused on the process of developing UOG regulations in each city.

All interviews were recorded and transcribed verbatim. Each interview was read at least three times, with themes and subthemes identified, coded, and verified using tests for interrater reliability. Field notes were taken, typed, and then revisited to code for themes common to our interview data. All historical and archival documents were consulted iteratively. These findings were revisited as we compared data from individual projects, and additional coding procedures and analyses were conducted to identify patterns across multiple data sources.

This research took place in several northern Colorado counties and communities, including Loveland and Fort Collins, neighboring cities on Colorado's Front Range, roughly sixty miles north of Denver. Fort Collins is about twice as large as Loveland in both area and population size. Legislatively, both of these Home Rule municipalities are governed under the same council-manager system. Other communities in the Niobrara Shale region were also study sites, including Greeley, Windsor, Kersey, and Ault.

Comparative Findings and Analyses

By the late 2000s, unconventional extraction methods dominated O&G extraction in Colorado. In early 2000, fewer than 22,000 wells operated in Colorado, but by January 2016, the COGCC reported 53,698 active wells (COGCC 2001, 2016a). State lawmakers and residents initially welcomed the impressive, if temporary, economic boom created by UOG production (Jaffe 2013). However, unconventional drilling methods posed a host of new technical, regulatory, and socio-environmental challenges for which many Colorado communities and citizens were largely unprepared. Social and environmental uncertainties became especially pernicious for our interviewees, as definitive data about air and water pollution, chemicals used in production, effects of living in close proximity to drilling, and many other issues remained elusive (Finkel et al. 2015). As the number of producing UOG wells increased (see Figure 8.1) and production sites crept closer to residential areas, schools, hospitals, and other community centers (see Figures 8.2a, b, c), citizen concerns and state and local regulatory efforts grew (Aguilar 2012). But citizens' power to influence UOG production did not keep pace.

State Control and Disempowered Communities. Public pressure in Colorado did help shape landmark O&G legislation, but not until 2007. That year, Governor Bill Ritter enacted three important policies: the Habitat Stewardship Act (HB 07–1298); the Reasonable Accommodation Act (HB 07–1252); and the Oil and Gas Commission Reform Act (HB 07–1341). Each of these added a new layer of state-level accountability for O&G operators, greatly diversified

Total Number of Actively Producing Colorado Oil and Gas Wells 2000–2016

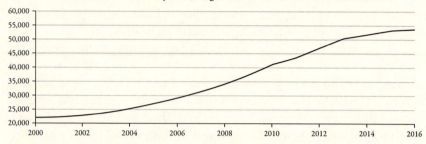

FIGURE 8.1 This figure shows significant increase in oil and gas wells in Weld County between 2000 and 2016., due in large part to the more widespread implementation of unconventional extraction methods such as hydraulic fracturing. Data source: Colorado Oil and Gas Conservation Commission.

the degree of environmental expertise on the COGCC, and helped balance these interests with industry influence in the regulatory process. Ritter was also a one-term governor.

The COGCC's subsequent regulations aimed to "toughen protections for communities, the environment, and wildlife and provide far greater opportunities for public and local government participation" (Randall 2012). They focused on chemical treatment disclosure, groundwater testing and monitoring, urban well setback distances,[3] spill reporting, targeted facility inspections, sensitive wildlife habitat, rule enforcement and penalties, and flood mitigation practices (COGCC 2016b). In 2014, Colorado became the first state to implement methane emission regulations for UOG operations. Still, the state's regulatory performance came under fire, due to incomplete permitting applications being granted and setback rules not being adequately enforced (University of Denver Review, DU Report 2015).

While these regulations offered landowners and communities greater protections than before, they did not alter basic procedural inequities and power imbalances. In effect, communities had little structural or political power compared to the state or UOG industry. To illustrate, one Loveland Council member observed:

> The decisive points are not with city and counties, it is with the state. The state allows for severed estates. With large swaths of properties being sold and the mineral rights retained by a preceding property owner, and then the mineral rights being sold off to other buyers, potentially even to anonymous shell company, a destructive practice. The structure and the rules that the Colorado Oil and Gas Conservation utilize are an improvement to what it was before 2008. But . . . a large majority of COGCC board members are still pretty openly and deliberately oil and gas industry operatives. They are to advance and protect the industry.

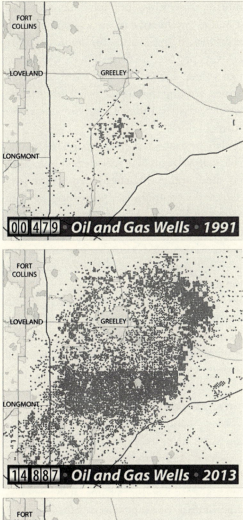

| 0 0 4 7 9 | Oil and Gas Wells · 1991 |

| 1 4 8 8 7 | Oil and Gas Wells · 2013 |

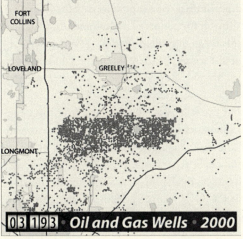

| 0 3 1 9 3 | Oil and Gas Wells · 2000 |

FIGURE 8.2A, B, C These maps show the location and relatively rapid increase in oil and gas wells in one part of the northern Colorado's Denver-Julesburg Basin. Figure 8.2a. shows production in 1991— with 479 conventional wells—as contrasted with the 14,887 wells operating by 2013.

Though state lawmakers and industry operatives publicly maintain that Colorado's rules were the strictest and most effective in the nation,[4] navigating this regulatory space has been uneven. Citizens and communities had to scramble to improve regulations as UOG production propelled forward. Even when local control was the wish of voters, UOG producers and Governor Hickenlooper opposed those measures, arguing that a patchwork of different community-based rules could thwart energy development and threaten rights of access for mineral owners (Ogburn 2014).

Community-led attempts to regulate or zone UOG production resulted in empty victories. Since 2012, the Colorado Oil and Gas Association (COGA), the industry's lobbying group, has successfully challenged democratic processes by filing lawsuits against Colorado communities with local bans or moratoria. State courts have supported UOG producers in these lawsuits and overturned communities' rights to control UOG activity on the grounds of preemption; this concluded with the May 2016 Colorado Supreme Court decision and dismissal of attempts in fall 2016 to introduce ballot initiatives allowing local control of UOG production.

Most municipal officials we interviewed felt disempowered by their inability to regulate UOG activity and believed their professional responsibilities to protect people's health and safety had been compromised. One Fort Collins city staff member expressed an acute awareness of staff members' structural limitations in determining and limiting socio-environmental risks:

> It's confusing to me that this industry is not regulated locally . . . especially that they do not have to comply with zoning. Traditionally, industrial uses . . . would be limited to industrial zoning areas to minimize the impact of . . . nuisance industries on people, their families, schools, and all those things that make a community work. . . . So the very idea that the oil and gas industry can drill a well, previously 150 feet from a home, now 350 feet from a home [at the time], is surprising to me. . . . The very idea that extraction industries can just go in wherever they want . . . I think is difficult for people to understand. And it compounds the risk of impacts to people through spills, or noise. . . . I'm not suggesting that I'm against the industry, it just seems like there should be a different way to plan it for communities.

Municipal lawmakers argued that the state was ill equipped to enforce regulations. The COGCC had to reconcile its conflicting mandates to promote O&G development, while simultaneously regulating the industry's environmental, health, and safety performance. Furthermore, the COGCC currently employs only 18 field inspectors to assess nearly 54,000 wells across the state (COGCC 2010; COGCC 2016a; COGCC 2016b), leading critics to question the organization's capacity to monitor UOG production sites. A City of

Fort Collins official noted: "the state doesn't fund nearly enough inspectors and other operation oversight to make sure that everything is happening the way the oil company says." The healthy skepticism of state officials has been further supported by recent reports on lax enforcement (DU Report 2015).

Community organizations continued to pursue procedural justice, as when Local Control Colorado attempted to put a measure on the 2014 ballot allowing local governments control over UOG activity. Power imbalances persisted, however. Back-room deals led to compromises in which ballot petitions were withdrawn in exchange for COGA dropping its lawsuit against Longmont, which had banned UOG activity. Instead, the 21-person Colorado State Oil and Gas Task Force was established in July 2014 and was mandated to provide recommendations to address UOG conflicts by February 2015.

In the end, the Oil and Gas Task Force did not facilitate procedural justice for communities. They proposed *40 nonbinding* recommendations, but *only nine* were approved.[5] Few stakeholders appeared particularly happy with the proposed changes, and industry supporters, such as the Weld County Commissioners, were able to work with the COGCC to opt out of the new regulations (Dunn 2016). Local community and conservation representatives were disappointed that several critical issues, especially enhanced local government control over siting operations, fell short of the required two-thirds support. By requiring a two-thirds vote instead of a majority, O&G industry representatives successfully blocked the task force from approving most of its major recommendations. The power disparities were obvious to many of our interviewees. As one Fort Collins Council member explained: "I think there was representation, but there was not empowerment. . . . I think they [environmental and community groups] have been involved, but [as in] 'well, we involved the environmental community in our discussion as we developed these lax regulations.' And none of them agreed with it. . . . The governor's commission this year was a complete waste of time. . . . Part of the compromise was to recommend regulations. Nothing came of that, to nobody's surprise. And a significant portion of the environmental community feels betrayed by that whole action."

While the task force's minimalist approach did not resolve any major regulatory issues, it did help reinvigorate activists' efforts to change Colorado's constitution. When the Colorado Supreme Court ruled against community rights in May 2016, concerned citizens introduced multiple potential ballot initiatives for fall 2016 to reassert procedural justice in the form of local control. For example, the Colorado Community Rights Network pursued a community rights amendment for the 2016 ballot, allowing local governments to regulate or ban hydraulic fracturing without threat of a lawsuit. Additional initiatives by Coloradans Resisting Extreme Energy Development (CREED) sought to increase local control and extend well setback distances to 2,500 feet from occupied structures. In this way, despite state preemption, Colorado community activists fiercely defended their rights to attain procedural justice through equal participation in decision-making. Yet, after the September 2016 decision by the Colorado secretary of state's office to reject all O&G-related ballot initiative petitions (due to allegedly forged signatures), even these attempts to realize procedural justice were effectively thwarted.

Municipal Government Approaches: Comparing Fort Collins and Loveland, Colorado. Municipal governments along Colorado's Front Range found themselves largely unprepared and unequipped to address the UOG boom and production problems. Despite most cities existing as Home Rule municipalities,[6] state preemption created serious regulatory challenges for local governments pulled between state laws, community members, industry demands, economic pressures, and public health and environmental concerns.

By late 2012, both Loveland and Fort Collins, Colorado had enacted various temporary controls on UOG activity to research UOG production before issuing drilling permits. In Loveland, this initiated a two-tiered application protocol for drilling permits. At the baseline tier, operators could choose not to sign voluntary enhanced user agreements, but were subject to the full development review process.[7] In the second tier, enhanced user agreements held industry operators to higher safety and mitigation standards, but then fast-tracked the operator's permitting process. Loveland could thus quickly negotiate relatively strict regulatory standards via industry operator Memoranda of Understanding (MOUs). Despite Loveland's new protocol, concerned citizens, particularly activists in Protect Our Loveland, developed a two-year hydraulic fracturing suspension initiative and collected enough signatures to place it on the November 2013 ballot. The initiative was contested, however, and 52% of voters rejected it during a June 2014 special election.

In Fort Collins, initial regulatory efforts mirrored Loveland's two-tiered system. Within a few months, however, the city began to change course. As one staff member noted, it became clear their approach would not meet Fort Collins citizens' strict standards for health, safety, and the environment: "There were community members who felt like even with what staff presented as best practice, that it didn't go far enough. There weren't strict enough protections. . . . There were community members who felt like there shouldn't be any oil and gas development here, period." Draft regulations then shifted to banning hydraulic fracturing in Fort Collins, but exempted operators who signed agreements mandating enhanced safety standards. In response, concerned community members established Citizens for a Healthy Fort Collins, which drafted a citizen-initiated referendum to implement a complete five-year moratorium on hydraulic fracturing. Despite official city opposition, the measure passed with 55% voter support. COGA immediately sued Fort Collins, and the Larimer County District Court ultimately ruled in favor of COGA by reversing the moratorium in August 2014. Following the City of Fort Collins's appeal, this became one of the cases heard by the Colorado Supreme Court in December 2015 and subsequently denied in May 2016.

Differences between Loveland and Fort Collins reflected broader struggles across Colorado to realize procedural justice and authentically participate in making decisions related to UOG production. Each community faced fundamentally different power dynamics among industry, citizens, and grassroots groups, which helped structure opportunities for or limitations to the goal of procedural justice. Importantly, Loveland city staff had few options outside their two-tiered permitting system, given that Loveland's city council mandated the city avoid lawsuits. The actions of Fort Collins lawmakers prioritized addressing local concerns over avoiding lawsuits, but only because they had more democratic parameters in which to act.

UOG producers also played more central roles in shaping regulations in Loveland than in Fort Collins. Though Fort Collins had a trusted relationship with Prospect Energy (a small, locally operated company), the city did not rely heavily on the business when drafting UOG regulations. Instead, Fort Collins retained staff members with O&G expertise. Loveland, however, did not retain any O&G experts on its staff but relied heavily on the COGCC and Anadarko, a large-scale UOG operator, to draft its UOG regulations. Loveland municipal employees' directive to adhere to state law, coupled with their limited technical expertise, meant that the state and industry possessed greater control over the process of regulatory development in Loveland than in Fort Collins, where such decisions were made with greater participation from key stakeholders.

Not all Niobrara Shale residents opposed UOG activity, of course, and some citizens and especially community leaders actively supported it. For instance, in the Home Rule township of Greeley, Colorado, town leaders trusted the industry to adequately self-regulate and many residents felt UOG production was vital for economic growth. But when the Colorado Supreme Court stripped municipalities of their rights to zone and regulate local UOG activity, all spaces for procedural justice collapsed and the democratic processes playing out in some communities stalled. Denied the ability to equally participate in the decision-making process, many citizens felt as though they had been forced to watch from the margins as the COGCC and UOG producers maintained oil production and controlled most related decisions.

Community Activism and Procedural Justice. Grassroots organizations mobilized in northern Colorado to challenge what they perceived as UOG producers' inequitable access to policy makers. Though some groups aligned with the precautionary principle—demanding that the industry show that their practices would not threaten environmental quality and public health before commencing production—others held less transformative, but perhaps more pragmatic, goals. Rather than rejecting UOG production outright, they targeted corporations for specific terms and protections from UOG activity. Community groups and activists were vulnerable to "divide and conquer" techniques, as some groups aligned with industry firms and regulators if their specific goals were met.

Patterned environmental inequalities limited the power that Colorado community organizations and citizens had to realize procedural equity or negotiate with UOG producers. A few examples illustrate this pattern. Despite local attempts to enact moratoriums, UOG activity continued to expand in Colorado until low oil prices statewide slowed production. Furthermore, split estate law institutionalized limited opportunities for procedural engagement and helped create gaps between citizens' desires and the decisions of community leaders. For instance, Greeley Colorado's City Council recently overturned the city Planning Commission's denial of a permit for Extraction Oil and Gas LLC's Triple Creek project, including 22 wells near homes and a school, despite hundreds of residents who protested its construction. Greeley mayor Tom Norton stated: "We have to protect the private property rights of many citizens of Greeley who have mineral rights and have the right to access them" (Aguilar 2016). Communities were caught in the middle as citizens faced disenfranchisement, due to split estate laws and uncertainty about how to respond to state preemption. In combination, these limited the power of citizens (or civil society) to counter the industry's dominance or achieve greater procedural equity.

Advocating for Moratoriums and Bans in Colorado. Some grassroots organizations sought transformative approaches to regulating UOG production, fighting to institutionalize precautionary and equitable approaches by reclaiming local control. They were often less likely to want UOG production anywhere in northern Colorado. The Colorado Community Rights Network (CCRN), for instance, worked to earn communities and citizens more power to locally zone, regulate, and even ban industrial production processes like UOG production. The CCRN's goal was to "protect communities by assisting them in creating local laws that outline and assert people's rights as that community defines them . . . to drive fundamental legal change upward, through the state and federal constitutions, so that corporate power can be curtailed or eliminated and new laws enacted reflecting, honoring, and promoting full democracy, sustainability, and community rights."[8]

These activists believed that the UOG producers' power could be meaningfully counteracted if communities could create local ordinances that controlled the pace and scale of UOG production, thereby helping to advance procedural equity. They believed that the socio-environmental risks related to

UOG production were too significant and unpredictable to allow complete state control. One member of the CCRN, after witnessing extensive and rapid changes to his environment, community, and daily life, completely rejected UOG production in Colorado. For him, drilling anywhere threatened environmental health and justice everywhere along the Front Range. He explained:

> I work with the Community Rights Network because I have witnessed the unpleasantness of having a multi-well pad constructed in close proximity to my home. For a couple months, there was around-the-clock noise from drilling equipment, compressors and other machinery. . . . Vibrations rattled the pictures on the wall in my office, we had to keep the windows closed on the side of the house facing the construction site, and at times there were turpentine-like smells in the air. There was heavy truck traffic along the only route that is the access to several neighborhoods and that road is now extremely rough and badly needs resurfacing. Our air quality is suffering. . . . I see a brown cloud hovering over in Weld County to the east that wasn't there four years ago. Neither were 22,000 active wells. . . . After witnessing this type of industrial activity in my neighborhood and how hard it is to stop the industry, I understand the dismay and frustration with our regulatory board [the COGCC]. . . . If our policy makers will not protect our health, safety and welfare from these drillers, then the citizens must take it upon themselves to drive change. Let us hope it works.

Another member of the CCRN noted UOG producers' power to set the pace of development and limit local democratic power of citizens. She expressed her frustration with activists who aided O&G industry goals once they were allowed some control over choosing locations for UOG facilities. She did not fault other activists, but felt community groups had been divided and conquered. She explained her support for transformative development goals: "Thomas Linzey and CELDF[9] had gone the regulatory route and found it only helped the industry sharpen up the regulations, all to the industry's good. It becomes obvious that regulation does not solve health or environmental issues. Fracking is inherently unsafe. . . . How do you regulate an industry that is totally out of control? . . . Some other groups that are opposed to fracking practice NIMBYism and think the industry is essential to the economy. . . . Community Rights gives local communities the right to self-regulate free of corporate dominance."

Still another northern Colorado resident explained that UOG activity near her home made her feel she lacked control over her daily life, health, and well-being, but that her activism was one way to combat disempowerment. One of the largest well pads in the state was slated to be drilled only 1,000 feet away from her back deck. Though this resident initially intended to work with the industry to resite its proposed drilling sites, she decided to found another group focused on banning UOG activity in Colorado altogether. In her mind, her initial inter-actions with the state had been inequitable and disrespectful, and she no longer trusted the industry to adequately protect public health. As she explained:

> Oil and gas will do whatever it takes to get what they want and make your life miserable. . . . What this does to neighborhoods, how it was once a friendly won-derful place to live but is divided because of the confusion, and they don't know who to believe. The stress it puts on you and the nights of no sleep and hearing fracking going on over a mile away. Oil and gas want to frack every square mile of Colorado and turn our beautiful state into one big industrial site. I feel I have to stand up and take action if I want to help save our state and planet. . . . We all do if we want a future for our children and grandchildren.

Groups with more state-directed tactics and transformative goals recognized the power the O&G industry had in setting the direction and scale of UOG development. Having been excluded from initial development decisions, they wanted to give communities greater agency to counter, and perhaps realign, the industry's agenda for shale gas production.

Advocating for Community-Led Development in Colorado. Other grassroots groups possessed less transformative but perhaps more realistic goals. They saw UOG production as insurmountable in Colorado and accepted the industry's power to set the agenda for energy production. They also believed community groups should develop better relationships with industry operators, mediated perhaps by legal counsel, to help determine the most practical, effective, and mutually beneficial method for siting UOG well pads. Disagreeing with coalitions like CCRN, such groups believed that the industry should be able to develop shale oil reserves, provided they also valued and incorporated the preferences of people living near production sites.

These strategic moves, while perhaps not optimal, created more demo-cratic opportunities for participation in decision-making and helped develop more open, communicative relationships with the industry, at least via the COGCC. For some citizens, these actions were the first, incremental steps toward realizing greater procedural equity and gaining some control, even if it led to shuffling around (potential) well pad sites. A founding member of one community group explained how her "middle-of-the-road path" created a greater sense of control, productivity, and reduced stress:

> Why don't I have a complete anti-industry approach? When you take a 100% stance against something, you will get tuned out. Regulators won't listen, no one will listen. You can't tell someone you don't want their product, they won't work with you. . . . And I knew the law. Changing the law takes too much time. . . .

And there are so many neighbors impacted without any benefits. . . . People listened to this middle-of-the-road path. I went from being up at 1:30 in the morning, crying, wondering if we should call the realtor and move. . . . Now, I'm in an office with Governor Hickenlooper, or [leaders] with COGCC, or any number of people with the industry who I can call and have lunch with.

While expressing mixed support for the industry, she was uncertain whether opposing them would lead to a more equitable outcome. She further observed: "I'm not totally sure that oil and gas development is a bad thing. We need it. . . . The mineral rights culture is so strong here and we knew it would prevail." While this activist wanted a greater voice in UOG activity, she did not find it worthwhile to quarrel with its fundamental existence.

Another northern Colorado resident, a recent retiree with a proposed well pad about 1,000 feet from his home, similarly advocated for middle-of-the-road approaches to regulating UOG activity. He acknowledged the industry's power and ample resources, despite compelling evidence that UOG activity could have substantial socio-environmental and public health impacts. He explained:

The O&G industry is strategically blind to the damage they cause to individuals' lives, environment and property. . . . At this point in time, neither the industry nor the state are remotely interested in the body of scientific data that suggest serious cause and effect of VOCs, noise, and light pollution, etc. All speakers appearing at the [recent public] meeting were immediately marginalized and universally ignored when discussing health issues. Decreased property values were also immediately dismissed. . . . So what is the right path for a citizen or citizen group to gain purchase? Learn the rules, the players and the jargon. Study and learn what they will listen to. . . . Our group was careful not to be sarcastic, hysterical, or distraught. We were always very respectful to everyone in all of the meetings. . . . It worked. At some point in time, health, environmental issues, common sense pad placement, and safety will be players at the table. Their time is yet to come.

While these observations showed how astute activists became in response to UOG production, they also illustrated how the industry retained the power to shape the terms of the discourse around development. Different tactics and goals divided grassroots efforts and created multiple groups whose energy and resources were often scattered, from which the industry certainly benefited.

Conclusion

Across the United States, the oil and gas industry expanded the pace and scale of unconventional production, while states, communities, and citizens scrambled to respond. Yet, few sociologists have attempted to examine procedural equity and justice in these processes. In states like Colorado, we need to know more about authentic participation in energy-related decisions, the roles communities and citizens play in policy formation and adoption, and "the decision processes distributed across complex social fields and bureaucratic arrangements" that uncover "how environmental policy and practices get shaped, are spread, are contested, and have consequences" (Hoffman and Ventresca 2002 p. xx).

In Colorado, UOG producers dominated decision-making processes—empowered by the legitimation of formal state support—even as Home Rule communities and citizens saw their voices diminished. Their backyards and playgrounds sat next to thousands of drilling rigs, but most citizens could do little more than wring their hands or hope for jobs. We have described and analyzed the key mechanisms structuring UOG producers' power over oil development in Colorado—namely, split estate laws and state preemption of local UOG regulations. We have illustrated how the state enabled UOG operators to retain control and effectively prevent communities and citizens from achieving procedural equity or local democratic control over energy production. Important environmental inequalities resulted, particularly in the form of compromised democratic governance for Home Rule communities, which decades ago passed ordinances to emphasize local control over land use decisions and prevent just this sort of state-corporate dominance. In the end, little has been done to ameliorate the structural disadvantages and inequalities that Colorado citizens have experienced, or to protect the rights of Home Rule communities to control local energy development.

Split estate legal provisions and the state's preference for mineral rights holders continue to institutionalize environmental inequalities in ways that have not been fundamentally challenged. Colorado privileges the ability of mineral owners to access their energy reserves regardless of the public sentiments, subdivisions, or schools on the surface. Instead, the desires of UOG producers have thus far trumped citizen concerns. One interviewee observed this inequality in decision-making power: "Towns [should] truly be able to exercise local control in setting up ordinances and procedures within their boundaries that really match the character of their community. The real sour apple in the barrel is oil and gas. Because as the deputy director of the COGCC told us in the first hearing that we had: 'I don't care what your local land use regulations are. If I want to drill at a certain location, I am going to approve the drilling. It doesn't matter what you want.'"

At stake are the rights of Colorado communities and citizens to (help) determine the scope of UOG activity near their homes, schools, and hospitals. Procedural equity demands that these stakeholders participate meaningfully and authentically in ways that allow them to shape criteria for UOG development and influence whether it occurs at all. Instead, profit margins and UOG producers' preferences have been given priority. Colorado communities and grassroots organizations have not yet experienced this level of inclusion. The proposed 2016 ballot initiatives held perhaps the last promise for local control, but those petitions were denied before making it onto the 2016 ballot.

To date, Colorado communities have lost ground. The distribution of costs and benefits associated with UOG production remain highly unequal: while state coffers and mineral rights owners benefited from the surging revenue during the UOG production boom, most Colorado communities saw only temporary benefits, if any. These communities also shouldered many of the post-boom costs associated with environmental remediation, the loss of local jobs, and the decline of economic activity in the region once prices fell for oil (Finley and Murphy 2015). The state Supreme Court's decision stripped municipalities of their Home Rule right to determine industrial activity within their jurisdictions, effectively overturning all attempts to realize Colorado communities' and citizens' constitutional rights to procedural equity. Even at the household level, where homeowners (particularly mineral rights holders) may sign leases and control development to an extent, these outcomes are often dissatisfying or alienating for lessees (Malin and DeMaster 2016). They also discourage more collective activism, as households and citizen groups with more pragmatic goals focus on their specific concerns.

Activists and communities need to be able to use legal mechanisms they have enacted, such as Home Rule ordinances, to challenge the awe-inspiring technology, infrastructure, and power held by the oil and gas industry (Wylie and Albright 2014). To balance UOG producers' power and technology, community organizations and leaders would benefit from encouraging, improving, and protecting higher-order tools that can enable people to network across spaces. For example, online tools like WellWatch have created virtual spaces where citizens affected by UOG production can compare their experiences with landmen, lease-signing, and environmental risks. These tools are necessary to connect citizens across physical locales and help counter the O&G industry's divide and conquer tactics.

The issue of local democratic governance over energy development extends beyond Colorado, as states like Pennsylvania, Texas, Oklahoma, Ohio, and Wyoming work to regulate and zone their own extensive UOG activity. Colorado has gained a reputation as a regulatory leader and other states pay close attention to Colorado's regulations and permitting processes. Yet, the political opportunities for Colorado communities and citizens to meaningfully and authentically participate in related decision-making have been seriously lacking. Recently, they have become even more limited by state decisions to reject community-led moratoriums on UOG production and even petitions for ballot initiatives to facilitate local control.

When considering the structural constraints and limitations placed on authentic public participation related to decisions about UOG production and its outcomes, Colorado's reputation as a regulatory and industry leader must be more critically examined and solutions more productively sought. This multilevel analysis offers initial evidence that more effective, collective, and institutionalized mechanisms for public participation must be created for Colorado communities and citizens to realize procedural equity. With Supreme Court decisions against them and ballot petitions rejected, perhaps community groups can now turn their attention to developing these institutions and, perhaps, a more thick democracy governing UOG production.

The challenges to this form of procedural justice remain deeply structured and embedded in Colorado's political economy, however. One closing example will suffice. Neighbors Affected by Triple Creek, one of many community groups that have formed in northern Colorado to contest UOG production near residential areas, worked to fight the proposed Triple Creek site. Triple Creek has record-setting dimensions: 22 wells extending horizontally for 2.5 miles (the current record in the state) and 22 oil tanks, plus other infrastructure and truck traffic, all directly in the middle of a large residential area with a neighborhood school. Neighbors Affected by Triple Creek and other citizens watched as the Greeley Planning Commission rejected Extraction Oil and Gas LLC's application to permit UOG production on the Triple Creek site, only to have the Greeley City Council overturn that decision and issue the permit. The group managed to hire legal counsel and file an appeal with the COGCC, based on the agency's new Large Urban Mitigation Area rules. Even with an effective case for appeal and a record 70 public comments supporting the appeal, however, the COGCC decided to support Extraction's application.

At this time, the longest horizontal wells ever drilled and fractured in Colorado will run under a highly populated residential area, school, and urban infrastructure. What this means for procedural equity and justice in the state remains a vital, continually unfolding issue. To wit, the legal counsel hired by Neighbors Affected by Triple Creek lamented the unenforceable nature of current regulations in a press release following the controversial permitting decision: "The rules state that large production facilities must be located 'as far as possible' from homes and the industry must use 'best available technologies.' Unfortunately, the state is allowing the oil and gas industry to determine whether it is as 'far as possible' from homes and what technologies they think are 'best' or even 'available.'"

Notes

1. The drilling boom in Colorado heavily impacted communities atop the Florence Basin and Green River shale formation along the state's western slope (Colborn et al. 2011; McKenzie et al. 2014, 2015).

2. The Niobrara Shale formation lies underneath portions of Colorado, between 3,000 and 8,000 feet below the surface. Drilling typically occurs at 7,000–8,000 feet. The Niobrara, like the Bakken, produces primarily oil.

3. A well setback is the minimum distance that must lie between the location of a well pad and any occupied buildings.

4. Still, O&G companies can apply for a variance in regulation when submitting any permit.

5. Approved recommendations encouraged collaborating with local government and comprehensive urban planning; increasing COGCC staff; introducing broader health initiatives; establishing an information clearinghouse; reducing truck traffic; approving new air quality rules; and developing an industry compliance assistance program (Keystone Policy Center 2015).

6. Over 100 Colorado cities have adopted a Home Rule charter (Bueche 2009). A Home Rule charter is a Colorado constitutional right that provides citizens the "freedom from the need for state enabling legislation and protection from state interference in both local and municipal matters" (Bueche 2009: 2).

7. Since 2013, only one permit has been issued for seismic testing in early 2015, attributed to a bust in regional UOG activity.

8. See http://cocrn.org/about-cocrn/.

9. CELDF, or the Community Environment Legal Defense Fund, provides legal counsel and education via Democracy Schools to municipalities that want to write their own ordinances to place moratoriums or bans on industrial activities deemed too risky to environmental and public health (see celdf.org).

References

Adgate, John L., Bernard D. Goldstein, and Lisa M. McKenzie. 2014. "Potential Public Health Hazards, Exposures and Health Effects from Unconventional Natural Gas Development." *Environmental Science & Technology* 48: 8307–8320.

Aguilar, John 2012. "Fracking Fury Reaches Fever Pitch in Erie." *Daily Camera*, January 7. Retrieved March 1, 2016. (http://www.dailycamera.com/ci_19696245).

Aguilar, John. 2016. "Greeley City Council Reverses Planning Commission's Oil and Gas Denial Verdict." *Denver Post*, March 8. Retrieved March 15, 2016. (http://www.denverpost.com/news/ci_29613923/greeley-wrestles-first-denial-oil-and-gas-permit)

Apple, Benjamin E. 2014. "Mapping Fracking: An Analysis of Law, Power, and Regional Distribution in the United States." *Harv. Envtl. L. Rev.* 38: 217–244.

Armstrong, Jonas. 2013. "What the Frack Can We Do: Suggestions for Local Regulation of Hydraulic Fracturing in New Mexico." *Natural Resources Journal* 53: 357–381.

Boudet, Hilary, Christopher Clarke, Dylan Bugden, Edward Maibach, Connie Roser-Renouf, and Anthony Leiserowitz. 2014. "'Fracking' Controversy and Communication: Using National Survey Data to Understand Public Perceptions of Hydraulic Fracturing." *Energy Policy* 65: 57–67.

Brady, William J., and James P. Crannell. 2012. "Hydraulic Fracturing Regulation in the United States: The Laissez-Faire Approach of the Federal Government and Varying State Regulations." *Vt. J. Envtl. L.* 14: 39–70.

Brasier, Kathryn J., Matthew R. Filteau, Diane K. McLaughlin, Jeffrey Jacquet, Richard C. Stedman, Timothy W. Kelsey, and Stephan J. Goetz. 2011. "Residents' Perceptions of Community and Environmental Impacts from Development of Natural Gas in the

Marcellus Shale: A Comparison of Pennsylvania and New York Cases." *Journal of Rural Social Sciences* 26(1): 32–60.

Bueche, Kevin G. 2009. *A History of Home Rule*. Denver, CO: Colorado Municipal League.

Carre, Nancy C. 2012. "Environmental Justice and Hydraulic Fracturing: The Ascendancy of Grassroots Populism in Policy Determination." *Journal of Social Change*, 4(1):1–13.

Clevenger v. Continental Oil Co., 149 Colo. 417, 369 P.2d 550 (1962).

Colborn, Theo, Carol Kwiatkowski, Kim Schultz, and Mary Bachran. 2011. "Natural Gas Operations from a Public Health Perspective." *Human and Ecological Risk Assessment: An International Journal*, 17(5): 1039–1056.

Colorado Energy Office. 2013. "Petroleum." Retrieved March 1, 2016. (https://www.colorado .gov/pacific/energyoffice/petroleum-0).

Colorado Oil and Gas Conservation Act 1951, §§ 34–60–101—129 (1951).

Colorado Oil and Gas Conservation Commission. 2001. "Staff Report." Retrieved March 1, 2016. (http://cogcc.state.co.us/documents/library/Staff_Reports/2k1/jan01/stats3.jpg).

Colorado Oil and Gas Conservation Commission. 2010. "Staff Report." Retrieved March 1, 2016 (http://cogcc.state.co.us/documents/library/Staff_Reports/2016/201603 _StaffReport.pdf).

Colorado Oil and Gas Conservation Commission. 2016a. "Staff Report." Retrieved March 1, 2016. (http://cogcc.state.co.us/documents/library/Staff_Reports/2016/201603 _StaffReport.pdf).

Colorado Oil and Gas Conservation Commission. 2016b. "Rules & Regulations." Retrieved March 1, 2016. (http://cogcc.state.co.us/reg.html#/rules).

Colorado Oil and Gas. (http://cogcc.state.co.us/about.html#/staffmaps).

Cook, Jeffrey J. 2015. "Who's Pulling the Fracking Strings? Power, Collaboration and Colorado Fracking Policy." *Environmental Policy and Governance* 25(6): 373–385.

Cowan v. Hardeman, 26 Texas 217 (1862).

Crowe, Jessica, Ryan Ceresola, and Tony Silva. 2015. "The Influence of Value Orientations, Personal Beliefs, and Knowledge about Resource Extraction on Local Leaders' Positions on Shale Development." *Rural Sociology* 80(4): 397–430.

Davis, Charles. 2012. "The Politics of 'Fracking': Regulating Natural Gas Drilling Practices in Colorado and Texas." *Review of Policy Research* 29: 177–191.

Davis, Charles. 2014. "Substate Federalism and Fracking Policies: Does State Regulatory Authority Trump Local Land Use Autonomy?" *Environmental Science & Technology* 48(15): 8397–8403.

Dunn, Sharon. 2016. "Oil, Gas Commission to Take Third Crack at New Rules on Monday; Weld Commissioners Go Their Own Way," *Windsor Now!*, January 23. Retrieved March 1, 2016. (http://www.mywindsornow.com/news/20277803-113/oil-gas-commission-to -take-third-crack-at).

Faber, Daniel. 2008. *Capitalizing on Environmental Injustice: The Polluter-industrial Complex in the Age of Globalization*. Lanham, MD: Rowman & Littlefield.

Finkel, Madelon L. (ed.). 2015. *The Human and Environmental Impact of Fracking: How Fracturing Shale for Gas Affects Us and Our World*. Santa Barbara, CA: Praeger.

Finley, Bruce. 2016. "Colorado Supreme Court Rules State Law Trumps Local Ban on Fracking." *Denver Post*, May 2. Retrieved May 3, 2016.

Finley, Bruce, and Joe Murphy. 2015. "Colorado Land Impact Oil and Gas Boom: Scars Spread and Stay." *Denver Post*, March 1. Retrieved March 1, 2016. (http://www .denverpost.com/environment/ci_27618385/colorado-land-impact-oil-and-gas-boom -scars).

Gerrity Oil and Gas Corp. v. Magness, 946 P.2d 913 (Colo. 1997).

Gregory, Kelvin B., Radisav D. Vidic, and David A. Dzombak. 2011. "Water Management Challenges Associated with the Production of Shale Gas by Hydraulic Fracturing." *Elements* 7: 181–186.

Gullion, Jessica Smartt. 2015. *Fracking the Neighborhood: Reluctant Activists and Natural Gas Drilling.* Cambridge, MA: MIT Press.

Heikkila, Tanya, Jonathan J. Pierce, Samuel Gallaher, Jennifer Kagan, Deserai A. Crow, and Christopher M. Weible. 2014. "Understanding a Period of Policy Change: The Case of Hydraulic Fracturing Disclosure Policy in Colorado." *Review of Policy Research* 31: 65–87.

Hoffman, Andrew and Marc Ventresca. 2002. *Organizations, Policy, and the Natural Environment: Institutional and Strategic Perspectives.* Stanford, CA: Stanford Business Books.

Howarth, Robert W., Renee Santoro, and Anthony Ingraffea. 2011. "Methane and the Greenhouse-gas Footprint of Natural Gas from Shale Formations." *Climatic Change* 106: 679–690.

Jacquet, J. B. 2014. "Review of Risks to Communities from Shale Energy Development." *Environmental Science & Technology* 48(15): 8321–8333.

Jaffe, Mark. 2013. "New Oil Boom Lurks in Denver-Julesburg Basin." *Denver Post*, December 15. Retrieved March 1, 2013. (http://www.denverpost.com/business/ci_24721419/new-oil-boom-lurks-denver-julesburg-basin).

Johnson, John Erich. 1998. "Gerrity Oil & Gas Corp. v. Magness: Colorado's Furtive Shift Toward Accommodation in the Surface-Use Debate." *Tulsa Law Journal* 33(3): 943–957.

Keystone Policy Center. 2015. Colorado Oil and Gas Task Force Final Report: Keystone, CO: Keystone Policy Center.

Ladd, Anthony E. 2013. "Stakeholder Perceptions of Socioenvironmental Impacts from Unconventional Natural Gas Development and Hydraulic Fracturing in the Haynesville Shale." *Journal of Rural Social Sciences* 28(2): 56–89.

Ladd, Anthony E. 2014. "Environmental Disputes and Opportunity-Threat Impacts Surrounding Natural Gas Fracking in Louisiana." *Social Currents* 1(3): 293–311.

Ladd, Anthony E. 2016. "Meet the New Boss, Same as the Old Boss: The Continuing Hegemony of Fossil Fuels and Hydraulic Fracking in the Third Carbon Era." *Humanity and Society* (http://dx.doi.org/10.1177/0160597616628908).

Lake, Robert W. 1996. "Volunteers, NIMBYs, and Environmental Justice: Dilemmas of Democratic Practice." *Antipode* 28(2): 160–174.

Malin, Stephanie A. 2014. "There's No Real Choice but to Sign: Neoliberalization and Normalization of Hydraulic Fracturing on Pennsylvania Farmland." *Journal of Environmental Studies and Sciences* 4(1): 17–27.

Malin, Stephanie A., and Kathryn Teigen DeMaster. 2016. "A Devil's Bargain: Rural Environmental Injustices and Hydraulic Fracturing on Pennsylvania's Farms." *Journal of Rural Studies* 47: 278–290.

Marvin v. Brewster Iron Mtn. Co., 55 NY. 538 (1874).

McKenzie, Lisa M., Ruixin Guo, Roxana Z. Witter, David A. Savitz, Lee S. Newman, and John L. Adgate. 2014. "Birth Outcomes and Maternal Residential Proximity to Natural Gas Development in Rural Colorado." *Environmental Health Perspectives (Online)* 122(4), 412.

McKenzie, Lisa M., Ruixin Guo, Roxana Z. Witter, David A. Savitz, Lee S. Newman, and John L. Adgate. 2015. "Birth Outcomes and Maternal Residential Proximity to Natural Gas Development in Rural Colorado." *Everyday Environmental Toxins: Children's Exposure Risks* 122(4): 412–417.

Minor, Joel. 2014. "Local Government Fracking Regulation: A Colorado Case Study." *Stanford Environmental Law Journal* 33(1): 59–120.

Mohai, Paul, David Pellow, and J. Timmons Roberts. 2009. "Environmental Justice." *Annual Review of Environment and Resources* 34: 405–430.

Negro, Sorell E. 2012. "Fracking Wars: Federal, State and Local Conflicts over the Regulation of Natural Gas Activities." *Zoning and Planning Law Report* 35(2): 1–13.

Nolon, John R., and Steven E. Gavin. 2013. "Hydrofracking: State Preemption, Local Power, and Cooperative Governance." *Case Western Reserve Law Review* 63(4): 994–1039.

Ogburn, Stephanie Paige. 2014. "Colorado's Local Control of Oil and Gas Is a Question of Degrees," *Colorado Public Radio*, August 4. Retrieved March 1 2016. (http://www.kunc.org/post/colorado-s-local-control-oil-and-gas-question-degrees#stream/0).

Opsal, Tara, and Tara O'Connor Shelley. 2014. "Energy Crime, Harm, and Problematic State Response in Colorado: A Case of the Fox Guarding the Hen House?" *Critical Criminology* 22(4): 561–577.

Perry, Simona L. 2012. "Development, Land Use, and Collective Trauma: The Marcellus Shale Gas Boom in Rural Pennsylvania." *Culture, Agriculture, Food and Environment* 34: 81–92.

Perry, Simona L. 2013. "Using Ethnography to Monitor the Community Health Implications of Onshore Unconventional Oil and Gas Developments: Examples from Pennsylvania's Marcellus Shale." *New Solutions: A Journal of Environmental and Occupational Health Policy*, 23(1), pp. 33–53.

Randall, Bob. 2012. "Colorado Oil and Gas Conservation Commission Rules and Regulations." Presented at the BLM Colorado Resource Advisory Councils Meeting, March 7, 2012. Retrieved March 1, 2016. (http://www.blm.gov/style/medialib/blm/co/resources/resource_advisory/2012_super_rac.Par.83149.File.dat/COGCC.pdf).

Reasonable Accommodation Provision, Ch. Colorado Oil and Gas Conservation Act, 314 Art. 60 Title 34 (2007).

Rinfret, Sarah, Jeffrey J. Cook, and Michelle C. Pautz. 2014. "Understanding State Rulemaking Processes: Developing Fracking Rules in Colorado, New York, and Ohio." *Review of Policy Research* 31: 88–104.

Schlosberg, David. 2007. *Defining Environmental Justice: Theories, Movements, and Nature.* New York: Oxford University Press.

Shapiro, Jennifer, and Barbara Warner. 2013. "Fractured, Fragmented Federalism: A Study in Fracking Regulatory Policy." *Publius: The Journal of Federalism* 43(4): 474–496.

Svaldi, Aldo. 2016a. "Oil under $30 a Barrel Carries Dangers for Colorado Economy." *Denver Post*, January 25. Retrieved May 3, 2016.

Svaldi, Aldo. 2016b. "Colorado Economy Facing Headwinds in Coming Years." *Denver Post*, June 20. Retrieved June 21, 2016.

Theodori, Gene L. 2009. "Paradoxical Perceptions of Problems Associated with Unconventional Natural Gas Development." *Southern Rural Sociology* 24(3): 97–117.

Toan, Katherine. 2015. "Not under My Backyard: The Battle Between Colorado and Local Governments over Hydraulic Fracturing." *Colorado Natural Resources Energy & Environmental Law Review* 26(1): 1–67.

Turner v. Reynolds, 23 Pa. St. 199 (1854).

Ubbelohde, Carl, Duane A. Smith, and Maxine Benson. 2006. *A Colorado History.* 9th ed. Portland, OR: WestWinds Press.

University of Denver, DU Report. 2015. "Review of Oil and Gas Industry and the COGCC's Compliance with Set-Back Rules." University of Denver Sturm College of Law. Retrieved on May 5, 2016. (http://www.law.du.edu/documents/student-law-office-clinical-programs/ELC-Form-2a-Executive-Summary.pdf).

U.S. Energy Information Administration. 2016. "Shale in the United States." Retrieved May 1, 2016. (https://www.eia.gov/energy_in_brief/article/shale_in_the_united_states.cfm).

Vidic, Radisav D., Susan L. Brantley, Julie M. Vandenbossche, David Yoxtheimer, and Jorge D. Abad. 2013. "Impact of Shale Gas Development on Regional Water Quality." *Science* 340(6134): 1235009-1–1235009-9.

Walsh, Patrick J., Stephen Bird, and Martin D. Heintzelman. 2015 "Understanding Local Regulation of Fracking: A Spatial Econometric Approach." *Agricultural and Resource Economics Review* 44(2): 138–163.

Weber, B. A., J. Geigle, and C. Barkdull. 2014. "Rural North Dakota's Oil Boom and Its Impact on Social Services." *Social Work*, swt068.

Wilber, Tom. 2015. *Under the Surface: Fracking, Fortunes, and the Fate of the Marcellus Shale.* Ithaca, NY: Cornell University Press.

Willow, Anna J. 2015. "Wells and Well-being: Neoliberalism and Holistic Sustainability in the Shale Energy Debate." *Local Environment* 21(6): 1–21.

Willow, Anna and Sara Wiley. 2014. "Politics, Ecology, and the New Anthropology of Energy: Exploring the Emerging Frontiers of Hydraulic Fracking." *Journal of Political Ecology* 21: 222–236.

Wiseman, Hannah Jacobs. 2009. "Untested Waters: The Rise of Hydraulic Fracturing in Oil and Gas Production and the Need to Revisit Regulation." *Fordham Environmental Law Review* 20: 115–170.

Wylie, Sara, and Len Albright. 2014. "WellWatch: Reflections on Designing Digital Media for Multi-Sited Para-Ethnography." *Journal of Political Ecology* 21: 320–348.

9

Citizen Resistance to Oil Production and Acid Fracking in the Sunshine State

PATRICIA WIDENER

Introduction

South Florida and the Greater Everglades are an emerging frontier for a form of unconventional oil and gas extraction known as acid fracturing or "acid fracking." Although South Florida has a history of low-volume oil production, new sites are being proposed for exploration and old sites for well stimulation that could use acid and/or hydraulic fracking and horizontal drilling to extract oil deposits from the region's porous limestone bedrock. As latecomers to the contested practices of fracking, a growing coalition of residents, activists, and local politicians have mobilized to support multiple town and county fracking moratoriums or bans. In addition, local and national conservation groups have united in a lawsuit to protect public lands. This broad network is both informed and inspired by impacted communities nationwide, regional campaigns of resistance, threats to conservation lands and endangered species, and increased scientific and journalistic scrutiny on unconventional oil and gas techniques.

Until now, Florida had never been considered part of the battleground over fracking. For journalists, fracking in the region conjures up the image of drilling into a "wet sponge" (Alvarez 2016: A10). The prospect alone has generated national headlines: "Oil Prospectors Seek Their Next Big Strike in South Florida's Everglades" (Bekiempis 2014); "Big Oil Eyes Florida's Public Lands, Plans to Drill in the Everglades" (Volz 2014); and "Could Florida Become the New Fracking Frontier?" (Drouin 2014). One reporter questioned the logic of drilling by asking: "Think Florida, and relaxing images come to mind of white sandy beaches . . . palm trees . . . oranges . . . sunny warm weather . . . oceanfront hotels . . . gentle waves lapping at the coastline . . . and oil production?" (Spencer 2013).

This chapter contributes to the sociological literature on unconventional energy development in three key respects. First, it explores the energy frontier of South Florida and the Greater Everglades, a nationally recognized wetlands ecosystem, endangered species habitat, and tourist destination. Second, it explores the extractive technique of acid fracking, a technology that has received little attention to date and is still trying to be understood in Florida. Third, it examines how a coalition of affected residents, environmental and climate groups, and local politicians developed multiple narratives of resistance by utilizing the firsthand accounts and mobilization efforts of affected communities nationwide in combination with their own regional knowledge of South Florida. In this manner, resident-activists of South Florida became part of a larger state- and nationwide effort that connected local concerns about the socioecological impacts of fracking with documented national accounts.

Based on qualitative fieldwork, this chapter demonstrates how three interlocking narratives of resistance emerged in the ensuing debate over acid and hydraulic fracking and expanded oil development in Florida. Despite their limited regional experience with oil extraction, residents adopted a narrative that replicated experiences in other U.S. communities regarding threats to public health, water quality, and water volume. The second narrative was a regional one that emphasized South Florida's distinctive features, including the Greater Everglades, Florida panthers and panther habitat, sinkhole-prone landscapes, a nature-based tourism economy, rising sea levels along the state's low-lying coastline, as well as the state's potential for expanded solar energy production. The third narrative focused on the overarching political debates, conflicts, and collaborations that developed between state and local officials, and between citizens and the state on energy and climate change. These narratives helped drive an antifracking campaign that served to energize and publicize a wider controversy in the state over how to protect its unique environmental amenities, expand its reliance on renewable energy sources, and mitigate the growing impacts of climate change.

Literature Review

Community Health

To date, groundwater contamination and public health have dominated community fears about hydraulic fracking (Hannigan 2014), supported in part by lived experience and scientific studies (Briggle 2015; Gullion 2015). With fracking, epidemiological studies have identified specific health risks such as preterm births, high-risk pregnancies, and infant congenital heart defects as associated with the presence, density, and proximity of shale gas wells (Casey et al. 2016; McKenzie et al. 2014). Though specific pathways have been difficult to identify, studies cite the entire multistage production process (including the use of heavy diesel equipment and trucks to prepare a site, as well as transport necessary water supplies and equipment to the well pad) as a threat to communities and the environment (Casey et al. 2016; McKenzie et al. 2014).

Fracking for natural gas has drawn attention to its health risks, but the socio-ecological risks associated with petroleum extraction were known well before the gas boom of the last decade (see Allen 2003; Horowitz 2007; Lerner 2005, 2010; O'Rourke and Connolly 2003; Ott 2005; Tamminen 2006). Indeed, scholars have long identified the important role of citizen science and local, lay, or popular knowledge in identifying the health risks of industrial toxins and challenging state and industry assurances (Brown 1992, 2007; Corburn 2005; Fischer 2000; Frickel and Gross 2005; Gunter and Kroll-Smith 2007). Often, community experiences and fears were ignored or downplayed. Even now with fracking, when companies submit their proposals for state decisions, citizen concerns and questions are generally given far less primacy than industry claims about the opportunities and benefits (Hannigan 2014).

In the past, states, counties, and communities with a history of production and employment in the extractive industries typically offered greater degrees of accommodation, license, and acceptance to expanded development (Boudet et al. 2016; Freudenburg and Gramling 1994; Widener 2011). Today, even residents in oil and gas states like Texas have become "reluctant activists" (Gullion 2015: 22). Although these activists might have mocked environmentalists and accepted oil and gas development in the past, many have now become motivated to organize against unconventional fracking and industry development near their backyards (Gullion 2015). Their activism is based on their own lived experiences, as well as the firsthand accounts of other citizens in impacted communities. The documentary *Gasland* (Fox 2010) in particular has served as an important source of resistance in major shale regions and represents perhaps the "single identifiable source responsible for elevating fracking to a broad level of public awareness and concern" (Hannigan 2014: 185). Research by Vasi and colleagues (2015) found *Gasland* to be extremely influential in generating public discussion and supporting community mobilization efforts against shale development (see also Vasi, this volume).

Regional Impacts

Beyond widespread concerns over fracking's impacts on local water quality, availability, and human health, communities have also organized around more generalized threats to regional farmlands, playgrounds, rural tranquility, livestock populations, and induced seismic activity and earthquakes from wastewater injection (e.g., Briggle 2015; Food and Water Watch 2014; Gullion 2015; Ladd 2014; Malin and DeMaster 2016). Despite the growth of regional frames of resistance, few studies have examined how community residents view the risks that fracking poses for iconic landscapes, wildlife, or tourist economies. The concept of socioecological inequalities serves as a frame to examine how humans, animals, and ecosystems "intersect to produce hierarchies" of the privileged and exploited (Pellow 2014: 7).

Climate change is another key issue. Although links between oil and gas development and global climate change are well established, research has not identified how local experiences of a changed climate inspire resistance to fracking. While Widener (2013) found that climate change represented a concern cited by citizens in their resistance to an oil refinery, Hilson (2015) and Neville and Weinthal (2016) found that antifracking campaigners were successful in linking the issue to opposing local siting plans. None of these works associated resistance to local climate-related impacts and local adaptation strategies. While national environmental campaigns have helped identify the connection between energy development and global climate change, it is in affected communities like South Florida where the actual experiences with climate change are most deeply felt. Climate change is no longer theoretical for many regions of the state.

Established regional economies—such as agriculture and tourism—also help determine whether fracking is met with acceptance or resistance. Farmers, for example, are often forced to accept production on or near their fields as a "devil's bargain," where they feel caught between the need for additional income and fracking's environmental risks (Malin and DeMaster 2016: 1). As for tourism, months before the BP *Deepwater Horizon* disaster in the Gulf of Mexico, a restaurant owner in the Florida Panhandle launched the first "Hands across the Sands" rally to demonstrate against offshore drilling along Florida's coastline, which was being debated in Washington, D.C. at the time. Since then, "Hands across the Sands" has become an annual, international event that includes "Hands across the Land" for those opposed to new drilling sites, fracking, or pipelines. Thus far, Florida's coastlines have been spared from offshore drilling due in large part to arguments based on the state's tourism and fishing industries. Yet, many tourist destinations also have a history of benefiting from the industry's sponsorships of music festivals, museums, aquariums, sporting events, and other activities (Barrett 2014; Jorgensen 2014; Stimeling 2014; Widener 2009). Following the *Deepwater Horizon* disaster, Florida was awarded a $3.25 billion settlement in 2015 to assist with the recovery of its marine and coastal environments, as well as to promote its tourism and fishing industries (Pittman 2015). Though counterintuitive, oil and tourism interests can occasionally coincide (Widener 2009).

Florida's Inclusion in a Protracted Age of Oil

One way to understand the expansion of fracking, despite its documented risks, is to consider the history and influence of the oil industrial complex—a privileged web of corporations, public offices, public agencies, and geopolitical interests (Widener 2011: 8).[1] This section organizes exploration and production into three waves of development to better understand the scale and impact of the political economy of oil today and to situate Florida within the wider context of recent fracking activities.

The first wave of commercial oil production began in Pennsylvania in the 1850s and helped make the United States the world's largest producer until the 1950s. The second wave, beginning in the 1950s and 1960s, was a period of expansion around the world in terms of new oil discoveries, as well as new private and national companies dedicated to energy development. Cable (2012) cites 1945 as the beginning of the petro-dependent period in the United States and documents the growth of federal and state policies and regulations that supported the goals of national energy production. This second wave was also a time of exploration and production throughout Latin America and Africa, as well as offshore drilling in North America and Europe. Routine disasters often resulted from the riskier technologies and industry practices in these energy frontiers, producing what Perrow (1984) referred to as an increasing number of "normal accidents."

The third wave, from the late 1990s onward, has involved considerably more aggressive and riskier projects in what has been called the "Third Carbon Era" (Ladd 2016), or the "protracted age of oil" (Widener 2013: 834). This period is one marked by proposals for large-scale projects involving infrastructure development (such as the Keystone Pipeline System), refinery expansion, new drilling and fracking technologies, and new locales, including farther offshore and deeper waters, tighter rock formations, and harsher climates (such as the Arctic). Yet, these contemporary extraction techniques are not supported by effective, advanced, or adequate risk reduction or cleanup procedures in the event of disasters or routine accidents (Freudenburg and Gramling 2011; Lustgarten 2012).

Although Florida has never been a top-tier destination for the oil industry in any of the phases of exploration and production, the state has a history of small-scale oil production. In the first wave, almost 80 dry holes were drilled between 1901 and 1939 before onshore oil production began in 1943 in Collier County (Spencer 2013). Florida reached peak production in 1978 (in the middle of the second wave), but has not experienced a new discovery since 1988 (Spencer 2013). For context, Florida produced approximately two million barrels of oil in 2013, equivalent to a day's production in Texas (Allen 2014a). It is not yet clear what will transpire in South Florida during this third phase of expansion.

In the event that onshore fracking were to be permitted in South Florida, the region may experience acid fracking of its softer limestone substrate and/or hydraulic fracturing at greater depths. In the current vernacular, acid fracking, fracture acidizing, and matrix acidizing are modified versions of pressurized fracturing, combined with the use of acid (primarily hydrochloric acid or a hydrochloric and hydrofluoric acid blend) to dissolve underground limestone or sandstone formations to release oil and gas deposits, rather than break open harder shale or sedimentary rock (see American Petroleum Institute 2014; Earthworks n.d.). As a distant precursor to today's fracking techniques, the first patent to "fracture" rock for oil was in 1866, while hydrochloric acid was first used in 1932 (Gold 2014). Acidizing and acid stimulation also predate contemporary fracking as tools to open wells or release stored hydrocarbons, while acid washing is a cleaning technique for oil wells (see American Petroleum Institute 2014).

Unlike hydraulic fracturing, acid fracking combines pressurized fracturing techniques with acid stimulation to dissolve soft rock formations to release oil and gas to the wellbore. The pressure with acid fracking also varies from hydraulic fracturing. In the case of matrix acidizing, acid is pumped into a well at low pressures, dissolving sediments and mud solids; while in fracture acidizing, acid is pumped at higher pressures, but still lower than those used during hydraulic fracking (Earthworks n.d.). In a fracking regulation bill passed in California in 2013 (SB4), both hydraulic fracturing and acid well stimulation "at any applied pressure" and/or "in combination with hydraulic fracturing treatments" will require groundwater monitoring.[2] Yet in Florida, there are no current state regulations specific to horizontal or directional drilling, or acid or hydraulic fracturing (Allen 2014b). In addition, this research also points to how many citizens in Florida are still confused about these new and diverse technologies.

With both acid and hydraulic fracturing, the water volume, water pressure, and acid and chemical compositions and concentrations vary by company and location. According to two Florida-based geophysicists, "At this point, there are more questions than answers. The specifics of proposed fracking in Florida are complicated by the very different regional geology of the peninsula" (Russo and Screaton 2016). As of this writing, it is not clear what would happen if the "wet sponge" that Florida represents were to be fracked, as there are no national comparisons. Little is known about the pressure thresholds, chemical concentrations, and waste products or gases that might be brought to the surface, nor has such information been disclosed to the public.

Opening the Door to Oil in South Florida

For the purposes of this research, South Florida includes the four southern-most counties of the peninsula (Broward, Collier, Miami-Dade, and Mon-roe). Each county has a coastline on either the Gulf of Mexico or the Atlantic Ocean, and to travel from coast to coast requires a drive along one of two roads through the Everglades. The region is recognized for its wetlands, resident and migratory wildlife, sandy beaches, and nature-based tourism. The Everglades ecosystem begins north of these southernmost counties and flows southward to the Gulf of Mexico and Florida Bay, the southern tip of the peninsula. Miccosukee tribal lands, the Big Cypress National Preserve, and the Everglades National Park also constitute a large portion of the region. Agriculture from the north and residential and business development along both coasts encroach on three sides.

Underground, South Florida's oil deposits are found in the Sunniland Trend that spreads onshore and offshore from Fort Meyers on the Gulf Coast to Miami (Bekiempis 2014; Reed 2015). According to the Florida Department of Environmental Protection (DEP), five permit applications for surveying, exploring, drilling, or wastewater injection (not specific to fracking) have been filed in two of the four southernmost counties, along with other applications throughout the state.[3] Depending on what the exploratory data reveal, these applications could signal the beginning of a deluge of drilling sites and/or a robust, southern campaign to ban fracking in the protected public areas, the region, or the state.

This study focuses on four proposed exploration or drilling sites that reflect a range of industry speculations and anti-drilling collaborations. The first two attempts were initiated in 2013 by Texas-based Dan A. Hughes Company near Naples, on the western edge of the Greater Everglades. In tandem, these two incidents (one involving horizontal drilling and one involving fracking) alerted the region to the possibility of unconventional technologies in the Greater Everglades. In the first location, residents learned of the company's application to begin drilling when they received a notice in the mail that they were now living in an area that required contingency plans for possible evacuation in the case of an explosion or leak (Allen 2014a; Bekiempis 2014). The well was to be approximately 1,000 feet from the nearest home and residential area, and closer than any previous horizontal drilling operation in South Florida history (Conlin and Grow 2013).

In the second location, the company began a fracking operation but did so without proper DEP permits. Although acid had been used before to loosen limestone, acid had not been used in pressurized applications that would constitute "acid fracking," thereby eliciting the DEP to produce a cease and desist order (Allen 2014b). While the company disputed whether they had technically fracked the well, a DEP consultant determined that a fracking operation and "some type of acid stimulation" or "an unspecified acid treatment" had indeed occurred at Collier-Hogan 20–3H well site (ALL Consulting 2014: 9, 29). Activists labeled this incident as an "acid fracking" operation, which galvanized residents to write letters, attend town hall meetings, organize, and become informed about the industry's newfound interest in the region. Eventually, the Dan A. Hughes Company withdrew its interests.

The third case began in mid-2014, but only became public knowledge a year later when Miami-based Kanter Real Estate proposed exploratory drilling on the eastern edge of the Everglades in a Water Conservation Area. Previously, Kanter had tried to develop a residential complex on its property, but was blocked by the South Florida Water Management District that controls the flow of surface water as part of a Greater Everglades restoration project (Fleshler 2015). As of June 2016, Kanter had applications pending with the DEP, Army Corps of Engineers, and South Florida Water Management District. In a county that has banned fracking, commissioners have indicated that they will not provide a zoning change permit (Fleshler 2016a). Also in opposition, the city closest to the proposed site passed a resolution urging the state or federal government to purchase the land for the Everglades National Park.[4] If oil is found, it is not clear whether acid or hydraulic fracking or horizontal drilling will be employed. This proposal linked communities, activists, and local politicians on the east coast with their counterparts on the west coast who mobilized against the first two production events.

The fourth proposal occurred in late 2014 in Big Cypress National Preserve, which represents more wilderness territory and wildlife habitat than the previous three locations. In Big Cypress, property owners and mineral rights holders are allowed to pursue their interests on public lands, including energy extraction (Reed 2015). Likewise, Native American communities of Miccosukee and Seminole descent maintain traditional land practices and access to the resources of Big Cypress. In this case, the Florida-based Collier Resources group, whose family wealth is in real estate, agriculture, and oil, retains mineral rights in parts of their namesake Collier County and Big Cypress (Bekiempis 2014; Drouin 2014). Given those rights, the Collier group leased access to Burnett Oil of Texas, which applied to the National Park Service (NPS) for a seismic survey permit for 70,000 acres. Currently, there are 10 active wells operated by California-based BreitBurn Energy in Big Cypress (Nielsen 2016). Although park stipulations indicate that exploration cannot exceed more than 10% of the preserve, it currently accounts for about 1% (Nielsen 2016). Whether the 10% limit reflects surface imprint or underground horizontal drilling is unclear.

For the seismic survey permit, NPS requested comments on whether the public wanted: (1) no change or no oil seismic survey; (2) a seismic survey using ground vibrations; or (3) a seismic survey using underground explosive charges. After the review of their own environmental assessment and public comments (which were not made publicly available), NPS decided that there would be no significant impact in conducting seismic surveys using "thumper trucks" to pound the surface to create underground vibrations. According to NPS (2016): "After extensive agency and tribal consultation and analysis of public comments, the agency found that the proposed seismic survey would pose no significant environmental impacts. The National Park Service considered the proposal because the federal law specifically grants owners of non-Federal oil and gas rights access to preserve lands to 'provide reasonable use and enjoyment' of those private interests." In response, six environmental groups filed a lawsuit to block NPS's approval (Fleshler 2016b), and some have called for a federal buyout of mineral rights in Big Cypress (Nielsen 2016).[5]

These four site proposals nearly encircle the Greater Everglades. Two were on its western flank, one on its eastern border, and the other was located in Big Cypress, near its center. As of June 2016, acid fracking was used once in Collier County in 2013 without proper approval, and hydraulic fracking was used twice in the Panhandle in 2003 (Eastman 2015). BreitBurn Energy began horizontal drilling in Collier County in 2010 (OGJ editors 2010), before the residential proximity and fracking operation of Dan A. Hughes alerted local residents and launched a campaign of resistance.

Data Collection

This study sought to understand how a region with little oil extraction or fracking experience would assess the potential impacts and craft a narrative of resistance. Data included participant observation of speeches, presentations, and public comments at eight public events attended between October 2014 and February 2016. Additional data were acquired from discursive documents, such as media reports, activist handouts, newsletters, websites, and emails. Public events included a rally and legislative meetings at the Florida capitol, a town hall meeting near a proposed site, a beachside public demonstration in front of the governor's private residence, an informational meeting to link local and state activists, a poverty and climate summit that included a round-table on fracking, and three community film screenings and discussions on *Gasland II*, *Groundswell Rising*, and *Great Frack Forward*, led by local environmental or climate groups.

I observed public gatherings, met or spoke with activists informally, took extensive notes at each event, and reviewed media accounts and available documents to assess the central arguments against fracking and oil development in South Florida. For example, at one town hall meeting, there were three formal presentations followed by 21 public comments in an audience of more than 200 people. Although only 13 men and eight women spoke during the comments period, they represented some of the ethnic and occupational diversity of the community. Fifth-generation residents of the Everglades and new residents for whom English was a second language spoke and were joined by regional representatives of national environmental groups. At the rally in Tallahassee, approximately 160 people took buses from across the state to meet in small groups with their local representatives or senators, as well as hear from experts and activists in the state and resident-activists from New York and Pennsylvania.

With the collected data, an assessment of how impacts were framed by citizens was conducted. Researchers of social movements, collective action, and environmental conflicts have used frame analyses to identify a campaign's key arguments or persuasive narratives, which are advanced to influence understanding, motivate action, and depict alternative solutions (see Benford and Snow 2000; Gunter and Kroll-Smith 2007; Hannigan 2014; Hilson 2015; Messer, Shriver, and Kennedy 2009; Robinson 2009; Widener and Gunter 2007). To be credible, frames should be consistent, originate from trustworthy claims-makers, and resonate with a person's lived experiences (Benford and Snow 2000). When frames are inconsistent, mobilization efforts may break down (Robinson 2009). For this study, data were revisited multiple times and reanalyzed as the analysis developed and new data were collected. Once initial frames were identified, data were reviewed and individual frames were organized into three scalar narratives.

Analysis and Discussion

Efforts to resist fracking started with the first fracking operation and developed into a network of at least 35 organizations representing citizen-led and community-based groups, as well as established climate and environmental organizations across the state. The broad campaign's initial and primary target was the state, followed by the NPS after the Big Cypress proposal was announced. The analysis found that several interrelated themes manifested into three narratives of resistance. With limited regional experience, the first narrative was a national one that utilized national lessons to argue for the protection of community health and the region's water supply. A second, regional narrative developed to emphasize South Florida's unique environmental qualities related to the Greater Everglades, including its panthers and nature-based tourism. The third narrative occurred within state boundaries and spoke to escalating strains between residents, local politicians, and the state on the interconnected and socioecological issues of oil development, climate change impacts, and opportunities for solar energy development.

National to Local Health Concerns

First and foremost, resident-activists feared that fracking and oil development would damage the health of the public and the greater ecosystem by unleashing hazardous and toxic chemicals into the water, air, and soil. Community film screenings, regional health experts, and out-of-state activists who toured the state with firsthand accounts informed resident awareness and galvanized resistance. At one rally to encourage the state legislature to reject fracking-friendly bills, speakers from New York and Pennsylvania spoke of skin lesions, migraines, tumors, and spontaneous nosebleeds. One activist told the crowd: "Fracking is a public health issue. Make it a national issue." A Tampa-based physician with Physicians for Social Responsibility added: "We now know without a doubt that fracking, like cigarettes, damages health. . . . [Let's] keep the fossil fuels in the ground. This is the best prescription for a healthy Florida" (see Ross 2016). After presenting a number of scientific studies, the physician added: "We don't need another study. [After tens of thousands of fracked wells,] they haven't been able to do it safely. There is no safe way."

In ongoing efforts to appeal to legislators, Urban Paradise Guild, a Miami-based organization, wrote to the Senate Appropriations Committee requesting them "to call on those who have lived with the consequences of fracking to testify. Their personal experiences qualified them to speak as experts on human impacts."[6] One visiting activist spoke to the House Democrats about the health and safety violations witnessed in Pennsylvania (Klas 2016a). When the group Floridians against Fracking was updating its Facebook page on county discussions to ban fracking (June 2016), it was also updating the national coverage on health risks, including comments by a Wyoming farmer who identified the family's experience of respiratory and neurological problems. Even without definitive lived experiences, resident-activists drew sufficient information from the growing body of citizen- and peer-reviewed science to doubt industry and regulatory assurances and to organize for prevention over prescription. Meanwhile, the Florida DEP was seeking to raise permissible levels on chemicals discharged into Florida's waters, a move suggested by the Florida chapter of Physicians for Social Responsibility as accommodating the fracking industry (Burlew 2016).

Protecting the Greater Everglades

The second narrative emphasized what was truly unique about South Florida: the Greater Everglades ecosystem. Despite its distinction, the

Everglades is also a long-running political and economic body of water (Grunwald 2006). Long ago a coalition of state elites put the region "on a path of self-destruction" in their efforts to control it for future agricultural production and population growth (Gramling and Freudenburg 2013: 652). Certainly, the Everglades has been filled in, built upon, diverted, canaled, and polluted by residents, developers, and farming interests. Yet, the ecosystem is also a national treasure, icon, and a critical freshwater source for approximately seven million people (Bekiempis 2014). In recognition of its value (and previous destruction), the U.S. Senate approved a $7.8 billion restoration plan in 2000. The plan is a collaborative state and federal effort to protect its 2.4 million acres of wetlands for ecological preservation, wildlife protection, and water security. To date, the plan remains a work in progress that has led to successes such as cleaner water in some areas, while failing to address other problems like agricultural runoff and runaway development (Grunwald 2006). If successful, the four proposals would lead to drilling in, near, and beneath the Greater Everglades.

Many in South Florida had concerns that fracking would contaminate the region's underground water labyrinth, deplete the region's fragile ground and surface waters, or result in the development of an extensive energy infrastructure (e.g., pipelines, roads, drill pads) that would impede the water's southward flow. A predominant question for many residents was: how much more human impact could this body of water take?

As shown in Figure 9.1, "It's all about the Water" was the impetus for a beachside demonstration in front of the governor's personal residence in October 2014. At this rally, the groups Stonecrab Alliance and ReThink Energy Florida, along with regional chapters of Food & Water Watch and Clean Water Initiative, asked residents to bring their "troubled water" that was threatened by the possibility of drilling for a traditional blessing from the Miccosukee Otter Clan. One of the organizers highlighted the essence of the rally in a protest poster: "Can't Clean Up an Aquifer: Don't Drill" (Dwyer 2014). Water tainted by fracking in the Marcellus Shale was also brought from Pennsylvania to be displayed on the steps of the old capitol building. These acts of displaying contaminated water from the Marcellus Shale and blessing local water reflected what Gullion (2015: 158) has termed "performative environmentalism," or environmental actions where the emotions, suffering, and fears of communities are put on public display, in contrast to the formal, "rational," and "scientific" presentations by industry or state officials.

The Florida Panther

The Greater Everglades is also home to the Florida panther and the Florida Panther National Wildlife Refuge. Unlike alligators, panthers are few in number and rarely seen. The South Florida Wildlands Association,

FIGURE 9.1 Antidrilling activists in front of the governor's private residence (Naples, October 2014). [Credit: P. Widener]

which joined the Big Cypress lawsuit, estimated that only about 160 of the endangered panthers still exist in the region, while those remaining could be startled from their nesting dens by seismic testing or killed by increased truck traffic. A spokesperson for the Florida Fish and Wildlife Commission disputed this position, arguing that "panthers have learned to coexist with all kinds of disruptions" (Allen 2014a; see also Volz 2014). Paradoxically, even the NPS (n.d.), which had approved seismic surveying in Big Cypress, promoted the value of the preserve for its protection of panther habitat: "Protecting over 729,000 acres of this vast swamp, Big Cypress National Preserve . . . [is] home to a diversity of wildlife, including the elusive Florida panther." Over time, antifracking groups, which initially identified panther protection as one justification for oil opposition, expanded their campaign in their lawsuit to embrace less charismatic species, including the Florida bonneted bat, eastern indigo snake, wood stork, and red-cockaded woodpecker (Fleshler 2016b).

Sinkholes

Another issue was the possibility of fracking-, wastewater-, or industry-induced sinkholes, similar to the seismic activity and earthquakes occurring in other parts of the country like Oklahoma (see Mix and Raynes, this volume).

While the process of fracking and/or wastewater injection into underground disposal wells does not trigger earthquakes in all locations, it has produced increased seismicity in many areas (Rubinstein and Mahani 2015). Inferring from Oklahoma's experiences, resident-activists of South Florida reasoned that industry-induced sinkholes could be a foreseeable risk worth avoiding. Although sinkholes are not new or specific to Florida, in recent years the state has made national news when areas of ground have collapsed quickly, swallowing homes, buildings, or parking lots. During one heavy withdrawal of water for agricultural purposes, 140 sinkholes occurred in neighboring communities (Barnett 2011: 5). Whether there was a geologically sound connection between Oklahoma's industry-induced earthquakes and the possibility of triggered-sinkholes in Florida, many believed that the two could be connected due to pressurized acid stimulation, intensive water extraction, or fluid injection or reinjection. In one Facebook post, someone wrote that with the use of hydro-chloric acid "you think parts of this State have Sink Hole problems now?" Another added: "A 3.3 magnitude earthquake near Perry, OK this morning... this is #861 this year!!! (W)hat would high pressure hydraulic fracking do to our limestone base???" The presumed answer was more sinkholes. In response, Floridians organized against becoming the next frontline geological test site for industry-triggered sinkholes.

Nature-based Tourism

Unlike communities already engaged in agriculture or extractive industries who accept oil and gas development for employment opportunities or leasing agreements (Boudet et al. 2016; Freudenburg and Gramling 1994; Malin and DeMaster 2016; Widener 2011), Florida's resident-activists expressed concerns that fracking or a related disaster would harm the state's tourism economy. In capturing the image of Florida as a nature-based destination, one journalist wrote: "Disney. The Everglades. Migrating Birds. Tourists. Fracking?" (Drouin 2014). Like the Everglades and endangered panthers, protection of the tourism economy was a discursive frame that linked regional interests to national ones, and vice versa. In one letter to the Senate Appropriations Committee, Urban Paradise Guild wrote: "What happens when tourists are scared to drink and wash in Florida's poisoned water supply? ... A vast part of Florida's economy is dependent on beautiful ecosystems... *if our environment is toxic, so is our economy* [italics in the original]."[7] South Florida Wildlands also made an argument to protect tourism: "The area of operations covers northern sections of the Florida Scenic Trail . . . as well as numerous recreational trails which access the Big Cypress backcountry. Noise, gasoline and diesel exhaust . . . cut landscapes, truck traffic, and low flying helicopters are hardly the reason people [visit Florida]."[8] Even the NPS reported that tourism to Big Cypress provided 1,255 jobs and $87 million to the local economy (Big Cypress 2016). It was two weeks after this report that NPS announced permission for seismic surveys.

State Disputes over Climate Impacts and Solar Alternatives

Fracking also fueled existing climate change and solar energy debates within the state. Given that climate change is clearly a significant, though still disputed, issue in Florida, resident-activists framed expanded oil production as amplifying the complex socioecological impacts of global warming already being experienced. While Hilson (2015) and Neville and Weinthal (2016) found that antifracking campaigners linked siting plans to global climate change, in Florida climate-related impacts were not spatially or temporally distant. Local governments were already initiating climate adaptation plans prior to the oil drilling applications.

At one town hall meeting, someone commented that there were "three elephants in the room: climate change, climate change, and climate change." The de facto argument was that any greenhouse gas–emitting project should be stopped, as sea levels were rising and saltwater intrusion and elevated storm surge were already new realities. From conferences, media reports, and announcements made by the Southeast Florida Regional Climate Change Compact, residents were learning about new and regionally relevant concepts such as "living with water," "nuisance flooding," "sunny day flooding," fresh and salt water "ponding," and living with "water in our landscapes."[9] Beyond issues of flooding, in Florida climate change signifies economic costs, higher temperatures, relocating and rebuilding public infrastructure, coral bleaching, ocean acidification, habitat loss, coastal erosion, tourism impacts, and tribal land deliberations (White House 2014).

Fracking aggravated existing tensions between citizens and public officials who supported climate action and those who denied anthropogenic climate change. Before reelection in 2014, the governor had repeated the Republican mantra that "I'm not a scientist," and hence could not be expected to know if the calculations of climate scientists were correct. Between this declaration and reelection, 30 state lawmakers organized a "Wake up Congress for Climate Action Rally," while residents, editors, and White House officials continued to write and comment on the state's lack of interest in climate change (Widener, Rowe, and Andreano 2015). Relations worsened when a news report claimed that state officials had imposed an unofficial ban on using the words "climate change" in policy discussions (Korten 2015).

At the same time, the state and the state's major utility were seen as acting to resist, slow, or control solar development. The coalition Floridians for Solar Choice mobilized in late 2014 for the right of citizens and businesses to buy and sell solar-generated electricity from a business other than a utility company. At one event, the Sierra Club's Miami chapter contrasted the risks of oil and solar energy in a sign that read: "Solar Energy Spill = A Nice Day!" The Floridians for Solar Choice campaign also aligned a number of expected environmental groups, such as the Sierra Club and Environmental Defense Fund, with state-based groups including ReThink Energy Florida and the Green Party of Florida—along with some unexpected allies, including Conservatives for Energy Freedom and the Libertarian Party of Florida. Environmental and conservative alliances, formed primarily around solar initiatives in the southern states, were being referred to as "green tea" coalitions. At the time of this writing, the green tea coalition from the solar campaign had discussed an alignment to oppose the state's efforts to ban local or county regulatory controls on issues like fracking.

As an added layer to the tug-of-war over energy development, the state's largest utility received state permission to invest its earnings from ratepayers into a fracking venture in Oklahoma with PetroQuest (Klas 2016b). Eventually, the State Supreme Court ruled such permission constituted an overreach. If the venture had not been overruled, the Florida utility would have had "the right to charge its customers up to $750 million a year for speculative natural gas fracking activities [in Oklahoma] without oversight from regulators [in Florida] for the next five years" (Klas 2016b). Even if Floridians blocked fracking in the state, they could have facilitated (unwillingly) a fracking operation in Oklahoma.

The Fracturing of Political Ties

Issues around fracking and climate change exposed cracks within the political system, stressed already difficult relations between climate activists and state leaders, intensified divisions between the two major political parties, and revealed splits between state and local politicians of the same party. In contrast, fracking and climate change concerns fortified socioecological alignments among environmental health, public health, climate action, conservation, and wildlife advocates, and local and state leaders from South Florida. The general sentiment across this broad coalition was that their voices on fracking, just like on climate change, were being ignored in the state capitol (see Figure 9.2).

At one town hall meeting, a local official declared: "Hell no!" to an oil drilling proposal. Another clarified: "This is personal, it's in our backyard, and we have told the state, no!" In an online thread, someone commented: "Calling them our 'representatives' is a misnomer. The local governments have made it clear that our community in Florida does not condone fracking. The elected state legislators are literally voting against what the vast majority of their constituents want."[10] In an opinion piece, a Republican state senator from Miami wrote: "Most leaders [from New York and around the country] have concluded that the dangers posed to communities by fracking far outweigh the benefits. . . . Protecting our communities from environmental risks should

FIGURE 9.2 Antidrilling activists in front of the old capitol building (Tallahassee, January 2016). [Credit: P. Widener]

never be a partisan issue. The flooding of coastal neighborhoods during king tides doesn't distinguish between households that vote Democrat and those that vote Republican" (Flores 2016). According to a local journalist for the *New York Times*, "strong local opposition has led to bipartisan misgivings in the State Senate, where a few prominent Republicans have broken ranks and publicly criticized fracking" (Alvarez 2016: A9). The diverse origins of these parallel comments convey how political and civic relations were fraying publicly over the issue of fracking and climate change.

By February 2016, 31 counties and 46 towns had passed resolutions for local bans, moratoriums, or determination (Flores 2016). Since then, more have followed suit. At the same time, the state legislature was deliberating on fracking. In early 2016, the House voted 73 to 45 in favor of an industry-friendly fracking regulation bill (HB191) with seven Republicans voting in opposition. The Senate Appropriations Committee rejected the bill (SB318) 10 to 9, with five Republicans participating in its defeat.[11] If the bill had been approved, it would have prohibited local or county bans retroactively and mandated a one-year environmental impact study to develop regulations. Bills to ban fracking statewide were not heard in committee.

HB191 also revealed the importance of understanding differences between unconventional technologies when applied regionally. HB191 was a bill to regulate "high-pressure well stimulation." Regulating this technology would have exempted the low-pressure acid fracturing or matrix stimulation that would most likely be used in South Florida from additional regulation beyond what was already in existence for conventional extraction technologies.[12] In other words, "acid fracking" was not debated. According to ReThink Energy, "The fracking regulatory bill only covers hydraulic fracturing. . . . It would allow this acidization to go unregulated" (Sullivan 2016). Future bills to regulate or reject "fracking," which is unclearly defined in Florida, are expected.

Conclusion

In this research, the antifracking efforts that surfaced in South Florida were built on the diffused awareness of how public health and environmental quality had been impaired by the expansion of fracking in other regions of the country, as well as a concentrated effort to protect public lands and resources. At the same time, debates over "fracking" became something of a double-edged sword. While the national controversy over hydraulic fracturing informed residents, mobilized resistance, and tarnished pro-industry arguments in Florida, it may have also undermined broad public and political understanding of low-pressurized acid fracking. In 2013, the first illegal fracking operation generated public awareness and media attention, but since then subsequent news stories have tended to reference "hydraulic fracking" or "oil development" in general (see Flores 2016 and Nielsen 2016 as examples). In other words, when extralocal images of "fracking" came to Florida, many in the state were still discovering the distinctions in technologies, impacts, proposals, and regulations between acid and hydraulic fracking, high-pressure and low-pressure water volume, horizontal and vertical drilling, and unconventional and conventional oil development. Florida's distinct geology and the more prominent national portrait of hydraulic fracking generated some confusion and may have weakened public understanding of the processes and appropriate legislation for the state.

As momentum built against "fracking," several interrelated narratives of protest came together to create a determined citizenry in South Florida. Resident-activists refused to become the next energy sacrifice zone (see Lerner 2010). The first narrative applied national experiences to argue for community and environmental health protections. The second lens emphasized the region's unique environmental amenities, risks, and opportunities. The Everglades, nature-based tourism, and the Florida panther were framed as deserving of protection, while sinkholes were pronounced to be a risk too great to take. Though regional in scope, the wetlands, wildlife, and tourism narratives were portrayed as an economic and ecological bridge between the region and the nation that would be destabilized by fracking or oil development. The third overarching narrative was a state-bounded frame that drew attention to the existing strains between residents, local politicians, and state leaders over the interlocking issues of climate impacts and energy development. The added climate change risks were framed as too great to absorb, and solar energy was portrayed as too important an opportunity to suspend for more oil. In a time of anthropogenic climate change, portions of the Sunshine State were debating whether to transition toward or away from expanded fossil fuel production.

Today, new extractive technologies, ongoing energy demand, and political acquiescence have enabled the oil and gas industry to enter into vulnerable communities that never expected to become industrial sites for energy production. Meanwhile, the risks of acid or hydraulic fracking continue to receive inadequate attention from industry and state regulators. Production and profit appear to supersede socioecological well-being, and scores of newly affected communities continue to realize (and resist) their marginal position in this sacrificial exchange. As of this writing, whether Florida will regulate or reject fracking remains uncertain. It also remains unclear whether a particular discursive narrative or combination of frames (protecting tourism, defending the Everglades, and/or containing greenhouse gas emissions) might prove successful in achieving support for a statewide ban or more stringent regulations. Which pockets of privilege will be protected (e.g., Big Cypress and Southeastern counties) and which portions of the state will become corridors of sacrifice (e.g., the rural, central, or northern communities) remain unanswered questions. Like other states and shale regions today, South Florida stands at the crossroads of determining which citizens will shoulder the social and environmental costs—as well as receive the benefits—of the energy path state officials select.

Notes

1. See Faber (2008) on the "polluter-industrial complex," and Watts (2004: 279–280) on the "oil complex."
2. California Senate Bill No. 4 (Section 2, Article 3, Section 3158). Retrieved June 27, 2016. (http://leginfo.legislature.ca.gov/faces/billNavClient.xhtml?bill_id= 201320140SB4).
3. Four Florida DEP permit requests were filed in Collier County since December 2012, and one in Broward County in July 2015. The one in Broward (Kanter Real Estate; permit #1366) is being reviewed. Of those in Collier, one was denied (December 15, 2014, permit #G169, for the operator Burnett Oil), and three were issued: to Burnett Oil (July 15, 2015, #G170); Century Oil (June 30, 2015, #1335C); and Dan A. Hughes (December 1, 2012, #1349H). See Florida DEP, last retrieved May 23, 2016. (http://www.dep.state.fl.us/water/mines/oil_gas/drill-apps.htm).
4. Pembroke Pines Resolution 2016-R-24: Reaffirming Opposition to Oil Drilling—Kanter, passed August 17, 2016.
5. The six environmental groups in alphabetical order with headquarters listed include: Center for Biological Diversity (Tucson, AZ), Conservancy of Southwest

Florida (Naples, FL), Earthworks (Washington DC), National Parks Conservation Association (Washington DC), Natural Resources Defense Council (New York), and South Florida Wildlands Association (Fort Lauderdale, FL).

6. Urban Paradise Guild. 2016. Letter, dated February 8, 2016.

7. Urban Paradise Guild. 2016. Letter, dated February 8, 2016.

8. South Florida Wildlands Association. 2015. Emailed letter, dated August 15, 2015.

9. Based on attendance at the annual Southeast Florida Regional Climate Leadership Summit in 2012, 2014, and 2015. See Widener, Rowe, and Andreano (2015).

10. Anonymous comment to Kaucheck's (2015) article, dated February 1, 2016. Retrieved April 20, 2016 (http://www.psr.org/environment-and-health /environmental-health-policy-institute/responses/fracking-around-the-everglades .html).

11. Based on a "Legislative Wrap-up" report by the Sierra Club, Broward County (received April 7, 2016). ReThink Florida Energy also provided extensive reports on the various bills for the legislative session.

12. Based on the report "Fracking Bill Up in Second Committee Wednesday" by the Sierra Club, Broward County (received November 30, 2015). ReThink Florida Energy also provided extensive reports on the various bills.

References

ALL Consulting. 2014. "Expert Evaluation of the D.A. Hughes Collier-Hogan 20–3H Well Drilling Workover." December. Retrieved September 30, 2016. (https://www .documentcloud.org/documents/1507525-allconsulting.html).

Allen, Barbara L. 2003. *Uneasy Alchemy: Citizens and Experts in Louisiana's Chemical Corridor Disputes*. Cambridge, MA: MIT Press.

Allen, Greg. 2014a. "Oil Industry Gets an Earful As It Eyes Florida's Everglades." National Public Radio, March 13. Retrieved June 4, 2016. (http://www.npr.org/2014/03/13 /289423090/oil-industry-gets-an-earful-as-it-eyes-floridas-everglades).

Allen, Greg. 2014b. "Florida County Goes to Court over 'Acid Fracking' near Everglades." National Public Radio, July 2. Retrieved February 19, 2016. (http://www.npr.org /2014/07/02/327373952/florida-county-goes-to-court-over-acid-fracking-near -everglades).

Alvarez, Lizette. 2016. "Unlikely Battle Intensifies in Florida After Company Is Found to Use Fracking." *New York Times*, February 24: A9–A10.

American Petroleum Institute. 2014. "Acidizing: Treatment in Oil and Gas Operators." Digital Media 2014–113. Retrieved June 2, 2016. (http://www.api.org/oil-and-natural -gas/wells-to-consumer/exploration-and-production/hydraulic-fracturing/acidizing -treatment-in-oil-and-gas-opera).

Barnett, Cynthia. 2011. "Our Water. Our Florida. A Water Ethic for Florida." Policy paper. Miami: Collins Center for Public Policy.

Barrett, Ross. 2014. "Picturing a Crude Past: Primitivism, Public Art, and Corporate Oil Promotion in the United States." Pp. 43–68 in *Oil Culture*, edited by R. Barrett and D. Worden. Minneapolis: University of Minnesota Press.

Bekiempis, Victoria. 2014. "Oil Prospectors Seek Their Next Big Strike in South Florida's Everglades." *Newsweek*, February 27. Retrieved November 9, 2015. (http://www .newsweek.com/florida-city-bans-fracking-do-not-run-embargo-embargo-embargo -untilsomething-353047).

Benford, Robert D., and David A. Snow. 2000. "Framing Processes and Social Movements: An Overview and Assessment." *Annual Review Sociology* 26: 611–639.

Big Cypress National Preserve. 2016. "Tourism to Big Cypress National Preserve Creates $124,524,400 in Economic Benefits." Press release, April 25.

Boudet, Hilary, Dylan Bugden, Chad Zanocco, and Edward Maibach. 2016. "The Effect of Industry Activities on Public Support for 'Fracking.'" *Environmental Politics* 25(4): 593–612.

Briggle, Adam. 2015. *A Field Philosopher's Guide to Fracking*. New York: Liveright.

Brown, Phil. 1992. "Popular Epidemiology and Toxic Waste Contamination: Lay and Professional Ways of Knowing." *Journal of Health & Social Behavior* 33(3): 267–281.

Brown, Phil. 2007. *Toxic Exposures: Contested Illnesses and the Environmental Health Movement*. New York: Columbia University Press.

Burlew, Jeff. 2016. "Florida DEP Seeks to Ease Restrictions on Discharged Chemicals." *Tallahassee Democrat*, May 15. Retrieved June 16, 2016. (http://www.tcpalm.com/news/state /florida-dep-seeks-to-ease-restrictions-on-discharged-chemicals-32e75fcf-3bc5-7056-e053 -0100007f2a9f-379587131.html).

Cable, Sherry. 2012. *Sustainable Failures: Environmental Policy and Democracy in a Petro-dependent World*. Philadelphia, PA: Temple University Press.

Casey, Joan A., David A. Savitz, Sara G. Rasmussen, Elizabeth L. Ogburn, Jonathan Pollak, Dione G. Mercer, and Brian S. Schwartz. 2016. "Unconventional Natural Gas Development and Birth Outcomes in Pennsylvania, USA." *Epidemiology* 27(2): 163–172.

Conlin, Michelle, and Brian Grow. 2013. "Special Report: U.S. Builders Hoard Mineral Rights under New Homes." *Reuters*, October 9. Retrieved November 9, 2015 (http://www.reuters.com/article/us-usa-fracking-rights-specialreport -idUSBRE9980AZ20131009).

Corburn, Jason. 2005. *Street Science: Community Knowledge and Environmental Health Justice*. Cambridge, MA: MIT Press.

Drouin, Roger. 2014. "Could Florida Become the New Fracking Frontier?" *Truth-Out*, February 4. Retrieved November 5, 2015 (http://www.truth-out.org/news/item/21642 -fracking-florida-could-florida-become-the-new-fracking-frontier).

Dwyer, Karen. 2014. "Groups Demand End of Everglades Oil Drilling, Fracking and Fix Water Problems." *GreenMedInfo*, October 15. Retrieved June 4, 2016 (http://www .greenmedinfo.com/blog/groups-demand-end-everglades-oil-drilling-fracking-and-fix -water-problems).

Earthworks. n.d. "Acidizing." Retrieved June 4, 2016. (https://www.earthworksaction.org /issues/detail/acidizing#.V1LzSuTmoiQ).

Eastman, Susan Cooper. 2015. "No Fracking Way." *Florida Field Notes: Northeast Florida's Outdoors Journal*. Retrieved June 18, 2016. (http://www.floridafieldnotes.com/2015/08 /27/no-fracking-way/).

Faber, Daniel. 2008. *Capitalizing on Environmental Injustice: The Polluter-Industrial Complex in the Age of Globalization*. Lanham, MD: Rowman & Littlefield.

Fischer, Frank. 2000. *Citizens, Experts, and the Environment: The Politics of Local Knowledge*. Durham, NC: Duke University Press.

Fleshler, David. 2015. "Oil Well Proposed for Everglades in Broward County." *Sun Sentinel*, July 12. Retrieved May 23, 2016. (http://www.sun-sentinel.com/local/broward/fl -broward-oil-drilling-20150710-story.html).

Fleshler, David. 2016a. "Controversial Everglades Oil Well Plan Moving Forward." *Sun Sentinel*, June 12. Retrieved June 16, 2016. (http://www.sun-sentinel.com/local/broward/fl -everglades-oil-update-20160612-story.html).

Fleshler, David. 2016b. "Suit Filed to Stop Big Cypress Oil Plan." *Sun Sentinel*, July 28. Retrieved September 27, 2016. (http://www.sun-sentinel.com/local/palm-beach/fl-big -cypress-oil-lawsuit-20160727-story.html).

Flores, Anitere. 2016. "Fracking Will Destroy Residents' Way of Life." *Miami Herald*, February 24. Retrieved February 25, 2016. (http://www.miamiherald.com/opinion/op-ed /article62179672.html).

Food and Water Watch. 2014. "The Urgent Case for a Ban on Fracking." Washington, DC: Food and Water Watch.

Fox, Josh. 2010. *Gasland* (documentary). New York: Wow Productions.

Freudenburg, William R., and Robert Gramling. 1994. *Oil in Troubled Water: Perceptions, Politics and the Battle over Offshore Drilling*. Albany: State University of New York Press.

Freudenburg, William R., and Robert Gramling. 2011. *Blowout in the Gulf: The BP Oil Spill Disaster and the Future of Energy in America*. Cambridge, MA: MIT Press.

Frickel, Scott, and Neil Gross. 2005. "A General Theory of Scientific/Intellectual Movements." *American Sociological Review* 70(2): 204–232.

Gold, Russell. 2014. *The Boom: How Fracking Ignited the American Energy Revolution and Changed the World*. New York: Simon & Schuster.

Gramling, Robert, and William R. Freudenburg. 2013. "The Growth Machine and the Everglades: Expanding a Useful Theoretical Perspective." *Society & Natural Resources* 26(6): 642–654.

Grunwald, Michael. 2006. *The Swamp*. New York: Simon and Schuster.

Gullion, Jessica Smartt. 2015. *Fracking the Neighborhood: Reluctant Activists and Natural Gas Drilling*. Cambridge, MA: MIT Press.

Gunter, Valerie, and Steve Kroll-Smith. 2007. *Volatile Places: A Sociology of Communities and Environmental Controversies*. Thousand Oaks, CA: Pine Forge Press.

Hannigan, John. 2014. *Environmental Sociology*, 3rd ed. New York: Routledge.

Hilson, Chris. 2015. "Framing Fracking: Which Frames Are Heard in English Planning and Environmental Policy and Practice?" *Journal of Environmental Law* 27(2): 177–202.

Horowitz, Joy. 2007. *The Poisoning of an American High School*. New York: Penguin Books.

Jorgensen, Dolly. 2014. "Mixing Oil and Water." Pp. 267–288 in *Oil Culture*, edited by R. Barrett and D. Worden. Minneapolis: University of Minnesota Press.

Kaucheck, Lynna. 2015. "Fracking around the Everglades?" Physicians for Social Responsibility. Washington D.C. Retrieved April 20, 2016. (http://www.psr.org/environment-and-health /environmental-health-policy-institute/responses/fracking-around-the-everglades.html).

Klas, Mary Ellen. 2016a. "Florida House Approves Bill to Authorize, Regulate Fracking." *Miami Herald*. January 27. Retrieved June 26, 2016. (http://www.miamiherald.com/news /politics-government/state-politics/article56938703.html).

Klas, Mary Ellen. 2016b. "Florida Supreme Court Rejects FPL Attempt to Have Customers Pay for Risky Investment." *Tampa Bay Times*, May 19. Retrieved May 23, 2016. (http:// www.tampabay.com/blogs/the-buzz-florida-politics/florida-supreme-court-rejects-fpl -attempt-to-have-customers-pay-for-risky/2278080).

Korten, Tristram. 2015. "In Florida, Officials Ban Term 'Climate Change.'" *Miami Herald*, March 8. Retrieved May 21, 2016. (http://www.miamiherald.com/news/state/florida /article12983720.html).

Ladd, Anthony E. 2014. "Environmental Disputes and Opportunity-Threat Impacts Surrounding Natural Gas Fracking in Louisiana." *Social Currents* 1(3): 293–311.

Ladd, Anthony E. 2016. "Meet the New Boss, Same as the Old Boss: The Continuing Hegemony of Fossil Fuels and Hydraulic Fracking in the Third Carbon Era." *Humanity & Society* (http://dx.doi.org/10.1177/0160597616628908).

Lerner, Steve. 2005. *Diamond: A Struggle for Environmental Justice in Louisiana's Chemical Corridor*. Cambridge, MA: MIT Press.

Lerner, Steve. 2010. *Sacrifice Zones: The Front Lines of Toxic Chemical Exposure in the United States*. Cambridge, MA: MIT Press.

Lustgarten, Abrahm. 2012. *Run to Failure: BP and the Making of the Deepwater Horizon Disaster*. New York: W. W. Norton.

Malin, Stephanie A., and Kathryn Teigen DeMaster. 2016. "A Devil's Bargain: Rural Environmental Injustices and Hydraulic Fracturing on Pennsylvania's Farms." *Journal of Rural Studies* 47: 278–290.

McKenzie, Lisa M., Ruixin Guo, Roxana Z. Witter, David A. Savitz, Lee S. Newman, and John L. Adgate. 2014. "Birth Outcomes and Maternal Residential Proximity to Natural Gas Development in Rural Colorado." *Environmental Health Perspectives* 122(4): 412–417.

Messer, Chris M., Thomas E. Shriver, and Dennis Kennedy. 2009. "Official Frames and Corporate Environmental Pollution." *Humanity & Society* 33(4): 273–291.

National Park Service. 2016. "National Park Service Signs Decision Document Related to Burnett Oil Company Nobles Grade 3-D Seismic Survey." Press release, May 6.

National Park Service. n.d. "Big Cypress: Freshwater to the Sea." Retrieved June 1, 2016. (https://www.nps.gov/bicy/index.htm).

Neville, Kate J., and Erika Weinthal. 2016. "Scaling Up Site Disputes: Strategies to Redefine 'Local' in the Fight Against Fracking." *Environmental Politics* 25(4): 569–592.

Nielsen, Kirk. 2016. "Bad Vibes in Big Cypress." *Sierra*, July/August, pp. 50–51.

OGJ editors. 2010. "Florida Sunniland Trend Well Wows BreitBurn." *Oil and Gas Journal* (June 8). Retrieved June 25, 2010. (http://www.ogj.com/articles/2010/06/florida-sunniland.html).

O'Rourke, Dara, and Sarah Connolly. 2003. "Just Oil? The Distribution of Environmental and Social Impacts of Oil Production and Consumption." *Annual Review of Environment and Resources* 28(1): 587–617.

Ott, Riki. 2005. *Sound Truth and Corporate Myths: The Legacy of the Exxon Valdez Oil Spill*. Cordova, AK: Dragonfly Sisters Press.

Pellow, David Naguib. 2014. *Total Liberation: The Power and Promise of Animal Rights and the Radical Earth Movement*. Minneapolis: University of Minnesota Press.

Perrow, Charles. 1984. *Normal Accidents: Living with High Risk Technologies*. New York: Basic Books.

Pittman, Craig. 2015. "Florida to Receive $3.25B from Gulf States' Deepwater Horizon Settlement with BP." *Tampa Bay Times*, October 5. Retrieved September 27, 2016. (http://www.tampabay.com/news/business/gulf-states-reach-187b-settlement-with-bp-over-deepwater-horizon-spill/2235887).

Reed, Jennifer. 2015. "The Truth about Our Oil." *Gulfshore Life* (April edition). Retrieved November 9, 2015. (http://www.gulfshorelife.com/April-2015/The-Truth-About-Our-Oil-fracking-Southwest-Florida/).

Robinson, Erin E. 2009. "Competing Frames of Environmental Contamination: Influences on Grassroots Community Mobilization." *Sociological Spectrum* 29(1): 3–27.

Ross, Kim. 2016. "Hundreds in Tallahassee Hold Largest Anti-Fracking Rally in State's History." *Sierra Club Florida News*. February 3. Retrieved June 4, 2016. (http://www.sierraclubfloridanews.org/2016/02/hundreds-in-tallahassee-hold-largest.html).

Rubinstein, Justin L., and Alireza Babaie Mahani. 2015. "Myths and Facts on Wastewater Injection, Hydraulic Fracturing, Enhanced Oil Recovery, and Induced Seismicity." *Seismological Research Letters* 86(4): 1060–1067.

Russo, Ray, and Elizabeth Screaton. 2016. "Should Florida 'Frack' Its Limestone for Oil and Gas? Two Geophysicist Weigh In." *The Conversation, Associated Press.* May 5. Retrieved June 16, 2016. (http://bigstory.ap.org/article/6ea5675acd694ee19c1921195b57f443 /should-florida-frack-its-limestone-oil-and-gas-two).

Spencer, Starr. 2013. "New Frontiers: A Once-Booming Florida Oil Industry Tries to Get Back to the Past." *Oilgram News,* January 7. Retrieved June 11, 2016. (http://blogs.platts .com/2013/01/07/florida_oil/).

Stimeling, Travis D. 2014. "Music, Place, and Gulf Coast Tourism since the BP Oil Spill." *Music & Politics* 8(2): 1–20.

Sullivan, Erin. 2016. "Florida House to Vote on Bill That Would Lay Groundwork for Fracking." *Orlando Weekly,* January 22. Retrieved June 26, 2016. (http://www.orlandoweekly .com/Blogs/archives/2016/01/22/florida-house-to-vote-on-bill-next-week-that-would -lay-the-groundwork-for-fracking-in-the-state-and-prevent-counties-from-banning-it).

Tamminen, Terry. 2006. *Lives per Gallon: The True Cost of Our Oil Addiction.* Washington, DC: Island Press.

Vasi, Ion Bogdon, Edward T. Walker, John S. Johnson, and Hui Fen Tan. 2015. "'No Fracking Way!' Documentary Film, Discursive Opportunity, and Local Opposition against Hydraulic Fracturing in the United States, 2010 to 2013." *American Sociological Review* 80(5): 934–959.

Volz, David. 2014. "Big Oil Eyes Florida's Public Lands, Plans to Drill in the Everglades." *Earth Island Journal,* April. Retrieved November 9, 2015. (http://ecowatch.com/2014 /04/30/big-oil-floridas-public-lands-the-everglades/).

Watts, Michael. 2004. "Violent Environments: Petroleum Conflict and the Political Ecology of Rule in the Niger Delta, Nigeria." Pp. 273–298 in *Liberation Ecologies,* edited by R. Peet and M. Watts. New York: Routledge.

White House. 2014. "What Climate Change Means for Florida and the Southeast and Caribbean." The White House Office of the Press Secretary. Washington, DC. (Released May 6).

Widener, Patricia. 2009. "Oil Tourism: Disasters and Destinations in Ecuador and the Philippines." *Sociological Inquiry* 79(3): 266–288.

Widener, Patricia. 2011. *Oil Injustice: Resisting and Conceding a Pipeline in Ecuador.* Lanham, MD: Rowman & Littlefield.

Widener, Patricia. 2013. "A Protracted Age of Oil: Pipelines, Refineries, and Quiet Conflict." *Local Environment* 18(7): 834–851.

Widener, Patricia, and Valerie Gunter. 2007. "Oil Spill Recovery in the Media: Missing an Alaska Native Perspective." *Society and Natural Resources* 20(9): 767–783.

Widener, Patricia, Carmen Rowe, and Celene Andreano. 2015. "Climate Awareness: A Media & Event Analysis of Locality, Social Problem & Institutional Amnesia." Paper presented at the Southern Sociological Society annual meeting, March 25–28, New Orleans, LA.

10

Public Participation and Protest in the Siting of Liquefied Natural Gas Terminals in Oregon

———————————————————●

HILARY BOUDET, BRITTANY
GAUSTAD, AND TRANG TRAN

Introduction

While most of the case studies in this volume explore communities facing unconventional oil and/or gas development and hydraulic fracking, the controversy surrounding the shale energy revolution today would not be complete without an examination of the conflicts over the siting of associated infrastructure projects involving pipelines and export terminals. Perhaps the most well known of these debates is related to the Keystone XL Pipeline that was vigorously and somewhat successfully opposed by a coalition of environmental organizations (McKibben 2013), but such conflicts are occurring in regions across the country. In Oregon, while communities have not faced proposals for fracking operations in their backyards, for more than a decade they have been involved in a contentious debate about the siting of liquefied natural gas (LNG) terminals and their related pipelines near the coastal cities of Astoria and Coos Bay. LNG is natural gas that has been cooled to cryogenic temperatures to reduce its volume to allow for easier and more economical transport.

While the recent national gas boom has made it difficult to fathom, in the early 2000s U.S. energy experts were predicting domestic gas shortfalls. As with many other coastal communities with deepwater ports, energy companies proposed to build LNG import terminals near both Astoria and Coos Bay. Import facilities receive LNG produced overseas and regasify it for domestic distribution. Almost a decade later, such proposals were switched to export facilities to receive natural gas produced domestically, largely via hydraulic fracturing and horizontal directional drilling, and liquefy it for transport on tankers overseas. These LNG proposals have stirred controversy and sparked protest among local citizens.

Using a comparative-case design and building on previous research on communities facing LNG import terminals (Boudet 2010; Boudet 2011; Boudet and Ortolano 2010; McAdam and Boudet 2012), in this chapter we examine how active participants (e.g., opponents, supporters, decision makers, agency officials, etc.) perceived the effectiveness of public participation processes in incorporating citizen concerns about the construction of local LNG terminals, as well as how those perceptions helped facilitate community mobilization efforts.

As in the case of communities facing unconventional oil and gas development and hydraulic fracturing, a proposed LNG terminal presents both potential risks and benefits to host communities (Anderson and Theodori 2009; Brasier et al. 2011; Ladd 2013, 2014). Such facilities may bring jobs, tax revenue, and associated business development, but also raise concerns about impacts to public safety, community character, and the surrounding environment. Broader economic and environmental concerns are also at play, including debates about fossil fuel dependence, the role of natural gas in our energy transition, and the viability of foreign markets for exported gas. Public perceptions of and response to such proposals has been wide-ranging, varying from ready acceptance to widespread opposition (Boudet 2015), as in the case of unconventional oil and gas development (Brasier et al. 2011; Eaton and Kinchy 2016; Kriesky et al. 2013; Schafft et al. 2013). The proposed projects in Oregon have been among some of the most controversial in the United States, apart from perhaps Maryland's Cove Point facility (Bengel 2015). With 27 other LNG proposals in development in the United States (Federal Energy Regulatory Commission [FERC] 2014), research into conflicts surrounding the infrastructure associated with the shale energy revolution is critical to understanding the entire context of such development. Indeed, the environmental community has taken note of this development: a 2012–2013 U.S. Department of Energy Environmental Impact Assessment about LNG export around the United States elicited almost 200,000 comments (U.S. Department of Energy 2014). The Sierra Club has established a "beyond natural gas" campaign that aims to "stop LNG exports" and intervenes in every proposal, recently (unsuccessfully) suing the Federal Energy Regulatory Commission (FERC) regarding related approvals (Sierra Club 2016).

Review of the Literature

Facility Siting and Public Participation

Two distinct literatures provide insights about perceptions of public participation processes and community mobilization: (1) the specialized scholarship on facility siting and public participation processes; and (2) the general literature on social movements. A long tradition of scholarship has explored proponent approaches to facility siting and how they can be improved to manage conflict. The Decide-Announce-Defend strategy, which relies on a technocratic, expert-based approach to determine an appropriate site that is then announced and defended in the public arena, has proven unsuccessful (Beierle and Cayford, 2002; Lesbirel and Shaw 2005). As a result, project proponents have been encouraged to move to more cooperative and collaborative approaches (Armour 1991; Lesbirel and Shaw 2005).

This shift in the academic literature to a more voluntary, cooperative approach to siting, which focuses on the importance of early communication, consultation, and public participation in decision-making, has been mirrored in regulation here and abroad (Armour 1991; Dietz and Stern 2008; Environment Division 1998; Freudenburg 2004; Kunreuther et al. 1993; Lesbirel & Shaw, 2005). Governments and financial institutions actively encourage project sponsors to consult the public (Environment Division 1998). Research on public participation has examined the effectiveness of different techniques for participation, such as public hearings, citizen advisory committees, and surveys (Bierle and Cayford 2002; Fiorino 1990). Part of the difficulty in analyzing the effectiveness of public participation is that effectiveness is often a subjective term. Depending on which lens the researcher chooses (typically project proponent or citizen), determining how effective a particular process is can yield very different conclusions. Here, we focus on perceptions of one measure of the effectiveness of public participation—incorporating citizen concerns into the decision-making arena (O'Faircheallaigh 2010).

Beliefs about fairness, both in terms of the process for citizen input and decision making, as well as the substantive choices about the facility location, have also proven important in determining community mobilization in response to a proposed facility (Boholm 2004; Hunter and Leyden 1995; Lesbirel and Shaw 2005; Kunreuther and Lathrop 1981). Distrust in government officials and project proponents has been shown to be a significant predictor of opposition (Boudet & Ortolano, 2010; Hunter and Leyden 1995;

Kraft and Clary 1991; Kunreuther and Lathrop 1981), although there is some debate on whether opposition to a proposed facility may actually spur distrust as opposed to the other way around (Smith and Marquez 2000).

Social Movements

Instead of analyzing the effectiveness of public participation processes, a social movements' perspective turns this question on its head to examine the conditions under which the siting of a facility results in significant community mobilization. Three broad factors are widely believed to shape the general prospects for emergent collective action: (1) new political opportunities or threats; (2) mobilizing structures; and (3) collective action framing. According to social movement theorists, under stable political/environmental conditions, movements are unlikely to develop. Movements, in this sense, should be seen as a response to environmental shifts that either grant new opportunities or leverage to potential challengers or pose new threats to some segment of the population (McAdam 1999; McAdam et al. 1996; Tarrow 1998; Tilly 1978). The announcement of a proposed terminal is precisely the kind of shift that has the potential to trigger collective action, as community groups come to define the proposal as posing either a significant new threat to, or an opportunity to advance, their interests. From the perspective of social movement scholars, public participation processes—instead of dampening conflict—may create political opportunities that spur additional mobilization. This implication is consistent with the general consensus among social movement scholars that open political systems are more prone to protest than are closed ones (Eisinger 1973; McAdam 1999).

While the announcement of a proposed LNG facility may create a motive for collective action, without an organizational basis for action, a sustained movement is unlikely to develop. Social movement scholars have consistently found that the majority of successful movements either emerge from similar, earlier struggles or develop within established social settings such as formal organizations and informal networks (Gould 1993; Gould 1995; Morris 1986; Snow et al. 1980). Though necessary, exogenous environmental shifts and the presence of mobilizing structures do not produce a social movement. Together, they only offer potential activists the possibility of collective action. Mediating between opportunity and action are people and the subjective meanings they attach to their situation. This process, which has come to be known as "collective action framing," must occur if organized collective action is to take place (Snow et al. 1986; Snow and Benford 1988; Snow and Benford 1992). One of the keys to movement emergence, then, is whether objective siting decisions are subjectively defined as sufficiently threatening or opportune by enough people to support and sustain organized collective action.

A few researchers have applied a social movements' lens to local action in response to a proposed infrastructure project (Carmin 2003; Sherman 2004; Walsh et al. 1997; Wakefield et al. 2006). Studying 20 U.S. communities facing large-scale energy facilities (of which 15 were LNG import terminal proposals), Wright and Boudet (2012) and McAdam and Boudet (2012) found that community mobilization was driven by a combination of aspects of the local community context. These aspects included economic need, prior community mobilization efforts, previous experience with similar siting proposals, the openness of the decision-making process to local voices, and the ability to access community resources. Mobilization levels were, however, unrelated to measures of the actual threat posed by the facility. Instead, the community's previous experiences and context provided a critical lens through which community members interpreted potential impacts of the facility and thus motivations to mobilize (Boudet 2010; McAdam and Boudet 2012; Wright and Boudet, 2012).

In this research, we build on these ideas to examine how active participants in the public participation processes surrounding LNG proposals in Astoria and Coos Bay perceived the effectiveness of public participation processes in incorporating citizen concerns and how these perceptions related to subsequent community mobilization.

Research Methods and Case Summaries

Using a similar approach as in previous research efforts related to LNG infrastructure (Boudet 2010; Boudet 2011; Boudet 2015; Boudet and Ortolano 2010), we first collected and coded information about the LNG proposals in relevant local newspapers to identify key players and events in the community, as well as outline the narrative of each case. Information from two local newspapers and one regional newspaper was retrieved from the Access World News database using the keywords "liquefied natural gas" and "LNG" from August 2004 to July 2015. We then conducted two site visits in August and September 2014 to each affected community to interview key players representing a wide variety of interests in the proposal (e.g., opponents, supporters, decision makers, agency officials). In Astoria, interviewees included a state agency representative, a (former) county commissioner, a county staff member, two city officials, a (former) port commissioner, two port officials, a reporter, three supporters, and 10 opponents. Despite repeated requests, we were unable to interview a project representative and therefore used Oregon LNG's website as a resource. In Coos Bay, interviewees included one state agency representative, three local decision makers, one federal agency representative, a reporter, a project representative, three supporters, 14 opponents (10 Coos Bay area residents, four landowners along the pipeline), and one neutral active citizen. (See the appendix for a complete list of interviewees by case.)

Semistructured interviews were used to encourage individuals to share their experiences with the LNG proposals in their community (Weiss 1994). Questions focused on: (1) previous political involvement in other issues that have historically driven local politics; (2) characteristics of the public participation process surrounding LNG terminal siting; and (3) activities undertaken by stakeholders around LNG issues. Interviews averaged one hour in length, ranging from 25 minutes to two hours. For analysis purposes, interviews were audio recorded and transcribed with detailed notes taken during the meetings as a guide. Additionally, one FERC public comment meeting on the Jordan Cove Energy Project was observed on December 9, 2014, in Roseburg, Oregon, for 4.5 hours, and transcripts from 10 FERC public hearings were coded for the stances of participants toward the proposal—four surrounding the release of the Draft EIS on the proposal to import LNG in October 2008 and six surrounding the release of the Draft EIS on the proposal to export LNG in December 2014.

Information collected from newspapers, interviews, and other sources (online or in regulatory documentation) was then analyzed and coded to identify general themes in terms of perceptions of public participation processes and to characterize community mobilization using standard approaches for analyzing qualitative data (Lofland et al. 1995). Table 10.1 provides additional information about our data collection strategies by case. Table 10.2 lists the emerging themes identified during this coding process.

A Brief Overview of the Regulatory Process

Initially, review of a LNG proposal in Oregon fell under the jurisdiction of Oregon's Energy Facility Siting Council. However, in August 2005, passage of the Energy Policy Act granted FERC, a five-member commission appointed by the president, exclusive authority to approve or deny the siting, construction, expansion, or operation of a LNG terminal located onshore or in state waters. The Coast Guard, Army Corps of Engineers, Fish and Wildlife Service, National Oceanic and Atmospheric Administration, and Department

Table 10.1
Data collection

Community	Newspaper	Articles	Letters and editorials	Interviews
Astoria	*The Daily Astorian*	1,410	1,046	21
Coos Bay	*The Coos Bay World*	1,035	514	22
Both	*The Oregonian*	154	118	2

Table 10.2
Emerging themes

Theme	Description	Examples
Regulatory / public participation process	Concerns that the LNG siting process was unfair, violated existing laws and regulations, and local comments and interests were not taken into consideration.	"FERC is dismissive of public input and does a poor job of incorporating even state concerns and concerns from other federal agencies, so I do not think it's a fair process" (interview 10).
Company's trustworthiness	Concerns that the company would pursue profit at the expense of public health and environmental integrity.	"Taxpayers get the risks and the oil companies get colossal profits" (Chasm 2008).
Decision maker's trustworthiness	Concerns that government officials could not be trusted to protect the public interest.	"On the whole, the city and the port will talk the talk but [will not] walk the walk. Once in a while somebody is responsible but a lot of times no. There was a lot of 'good old boy' politics in Astoria, and everyone is connected to somebody that wants something" (interview 6).
Potential risks	Concerns that the LNG facility would bring about negative impacts to public safety and security, the economy, and environmental quality.	"We are becoming a staging area for the LNG speculators to reap huge financial rewards, while we assume the very real risks, and the costs, of security, safety, and environmental and economic degradation" (Caplan 2007).
Potential economic benefits	Comments about the potential financial and employment benefits associated with a LNG facility.	"I'm in favor of LNG, a more modern port and economic growth" (Mann 2006).
Community context	References to past failed projects in the community.	"It is implied that the millions of dollars we have spent chasing [projects like] Daishowa Paper, Nucor Steel, Jordan Cove LNG, ad infinitum was money well spent, and we ought to continue dumping money down the chute until we hit the jackpot" (Sadler 2008).
Irrational, emotional nature of comments by opponents	Statements made by supporters that the concerns of opponents are not based on facts.	"All I've heard about the risk is the fear-mongering, fictitious conjuring, outright lies, and wild speculations of what could happen" (Penny 2008).

of Transportation assist FERC in concurrently integrating requirements of Waterway Suitability Assessment, the National Environmental Policy Act, Clean Water Act, Rivers and Harbors Act, and National Fire Protection Association in the review process (U.S. Department of the Army 2004). In terms of public participation, the National Environmental Policy Act, which requires that federal agencies prepare an environmental impact statement (EIS) for major federal actions that significantly affect the quality of the human environment, is particularly important.

At the state level, the Oregon Department of Energy was designated by the governor as the lead state agency to ensure that state interests, as well as community safety and security, were adequately addressed in the federal review. It also coordinates state permitting. Other state agencies, such as the Oregon Department of Environmental Quality, were involved in drafting, issuing, and ensuring compliance with environmental permits. Under the Coastal Zone Management Act, Oregon—through the Oregon Department of Land Conservation and Development—has the authority to review all federal actions, licenses, and permits affecting the coastal zone for consistency with "national standards implemented by states" (Oregon Department of Energy n.d.).

At the local level, counties and local ports have authority over land use decisions related to the terminal site and associated pipeline. However, these decisions can also be challenged in Oregon's Land Use Board of Appeals and the Ninth Circuit Court of Appeals. Pipeline developers generally negotiate directly with landowners along the pipeline route to secure access and determine appropriate levels of compensation. However, the power of eminent domain (with just compensation) may be granted to private companies if construction of a project is deemed for public use or benefit.

LNG in Astoria: Oregon LNG and Bradwood Landing

At one point, four LNG proposals were under consideration in and around Astoria. We focus on the two projects that went the farthest in the review process: Oregon LNG and Bradwood Landing.

Oregon LNG started off as the Skipanon Natural Gas Facility, originally proposed as an import facility by Calpine, a California-based power company, for 96 acres of state-owned land at the tip of the Skipanon Peninsula. The Port of Astoria quickly and controversially signed a lease with Calpine for the project in November 2004 after only one day of public notice. Just a few months later, in February 2005, Northern Star Natural Gas LLC announced plans to build the Bradwood Landing LNG import facility on 55 acres of a 420-acre site located between Astoria and Clatskanie on the Columbia River. Northern Star swiftly initiated the prefiling process with FERC in March 2004 and completed scoping of the project's environmental review in September 2005, conducting meetings in Knappa, Oregon, and Cathlamet, Washington.

While FERC was putting together the Draft EIS for the Bradwood Landing project, Calpine went bankrupt and sold its lease to the New York–based Leucadia National Corporation in January 2007. The Skipanon Natural Gas Facility was renamed Oregon LNG, and Leucadia initiated FERC's prefiling process in June 2007. Meanwhile, the Bradwood Landing proposal continued to move forward with the release of a Draft EIS in August 2007 and the associated comment period until December 2007, with six public hearings held in November 2007. In June 2008, FERC issued the final EIS for Bradwood Landing and subsequently approved the project in September 2008, making it the first LNG import facility approved on the West Coast of the United States. The states of Washington and Oregon filed appeals of FERC's decision shortly thereafter, arguing that FERC's approval could only come after clean air, clean water, and coastal zone permits had been approved at the state level. Controversy about Clatsop County's local land use permit for the terminal continued. In May 2010, Northern Star declared bankruptcy and officially suspended the Bradwood Landing LNG project, citing delays in state and federal permitting and financial challenges.

Oregon LNG did not get very far in its environmental review as an import project: FERC conducted a few scoping meetings about the proposal in late 2007 and early 2008. Instead, in mid-2011, citing changing market conditions, the company announced plans to retrofit its proposal to export LNG and initiated prefiling with FERC in July 2012. Scoping for the project was conducted in the fall of 2012, and FERC issued a draft EIS for Oregon LNG in August 2015 with public comment and hearings in fall 2015. Facing controversies surrounding the company's control of the proposed site, including a series of unsuccessful lawsuits with the U.S. Army Corps of Engineers and the denial of a local land use permit, the proposing company officially withdrew in April 2016, before FERC issued a final EIS.

LNG in Coos Bay: The Jordan Cove Energy Project

Initial discussions on the Jordan Cove Energy Project between the port and the company in late 2003 were followed by a public announcement in August 2004. In September 2004, the company finalized a purchase agreement for 90 acres of industrial land on Coos Bay's North Spit, initiating the permitting process just two months later by submitting a notice of intent to the Oregon Energy Facility Siting Council. With the shift in siting authority to FERC in 2005, the proposing company initiated the prefiling process with FERC in April 2006. Scoping meetings were held during the summer of 2006 and continued into early 2007. In late 2008, FERC issued a draft EIS for the import proposal, and one year later FERC approved the project. However, in early 2011, like Oregon LNG, Jordan Cove also proposed a change to export LNG.

Environmental review of the proposed export project began in May 2012, with scoping meetings conducted in August and October 2012. The draft EIS for the export proposal was released in November 2014 with public meetings held at the end of 2014. The final EIS was released in late September 2015. In March 2016, in a move the shocked both supporters and opponents, FERC rejected the proposal, citing the lack of a demonstrated market for proposal's natural gas and landowner concerns along the pipeline route. The proposing companies have since requested a rehearing but FERC has yet to decide whether to grant this request. In both cases, our data collection efforts occurred largely before these latest decisions, thus we provide a glimpse of perceptions during ongoing decision-making processes.

Results

Active participants in the public participation processes surrounding the two proposals perceived the effectiveness of such processes differently depending on their stance toward the proposal. For this reason, we discuss opponents and supporters separately.

Opponent Perceptions of Public Participation Efforts and Mobilization

We found that while opponents in both siting processes agreed that there had been many opportunities for citizens to attend and be informed about the proposals, they did not feel that public input had been adequately incorporated into decisions. As stated by one opponent at a hearing along the proposed Pacific Connector pipeline route: "We make comments, we come up here . . . we put our heart on the line, we stay up nights, we travel all over the place to put our comments into this process and you just throw them into the garbage, I guess, because none of them ever get addressed. It's a damned dog and pony show by the industry. You're owned by the industry, you're a puppet for the industry!" (as quoted in Federal Energy Regulatory Commission 2014). Echoing similar concerns in an interview, one Coos Bay opponent lamented: "You can express your concerns, but they won't listen. The purpose is not for the citizen to get heard, it's to check off the boxes in the process" (interview 9). In fact, this sentiment was echoed in 21 out of the 25 interviews we conducted with opponents.

In particular, the initial decisions by the ports in both locations to lease land to the proposing companies were viewed by eventual opponents as occurring in secret with little public input. This secrecy not only motivated eventual opponents but also contributed to feelings of distrust toward both the proposing companies and local decision makers. In Astoria, all opponents interviewed mentioned the property lease to Calpine in 2004, which was agreed upon and voted by the former port director and commissioners after only one day of public notice, as a motivating event. And, in letters to the editor, opponents revealed that the secrecy of the lease played a key role in shaping their decision to mobilize against LNG. This sentiment was repeatedly mentioned; for example, one opponent wrote in a letter to the editor: "Oregon LNG, the recipient of the benefits of a bonehead deal between the now defunct Calpine Energy Group and our own Port of Astoria, is already flaunting a facility on the Skipanon Peninsula. . . . We, the citizens of Clatsop County cannot afford to sit home on this one" (McGee 2007). One opponent who attended the meeting that the port commissioners held prior to approving the Calpine lease described it as "an illegal meeting, making an illegal decision because they did not fulfill the requirement for public notice of the meeting" (interview 5).

Similarly, in Coos Bay, opponents viewed the process of making the initial decision regarding the lease in an unfavorable light. According to one opponent, who learned of the Jordan Cove Energy Project proposal in May of 2004, "We were told it was already a done deal that had been decided in the back room" (interview 25). Although an interview participant mentioned that focus groups were organized before moving forward with plans to site Jordan Cove, those invited were "political leaders, business leaders and primarily citizens with an interest in economic development" (interview 28). Jordan Cove's Public Affairs representative concurred, "The public doesn't necessarily have a say in Jordan Cove's business. However, elected officials were consulted. The public can kind of come after when the site is decided" (interview 26). Thus, it is not surprising that the initial lease decisions in Astoria and Coos Bay, which in many ways correspond to the discredited Decide-Announce-Defend strategy of facility siting, created feelings of mistrust and motivated many eventual opponents.

In addition, our interviews with opponents revealed that these feelings of mistrust of local decision makers actually predated the LNG proposals and were often rooted in past experiences with other industrial development projects (interviews 1, 3, 4, 5, 6, 7, 10). As explained by one eventual Coos Bay opponent, Richard Knablin, founder of Coos County Citizens for Responsible Government, a precursor to the main group opposed to Jordan Cove, Citizens Against LNG, in an initial newspaper article about the proposal, "The history here has been that jobs at any price prevails and all these plans, from Nucor to (Daishowa Paper Products), they never involve citizen involvement. . . . All they do is go to the same people—the South Coast Development Council, Friends of New and Sustainable Industry—and here they are making decisions for everyone" (as quoted in Sirocchi 2004).

Opponents repeatedly lamented the unresponsiveness of local decision-making circles in our interviews. All seven opponents in Coos Bay who lived in the community prior to the Jordan Cove Energy Project proposal stated that local decision makers had been unresponsive to concerns of the locals. As explained by Wim de Vriend, who wrote a book about the history of failed industrial development projects in Coos Bay entitled *Job Messiahs*, explained: "Those who get elected get elected because they have more funds and resources. . . . They blame all past failures on environmentalists and use stereotypes to label people" (interview 27).

In Astoria, when asked about the responsiveness of elected officials to community concerns prior to the LNG proposals, 9 out of 10 opponents interviewed suggested that their local officials had not been responsive. For example, one resident reported: "The county council had a very narrow focus and was originally very pro-industrial and pro-development; [they were] less interested in environmental issues and quality of life issues. The Port of Astoria was originally also very pro-industrial, always looking for the brass ring, willing to bring in anything promised to them that would bring in money without consideration for how realistic, impacts to quality of life and the river" (interview 1). As evidenced by this quote, the lack of trust in decision makers who played a key role in the siting of the proposals clearly motivated opponents.

Opponents also expressed feelings of distrust toward the proposing companies. During interviews in Astoria, all opponents stressed concerns about the proposing company's trustworthiness, with 7 of the 10 opponents interviewed mentioning the example of Northern Star Natural Gas's May 2010 bankruptcy. The company's bankruptcy, which marked the end of the Bradwood Landing proposal, left an unpaid debt of roughly $200,000 to the county, local nonprofit organizations, and businesses (*Daily Astorian* 2010). For many opponents in Astoria, this past experience was an important motivator for their opposition to LNG. The switch from import to export in both cases only cemented these concerns about company intentions and reliability. After years of shying away from public questions about export, project backers officially announced their export plans in 2011. Opponents in both cases suspected that this move would eventually occur. They used what they called this "bait and switch" event as powerful evidence to allege the companies' lack of credibility, asking others: "Are you going to blindly trust these company men?" (interview 4). A notable quote exemplifying these feelings about the change from import to export follows:

Exporting natural gas has hardened opposition to the project. . . . When you're talking about shipping fracked gas to overseas markets; this shifts the public toward our [the opposition's] direction because the projects are so different than what they originally proposed. [The LNG companies] have supported LNG for years on end, they said they would never export LNG and called us conspiracy theorists and then, all of a sudden, it flipped to exporting LNG. People in Oregon who were against the pipeline really noticed that. Opponents were anticipating this twist even before it happened. This inspired people to stay connected to this project. (interview 10)

In sum, the opponents' distrust of the proposing companies in each case provided a powerful motivating force for opposition.

Regarding the degree to which opponents felt citizen concerns were reflected in the federal public participation process, opponents in both cases perceived FERC's overall approach to environmental review to be biased in favor of approval (interviews 10, 25, 27, 29, 35, 40, 41, 42, 43). Additionally, many opponents expressed frustration because they felt FERC not only disregarded public comments, but also those from other federal and state agencies (interviews 1, 2, 5, 8, 9, 10). Said one opponent: "FERC is dismissive of public input and does a poor job of incorporating even concerns from state and other federal agencies. It's as close to a rubber stamp as you can get. The people who have been to the hearings get the impression that their input is not taken seriously, and we get the impression that FERC has already made up their minds. And, when you look at their record, FERC very, very seldom rejects an LNG proposal" (interview 10). A frequently mentioned example came from the Bradwood Landing proposal in Astoria. In September 2008, despite the expressed misgivings of other federal and state agencies, FERC conditionally approved the proposal—a decision that the states of Oregon and Washington appealed in court. In fact, opponents in both cases were most complimentary of the participation processes implemented by state officials. For example, one interviewee said that Oregon state agencies have "the reputation of being responsive to citizens and responsible in an environmental way that affects quality of life, . . . [Oregon] is known for having very good land use laws that limit development" (interview 5). This is likely why LNG opponents have repeatedly targeted state representatives and involved agencies to challenge FERC's decisions.

Our coding of comments by opponents in letters to the editor in local newspapers underlines the importance of opponents' concerns about the regulatory process, decision maker trustworthiness, and company trustworthiness in the framing of LNG opposition in both cases (see Figures 10.1a, b). This combination accounts for roughly a quarter of opponent comments in letters to the editor in *The Daily Astorian* and one-third in *The Coos Bay World* and is second only to comments about the potential risks posed by the proposal. The relationship between perceptions of trustworthiness of decision makers and project proponents *and* subsequent acceptance of siting decisions is also well recognized in the literature (Terwel et al. 2010).

While opponents in both cases were able to launch significant mobilization efforts that included extensive letter-writing campaigns, protests, and appeals (see Table 10.3), opponents differed in the two cases in their ability to influence local decision-making bodies. Opponents in Astoria, building on an existing network of politically active groups and individuals in the community, successfully recalled county officials and replaced county and port commissioners with allies sympathetic to their cause, creating a more politically opportune

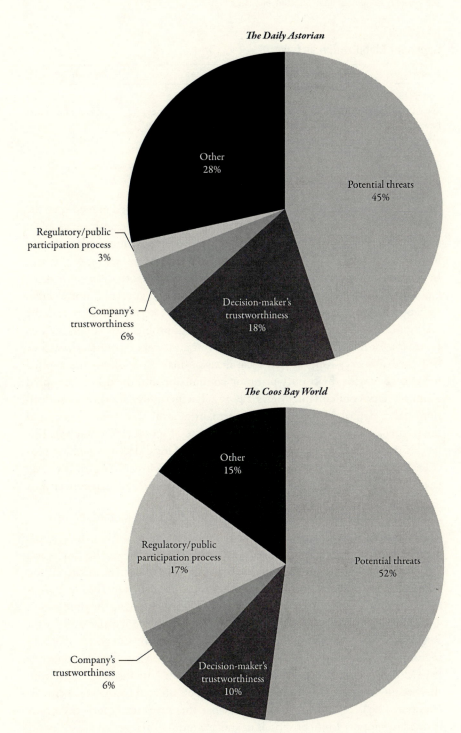

FIGURE 10.1A, B Topics of comments in letters to the editor by case. (a) *The Daily Astorian*. (b) *The Coos Bay World*.

Table 10.3
Opponent Mobilization by Case

Activity	Astoria	Coos Bay
Letters to the editor	872	326
Speakers at EIS hearings	218 (9 hearings)	452 (10 hearings)
Lawsuits / appeals (Oregon Department of Energy, n.d.).	6	12
Protest events	20	23

environment for opposition. The opponents we interviewed agreed that the new county commission (elected in May 2010) marked a "transformation" in local decision making after which citizen input was taken more seriously and incorporated into the decision-making process (interviews 1, 2, 5, 6, 7, 9, 10). In contrast to Astoria, where opponents had more access to decision makers, Coos Bay's port commissioners are appointed by the governor, not elected, a change that was instituted by referendum in the late 1980s and repeatedly mentioned by Coos Bay opponents interviewed as problematic (interviews 27, 31, 32, 35). Unlike in Astoria, two Coos Bay interviewees recounted how they had tried unsuccessfully to join local decision-making bodies—one applied for a vacancy with the county planning commission and the other attempted a run for city council (interviews 24, 27). Furthermore, another local decision maker described the pressure to remain neutral about the project and the stigma attached to being openly against the proposal (interview 35). This reveals that Coos Bay opponents felt shut out of local decision-making processes and were driven toward other methods of participation (e.g., protesting, lawsuits) in response to a lack of local elite allies to represent their concerns.

Supporter Perceptions of Public Participation Efforts and Mobilization

The two cases also differed in terms of supporter mobilization. In Astoria, most supporters faded from public view after the demise of the Bradwood Landing proposal. Supporters in Astoria sent only 95 letters to the editor, accounting for only 9% of all letters to the editor published in *The Daily Astorian*, and only 52 supporters testified at FERC hearings during the time period analyzed—mostly representatives from unions like the Pacific Northwest Regional Council of Carpenters, Union Carpenters for Oregon and Washington, and Carpenters Local 156. According to interviews, local supporters have also been less visible: As one supporter said: "They stayed away from this because they are afraid their business will be affected by the vocal crowd" (interview 20). Thus, it is unsurprising that there was no group specifically formed to support LNG proposals in and around Astoria.

We noted the importance of Bradwood Landing's bankruptcy and the differences in company strategy between Northern Star Natural Gas (of

Bradwood Landing) and Oregon LNG. Indeed, all supporters interviewed in Astoria recalled how Northern Star's representatives had actively cultivated support in Astoria. They established an office front in downtown Astoria, spoke regularly to the media, signed a memorandum of understanding with the Columbia Pacific Building Trades Council, and sponsored union worker attendance at public meetings (interviews 11, 12, 13, 20). In general, Northern Star was known to "financially support various things and attempted to be engaged in the community" (interview 11).

Compared to Northern Star and perhaps because of the negative impression left by Northern Star's bankruptcy, Oregon LNG took a different approach. As noted by the Oregon LNG's CEO, "Our strategy has been based on making [sure] that the state of Oregon follows the rules" (as quoted in Magill and Rubin 2015). The company has thus downplayed many decisions made by the county and state boards, citing FERC's exclusive authority over LNG siting, and has not spent much time cultivating support from the local community.

In contrast, in Coos Bay, a more concerted effort to mobilize support has taken place, in some ways akin to Northern Star's initial efforts in Astoria, especially since the change from import to export. As revealed in letters to the editor and interviews, supporters in Coos Bay felt that public discussions were being dominated by opponents. Indeed, our analysis of stances of letters to the editor in *The Coos Bay World* shows that, of the 514 letters collected, 326 (63%) took oppositional stances, with 112 (22%) supportive—one indication that opponents were more vocal in public discussions about the project.

Whether these oppositional views did represent the larger community's assessment of the project is unclear; what is clear is that some supporters believed opponents to be a vocal minority and wanted to mobilize the "large, silent majority" (interviews 26, 34, 36). According to one supporter, "Opponents were pushing fear rhetoric. They're a minority; we wanted to give everyone else a voice" (interview 34). Another supporter wrote in a letter to the editor that opponents' arguments were "based on emotions and fear" (Mann 2006). Indeed, our coding of supportive comments in letters to the editor in *The Coos Bay World* shows that comments about the irrational and emotional fears of opponents was the third-most commonly mentioned issue (17%), trailing only references to the proposal's potential economic benefits (48%) and community context (19%).

As a result of these concerns, local supporters and the company joined forces in 2013 to create Boost Southwest Oregon (BSWOR), an organization whose primary focus was to gather support for Jordan Cove, mainly by emphasizing the project's potential economic benefits to the economically depressed region (interviews 34, 36). According to one BSWOR member, "The whole organization took shape because . . . we wanted to show (the decision makers and congressional delegation) our community support for the project. . . . The [BSWOR] website was developed to combat some misinformation about the project" (as quoted in Novotny 2014). In addition to the group's website, BSWOR purchased radio, newspaper, and TV ads promoting the project and organized rallies at public meetings about the project (Davis 2014; Novotny 2014), with funding and other types of support, including educational and training materials supplied by the proposing company (Davis 2014; Novotny 2014). BSWOR provided training for writing effective letters to the local newspaper and representatives, tips for speaking at public hearings, and pre-prepared presentations on the project's potential benefits. In one interview, a BSWOR member explained that it was "helpful to train the blue collar workers who are local to know about the reality [so they are able to] speak intelligently with the same language. This is what BSWOR is working on. Community members don't feel confident, so we should know how to empower this population . . . get more turnout and bring out the workers" (interview 36).

Our coding of the stances of speakers at FERC public meetings suggests that BSWOR was effective in its efforts (see Figure 10.2). While only 11 people spoke in support of the Jordan Cove project in 2008 during hearings about the draft EIS for LNG import, 110 spoke in support in 2014 during hearings about the draft EIS for LNG export. Photographs of buses bringing supporters to meetings, many of whom belong to unions associated with BSWOR, highlight the organization's role in increasing turnout of project supporters (Davis 2014; interviews 25, 27, 30, 33, 36, 40, 41, 42, 43).

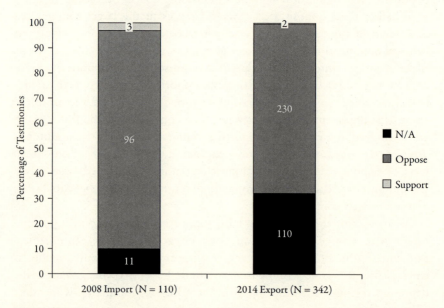

FIGURE 10.2 Change in stances of speakers at FERC public meetings about the Jordan Cove project.

Discussion and Conclusions

Our results clearly show that recent developments in unconventional oil and gas development not only affect adjacent communities, but also those far afield, who face proposals for related infrastructure. Moreover, the current regulatory structure geared toward incorporating public participation has created frustration on both sides. In sum, we found that while opponents in both communities agreed there were many opportunities for citizens to attend meetings and be informed about the proposals, they did not feel that their input was adequately incorporated into the decisions. As a result, they turned to other tactics to express their opinions, including protests and lawsuits. In contrast, supporters, most especially those in Coos Bay, perceived opponents to be a vocal minority and organized a counter-movement in response.

In many ways, our results confirm much of the reviewed literature about siting and social movement emergence. Initial lease decisions in both locations mimicked the discredited Decide-Announce-Defend strategy of facility siting, created feelings of mistrust, and motivated many eventual opponents. Structural political opportunities to recall and replace local elected officials, particularly in Astoria, allowed a well-organized local opposition to put allies in positions of power and overturn important permits previously granted.

Our findings also echo the findings of other writers in this volume in terms of the fracturing that can occur within communities during such debates as residents weigh the opportunities and costs associated with energy development. Astoria and Coos Bay are not alone. Many other communities throughout the United States face proposed LNG development, often with debates spanning more than a decade, first as import and now export terminals. Moreover, deliberations about export are not limited to LNG. Several coal and oil export terminals have been proposed in the Pacific Northwest and have faced fierce opposition, as have associated rail and pipeline transport projects. This flurry of proposed export activity and the associated, and largely successful, opposition has led the Sightline Institute—a Seattle-based progressive think tank—to call for the region to hold the "The Thin Green Line" separating North America's energy-producing regions and energy-hungry Asian regions to limit fossil fuel development (de Place 2014).

For Astoria, the LNG debate is essentially over—and Oregon LNG has withdrawn. The fate of the Jordan Cove Energy Project in Coos Bay, however, remains an open question. Since FERC's rejection, the proposing companies have secured preliminary agreements with several potential buyers and have applied for a rehearing with FERC. As of this writing, FERC has yet to issue a decision as to whether to grant the companies' request.

Appendix

Number	Name	Description	Date	Case
1	C. Johnson	opponent	8/7/14	Astoria
2	C. Newman	opponent	8/5/14	Astoria
3	C. Dominey	opponent	8/6/14	Astoria
4	J. Dominey	opponent	8/6/14	Astoria
5	L. Caplan	opponent	8/7/14	Astoria
6	L. Durheim	opponent	8/5/14	Astoria
7	T. Durheim	opponent	8/5/14	Astoria
8	P. Corcoran	Oregon State University Extension	8/7/14	Astoria
9	S. Skinner	opponent	8/6/14	Astoria
10	Anonymous	opponent	7/26/14	Both
11	J. Dunzer	supporter	8/28/14	Astoria
12	R. Balkins	supporter	9/8/14	Astoria
13	E. Nyberg	supporter	9/8/14	Astoria
14	K. Fritsch	Warrenton city manager	9/9/14	Astoria
15	S. Urling	Warrenton planning director	9/9/14	Astoria
16	J. Bunch	Clatsop County planning staff	8/7/14	Astoria
17	K. Sanders	former port commissioner	8/27/14	Astoria
18	M. Weston	port official	8/7/14	Astoria
19	P. Trapp	port official	8/6/14	Astoria
20	A. Samuelson	local decision maker	9/9/14	Astoria
21	S. Shribbs	state agency staff	9/4/14	Astoria
22	T. Shorack	newspaper reporter	8/7/14	Astoria
23	R. Sadler	active citizen	7/30/14	Coos Bay
24	R. Knablin	opponent	7/30/14	Coos Bay
25	J. McCaffree	opponent	7/31/14	Coos Bay
26	M. Hinrichs	company representative	7/31/14	Coos Bay
27	W. de Vriend	opponent	8/1/14	Coos Bay
28	Anonymous	local decision maker	8/1/14	Coos Bay
29	J. Clark	opponent	7/31/14	Coos Bay
30	M. Geddry	opponent	8/1/14	Coos Bay
31	J. Jones	opponent	8/2/14	Coos Bay
32	J. Jones	opponent	8/2/14	Coos Bay
33	J. Dilley	opponent	8/2/14	Coos Bay
34	M. Wall	supporter	8/5/14	Coos Bay
35	Anonymous	local decision maker	9/3/14	Coos Bay
36	D. Granger	supporter	9/4/14	Coos Bay
37	Anonymous	local decision maker	9/4/14	Coos Bay
38	L. Campbell	newspaper reporter	9/25/14	Coos Bay
39	Anonymous	decision maker	11/10/14	Both
40	C. Adams	opponent along pipeline	12/12/14	Coos Bay
41	L. Hyde	opponent along pipeline	1/19/15	Coos Bay
42	P. Hyde	opponent along pipeline	1/19/15	Coos Bay
43	B. Hyde	opponent along pipeline	1/19/15	Coos Bay
44	Anonymous	agency staff	1/14/15	Coos Bay
45	L. Clausen	opponent	8/1/14	Coos Bay

References

Anderson, Brooklynn J., and Gene L. Theodori. 2009. "Local Leaders' Perceptions of Energy Development in the Barnett Shale." *Southern Rural Sociology* 24(1): 113–129.

Armour, Audrey M. 1991. "The Siting of Locally Unwanted Land Uses: Towards a Co-operative Approach." *Progress to Planning* 35(1) (http://dx.doi.org/10.1016/0305-9006 (91)90007-O).

Beierle, Thomas C., and Jerry Cayford. 2002. *Democracy in Practice: Public Participation in Environmental Decision Making*. Washington, D.C.: Resources for the Future Press.

Bengel, Erick. 2015. "Wyden, Merkley, Bonamici ask for extension of public comment period on LNG." *Daily Astorian*, September 18, News Section.

Boholm, Asa, ed. 2004. *Facility Siting: Risk, Power and Identity in Land Use Planning*. Toronto: Earthscan Canada.

Boudet, Hilary S. 2010. *Contentious Politics in Liquefied Natural Gas Facility Siting*. Ph.D. Thesis, Stanford University: Stanford, CA.

Boudet, Hilary S. 2011. "From NIMBY to NIABY: Regional Mobilization Against Liquefied Natural Gas in the United States." *Environmental Politics* 20: 786–806. (http://dx.doi .org/10.1080/09644016.2011.617166).

Boudet, Hilary S. 2015. "An 'Insiteful' Comparison: Contentious Politics in Liquefied Natural Gas Facility Siting in the U.S." *MIT Projections* 11: 47–76.

Boudet, Hilary S., and Leonard Ortolano. 2010. "A Tale of Two Sitings: Contentious Politics in Liquefied Natural Gas Facility Siting in California." *Journal of Planning Education and Research* 30: 5–21. (http://dx.doi.org/10.1177/0739456X10373079).

Brasier, Kathryn J., Matthew R. Filteau, Diane K. McLaughlin, Jeffrey Jacquet, Richard C. Stedman, Timothy W. Kelsey, and Stephan J. Goetz. 2011. "Residents' Perceptions of Community and Environmental Impacts from Development of Natural Gas in the Marcellus Shale: A Comparison of Pennsylvania and New York Cases." *Journal of Rural Social Sciences* 26(1): 32–61.

Caplan, Laurie. 2007. "Ripe for Speculators." *Daily Astorian*, January 12, Opinion Section.

Carmin, Joann. 2003. "Resources, Opportunities and Local Environmental Action in the Democratic Transition and Early Consolidation Periods of the Czech Republic." *Environmental Politics* 12: 42–64. (http://dx.doi.org/10.1080/09644010412331308274).

Chasm, Richard. 2008. "LNG Will Hurt up to 400 Landowners." *Coos Bay World*, October 11, Opinion Section.

The Daily Astorian. 2010. "Let Us Now Sober Up: Bradwood's Demise Echoes the Crash of Enron and WPPSS Nuclear Plants." Newspaper editorial. May 6, Opinion Section.

Davis, Chelsea. 2014. "Union Members Crowd Jordan Cove Hearing in Effort to Drown Out LNG Opponents." *Coos Bay World*, June 26, News Section.

de Place, Eric. 2014. *Northwest Fossil Fuel Exports: Planned Facilities Would Handle Five Times as Much Carbon as the Keystone XL Pipeline*. Seattle, WA: Sightline Institute. Retrieved from http://www.sightline.org/research_item/northwest-fossil-fuel -exports-2/.

Dietz, Thomas, and Paul Stern. 2008. *Public Participation in Environmental Assessment and Decision Making*. Washington, D.C.: National Academies Press.

Eaton, Emily, and Abby Kinchy. 2016. "Quiet Voices in the Fracking Debate: Ambivalence, Nonmobilization, and Individual Action in Two Extractive Communities." *Energy Research and Social Science* 20: 22–30.

Eisinger, Peter K. 1973. "The Conditions of Protest Behavior in American Cities." *American Political Science Review* 67(1): 11–28 (http://dx.doi.org/10.2307/1958525).

Environment Division. 1998. *Doing Better Business Through Effective Public Consultation and Disclosure: A Good Practice Manual.* Washington, D.C.: International Finance Corporation.

Federal Energy Regulatory Commission. 2014. *Public Hearing Transcripts for the Draft Environmental Impact Statement for the Jordan Cove-Pacific Connector Pipeline Project.* Docket Nos. CP13–483–000 and CP13–492–000. Retrieved from www.ferc.gov on Jan. 15, 2015.

Fiorino, Daniel J. 1990. "Citizen Participation and Environmental Risks: A Survey of Institutional Mechanisms." *Science, Technology and Human Values* 15: 226–243 (http://dx.doi.org/10.1177/016224399001500204).

Freudenburg, William R. 2004. "Can We Learn from Failure? Examining U.S. Experiences with Nuclear Repository Siting." *Journal of Risk Research* 7(2): 153–169.

Gould, Roger V. 1993. "Collective Action and Network Structures." *American Sociological Review* 58: 182–196.

Gould, Roger V. 1995. *Insurgent Identities: Class, Community and Protest in Paris from 1848 to the Commune.* Chicago, IL: University of Chicago Press.

Hunter, Susan, and Kevin M. Leyden. 1995. "Beyond NIMBY: Explaining Opposition to Hazardous Waste Facilities." *Policy Studies Journal* 23(4): 601–604 (http://dx.doi.org/10.1111/j.1541-0072.1995.tb00537.x).

Kraft, Micheal E., and Bruce B. Clary. 1991. "Citizen Participation and the NIMBY Syndrome: Public Response to Radioactive Waste Disposal." *Western Political Quarterly* 44(2): 299–328. (http://dx.doi.org/10.1177/106591299104400204).

Kriesky, Jill, Bernard D. Goldstein, Katrina Zell, and Scott Beach. 2013. "Differing Opinions about Natural Gas Drilling in Two Adjacent Counties with Different Levels of Drilling Activity." *Energy Policy* 58: 228–236.

Kunreuther, Howard, and John W. Lathrop. 1981. "Siting Hazardous Facilities: Lessons from LNG." *Risk Analysis* 1(4): 289–302.

Kunreuther, Howard, Kevin Fitzgerald, and Thomas Aarts. 1993. "Siting Noxious Facilities: A Test of the Facility Siting Credo." *Risk Analysis* 13(3): 301–318.

Ladd, Anthony E. 2013. "Stakeholder Perceptions of Socioenvironmental Impacts from Unconventional Natural Gas Production and Hydraulic Fracturing in the Haynesville Shale." *Journal of Rural Social Sciences* 28(2): 56–89.

Ladd, Anthony E. 2014. "Environmental Disputes and Opportunity-Threat Impacts Surrounding Natural Gas Fracking in Louisiana." *Social Currents* 1(3): 293–311.

Lesbirel, S. Hayden, and Daigee Shaw, eds. 2005. *Managing Conflict in Facility Siting: An International Comparison.* Cheltenham, UK: Edward Elgar.

Lofland, John, David A. Snow, Leon Anderson, and Lyn H. Lofland. 1995. *Analyzing Social Settings: A Guide to Qualitative Observation and Analysis.* Belmont, CA: Wadsworth.

Magill, Jim, and Richard Rubin. 2015. "Oregon LNG to Move Ahead, Despite Adverse Pipeline Decision." May 1. Retrieved from http://www.platts.com/.

Mann, Jim. 2006. "Arguments Based on Emotion, Fear." *Coos Bay World*, December 9, Opinion Section.

McAdam, Doug. 1999. *Political Process and the Development of Black Insurgency, 1930–1970.* Chicago, IL: University of Chicago Press.

McAdam, Doug, and Hilary S. Boudet. 2012. *Putting Social Movements in Their Place: Explaining Opposition to Energy Projects in the United States, 2000–2005.* Cambridge, UK: Cambridge University Press.

McAdam, Doug, John D. McCarthy, and Mayer N. Zald. 1996. *Comparative Perspectives on Social Movements.* Cambridge, UK: Cambridge University Press.

McGee, Patrick. 2007. "Strength in Unity." *Daily Astorian*, October 12, Opinion Section.

McKibben, Bill. 2013. *Oil and Honey: The Education of an Unlikely Activist*. New York, NY: Times Books.

Morris, Aldon D. 1986. *The Origins of the Civil Rights Movement: Black Communities Organizing for Change*. New York, NY: Free Press.

Novotny, Tim. 2014. "Jordan Cove Battle Waged on Different Fronts." *Coos Bay World*, January 28, News Section.

O'Faircheallaigh, Ciaran. 2010. "Public Participation and Environmental Impact Assessment: Purposes, Implications, and Lessons for Public Policy Making." *Environmental Impact Assessment Review* 30(1): 19–27. (http://dx.doi.org/10.1016/j.eiar.2009.05.001).

Oregon Department of Energy. N.d. *LNG in Oregon*. Retrieved from http://www.oregon.gov/.

Penny, Ray. 2008. "LNG Reward Far Outweighs Risks." *Coos Bay World*, October 20, Opinion Section.

Sadler, Ron. 2008. "Longtime Investment Hasn't Paid Off." *Coos Bay World*, December 15, Opinion Section.

Schafft, Kai A., Yetkin Borlu, and Leland Glenna. 2013. "The Relationship between Marcellus Shale Gas Development in Pennsylvania and Local Perceptions of Risk and Opportunity." *Rural Sociology* 78(2): 143–166.

Sierra Club. 2016. *Beyond Natural Gas*. Retreived March 15, 2016. (http://content.sierraclub.org/naturalgas/).

Smith, Eric R.A.N., and Marisela Marquez. 2000. "The Other Side of the NIMBY Syndrome." *Society and Natural Resources* 13(3): 273–280. (http://dx.doi.org/10.1080/089419200279108).

Sherman, Daniel J. 2004. "Not Here, Not There, Not Anywhere: The Federal, State and Local Politics of Low-Level Radioactive Waste Disposal in the United States, 1979–1999." PhD Thesis. Ithaca, NY: Cornell University.

Sirocchi, Andrew. 2004. "Citizen Group to Review Dangers, Impact of Liquid Natural Gas Plan." *Coos Bay World*, September 14, News Section.

Snow, David A., E. Burke Rochford Jr., Steven K. Worden, and Robert D. Benford. 1986. "Frame Alignment Processes, Micromobilization, and Movement Participation." *American Sociological Review* 51: 464–481.

Snow, David A., and Robert D. Benford. 1988. *From Structure to Action: Social Movement Participation Across Cultures*, eds. Bert Klandermans, Hanspeter Kriesi, and Sidney Tarrow, 197–217. Stamford, CT: JAI Press.

Snow, David A., and Robert D. Benford. 1992. *Frontiers in Social Movement Theory*, eds. Aldon D. Morris and Carol McClurg Mueller, 133–155. New Haven, CT: Yale University Press.

Snow, David A., Louis A. Zurcher, and Sheldon Ekland-Olson. 1980. "Social Networks and Social Movements: A Microstructural Approach to Differential Recruitment." *American Sociological Review* 45(5): 787–801. (http://dx.doi.org/10.2307/209489).

Tarrow, Sidney. 1998. *Power in Movement: Social Movements and Contentious Politics*. Cambridge, UK: Cambridge University Press.

Terwel, Bart W., Fieke Harinck, Naomi Ellemers, and Dancker D. L. Daamen. 2010. "Voice in Political Decision-Making: The Effect of Group Voice on Perceived Trustworthiness of Decision Makers and Subsequent Acceptance of Decisions." *Journal of Experimental Psychology: Applied* 16(2): 173–186. (http://dx.doi.org/:10.1037/a0019977).

Tilly, Charles. 1978. *From Mobilization to Revolution*. New York, NY: McGraw-Hill College.

U.S. Department of the Army. 2004. *Regulatory Standard Operating Procedures for Processing Liquefied Natural Gas Projects*. Washington, D.C.

U.S. Department of Energy. 2014. *Order Conditionally Granting Long-Term Multi-Contract Authorization to Export Liquefied Natural Gas by Vessel from the Jordan Cove LNG Terminal in Coos Bay, Oregon to Non-Free Trade Agreement Nations.* Washington, D.C.: Office of Fossil Fuels.

Wakefield, Susan E. L., Susan J. Elliot, John D. Eyles, and Donald C. Cole. 2006. "Taking Environmental Action: The Role of Local Composition, Context, and Collective." *Environmental Management* 37(1): 40–53. (http://dx.doi.org/10.1007/s00267-004-0323-3).

Walsh, Edward J., Rex H. Warland, and Douglas C. Smith. 1997. *Don't Burn It Here: Grassroots Challenges to Trash Incinerators.* State College: Pennsylvania State University Press.

Weiss, Robert S. 1994. *Learning from Strangers: The Art and Method of Qualitative Interview Studies.* New York, NY: Free Press.

Wright, Rachel A., and Hilary S. Boudet. 2012. "To Act or Not to Act: Context, Capability, and Community Response to Environmental Risk." *American Journal of Sociology* 118: 728–777. (http://dx.doi.org/10.1086/667719).

Conclusion

Standing at the Energy
Policy Crossroads

ANTHONY E. LADD

As the preceding chapters and case studies featured in *Fractured Communities* make clear, the controversy over unconventional oil and gas development and hydraulic fracturing represents one of the most contentious arenas of social and environmental conflict in the United States today. Proponents trumpet fracking as an economic "game changer" that creates jobs, personal wealth, tax revenues, rural revitalization, and energy independence from foreign oil. Other supporters, including some environmental groups, view natural gas fracking as providing a cleaner, "bridge fuel" to an alternative energy future that will lower carbon emissions, as well as reduce the planetary risks posed by climate change. Conversely, opponents attack fracking for its myriad threats to water quality, aquifers, public health, rural landscapes, roads, property values, animals, farm communities, and nature-based tourism, among other problems. Critics see natural gas fracking as not just another risky and inefficient energy source that keeps the nation chained to the fossil fuel treadmill driving climate change, but a dangerous and invasive technological process that is linked to "every part of the environmental crisis, from radiation exposure to habitat loss" (Steingraber 2012: 175).

In major shale regions and states across the country, communities have become deeply divided—indeed *fractured*—over the differential opportunity-threat impacts posed by oil and gas fracking, the rise in induced seismicity and earthquakes from underground wastewater injection, and the power of the industry to override local bans or municipal regulations on shale production. Given the siting of many fracking wells in close proximity to residential homes, schools, farms, and businesses, a growing body of citizens have become fearful of what these new industrial neighbors mean for their health, safety, and well-being, as well as the risks posed by the labyrinth of diesel trucks, pipelines, compressor stations, export terminals, and related infrastructure required to transport the energy to market (Finkel 2015). Indeed, in an era of risk and extreme energy development, the increasing use of high-volume hydraulic fracking methods to extract shale deposits from miles below the surface of the earth and oceans signals the emergence of a "New Species of Trouble" for the community, nation, and planet (Beck 1995; Erikson 1994; Ladd 2016).

While the word "fracking" has clearly entered the popular lexicon of cultural discourse (Evensen et al. 2014), its substantive meaning represents a relatively new system of technological innovation within the energy industry that most natural and social scientists, lay citizens, and college students know little about. Given the cursory understanding of industrial fracking techniques in both public and academic circles, this volume of original sociological research provides readers with a descriptive and critical assessment of some of the history, benefits, risks, impacts, policy questions, community initiatives, and collective protests surrounding the controversy today.

Three important features of *Fractured Communities* set this book apart from other works. First, the volume brings together a diverse set of ten comparative case studies of local fracking disputes that have emerged in U.S. shale plays and states nationwide: from the Pacific Coast of Oregon to the Bakken Shale of North Dakota, the Niobrara Shale of Colorado, the Woodford Shale of Oklahoma, the Barnett Shale of Texas, the Haynesville/Tuscaloosa Shale of Louisiana, the Utica Shale of Ohio, the Marcellus Shale of New York and Pennsylvania, and the Sunniland Trend of South Florida. Second, it constitutes the only edited book in sociology that offers a synthesis of original, previously unpublished research by many nationally known and emerging scholars in environmental sociology, rural sociology, and social movements on fracking-related conflicts and impacts in their nearest shale regions, if not their own communities. Third, the book has been designed to be academically suitable for use in undergraduate- and graduate-level courses in sociology, environmental studies, political economy, energy policy, and natural resource management issues, as well as written in an engaging and accessible style for nonacademic readers, journalists, or other interested researchers. While a number of good books on fracking have been published in recent years (see, for example, Bamberger and Oswald 2014; Finkel 2015; Gullion 2015; Hauter 2016; Wilber 2015), none to date have the diverse academic features of *Fractured Communities*.

New Insights and Contributions to the Literature

The ten case studies in this collection make several important contributions to the rapidly evolving sociological literature on unconventional oil and gas development and hydraulic fracking. All of the authors explore relatively new topics that have not been examined in previous fracking research, as well as address these issues in shale regions or states that have received little attention. Indeed, the work contained here stimulates our sociological imaginations by drawing connections between the history of this nascent form of energy development and its impacts on social life and citizen agency in extractive communities. More than just "new wine in old bottles," these chapters offer some novel insights, directions, and questions for future research. At the same time, they bring together some of the useful analytical lenses of environmental sociologists with those employed by scholars of social movements and collective protest.

These case studies also serve to effectively underscore the guiding conceptual framework behind the title of the volume, *Fractured Communities: Risk, Impacts, and Protest against Hydraulic Fracking in U.S. Shale Regions*—one intended to convey both a metaphorical and substantive image of the parallel fault lines, cracks, and prominent divisions that tend to develop in local shale communities targeted by industry for unconventional oil and gas exploration and drilling. Collectively, the research illuminates the book's central thesis that the fracturing of a shale region's biophysical landscape for energy development inherently creates a system of differential risks and opportunity-threat impacts for local stakeholder groups and residents. Depending on how these socioenvironmental impacts play out over time, place, and audience, mediated by conflicting discursive frames of assorted institutional and residential voices, these synergistic forces can produce the fracturing of a community's social fabric and social capital, as well as its patterns of political mobilization and democratic governance. Taken together, the probable outcomes of these complex processes for the community constitute a continuum ranging from exuberant support for fracking and energy development, to varying degrees of acceptance or ambivalence, to quiescence, opposition, or collective protest. As the section below summarizes in more detail, the research presented in *Fractured Communities* illustrates a diverse pattern of citizen responses to various anticipatory, perceived, and documented objective impacts associated with oil and gas fracking operations, wastewater injection practices, or proposed export terminals in a dozen different shale formations and states across the country.

While environmental sociologists have long explored the social impacts of energy development and resource extraction on communities, Freudenburg and Gramling (1992) were among the first to call for more systematic attention to the predevelopment phases of large-scale industrial projects and their anticipatory impacts on rural or metropolitan populations. Indeed, controversies over state-industry proposals to develop shale oil and gas reserves near water sources, parks, forests, farms, small towns, and even cities have been at the heart of the "fracking wars" over the last decade. In these scenarios, key stakeholder groups and citizens, acting on what they view as the expected benefits and/or costs of development, often begin to mobilize around the issue before the project's "first shovel of dirt is turned" (Freudenburg and Gramling 1992: 938).

Struggles over anticipatory impacts are well illustrated in the majority of the case studies here, including Sherry Cable's research on fracking on public lands in the Utica; Suzanne Staggenborg's research on the success of movement-building in the Marcellus to ban fracking in the city of Pittsburgh; Carmel Price and James Maples's study of disputes over fracking below a historic veterans' cemetery in the Utica; Patricia Widener's research on the mobilization of citizens in South Florida to stop oil development and acid fracking in the Sunniland Trend; and Hilary Boudet, Brittany Gaustad, and Trang Tran's study of citizen resistance to proposals to build liquified natural gas (LNG) export terminals in Astoria and Coos Bay, Oregon. Additionally, the issue of anticipatory impacts is also highlighted in Cameron Whitley's examination of the research assessing the impacts of fracking on wild and domestic animals in the Marcellus Shale and beyond. Addressing the issue of impacts from a more ecological and less anthropocentric perspective, Whitley's chapter serves to remind us how nonhuman species still function today—perhaps more than ever—as sentinels that can warn us about the future risks of an environment polluted by the hazardous by-products of energy extraction and consumption.

While the ambiguity of harm in environmental disputes almost always favors the interests of industry, these chapters demonstrate how concerned residents and activists have become increasingly proficient at organizing "Not in My Backyard" (NIMBY) responses to fracking long before the wheels of development begin to turn and natural habitat is destroyed. Rather than waiting to draw on their own personal experiences once resource extraction is underway, many citizen groups today are utilizing preexisting resources, organizational networks, media accounts, scientific reports, and protest strategies from antifracking movements outside their regions who have faced similar industry threats. By the same token, when economically depressed communities anticipate that shale development will bring them badly needed jobs and other financial benefits, local residents have also organized "Please in My Backyard" (PIMBY) campaigns to recruit oil and gas companies to frack in their vicinity (Jerolmack 2014).

Three of the case studies included in *Fractured Communities* also serve to demonstrate the role that perceived opportunity-threat impacts play in generating varying degrees of support, ambivalence, or opposition to oil and gas fracking, particularly once boomtown conditions have peaked, declined, and stabilized (Ladd 2013). Illustrating the familiar sociological maxim that "what is perceived as real becomes real in its consequences," citizens typically construct risk perceptions of fracking based on a wide range of past and present lived experiences, which, in the context of other discursive messages, beliefs, and events, become accepted as relatively "factual" characterizations of the technology. This impact scenario is readily exemplified by Stephanie Malin, Stacia Ryder, and Peter Hall's research on citizen efforts to ban oil fracking in several front-range communities in Colorado's Niobrara Shale; Ion Vasi's analysis of the role of film documentaries in creating discursive opportunities for antifracking protest in the Marcellus Shale, as well as other shale communities and states; and Anthony Ladd's research on the motivational frame disputes surrounding the perceived benefits and threats of natural gas fracking in the Haynesville Shale of Louisiana. In all these cases, citizens mobilized on both sides of the fracking debate based largely around their own local observations and perceived economic interests, as well as the different types of popular media and partisan discourse they encountered. While residents took proactive and successful steps to organize an assortment of municipal bans, zoning regulations, and statewide moratoriums against oil and gas fracking in New York, Pennsylvania, and Colorado, the outcomes were quite different in Louisiana. Despite the significant skepticism among a significant minority of Louisiana stakeholders about the benefits of natural gas fracking in the Haynesville, the overwhelming response of citizens in that state's oil- and gas-friendly culture was one of general acceptance, ambivalence, or quiescence about the impacts of the natural gas boom on the region. Though community perceptions were relatively divided, there was a complete reluctance of residents to engage in any kind of mobilization efforts in support of fracking operations, nor take any collective actions that would openly oppose or challenge the Louisiana gas industry's practices, impacts, or hegemonic status. Such responses are emblematic today of what has been described as the "quiet voices" behind many fracking debates (Eaton and Kinchy 2016).

Controversy over oil and gas fracking is also a product of the contested terrain concerning expert knowledge claims, scientific facts, and peer-reviewed research publications (Gullion 2015). In most fracking battlegrounds today, concerned citizens and activists are significantly disadvantaged in their struggles to challenge the policies of state regulatory agencies that generally favor the growth machine revenues provided by energy development. Time and time again in public hearings or court proceedings, the arguments of fracking opponents are typically undercut by industry claims that the data they present are not "scientifically credible"—or fail to demonstrate that any significant social or environmental impacts occurred. Citizen groups are then forced to raise the question: how often must these impacts be physically observed—and in how many places—before they are seen as constituting valid threats that can no longer be ignored, dismissed, or denied by state officials, industry spokespeople, or other local proponents?

Nowhere is this scenario concerning objective impacts better illustrated than in Tamara Mix and Dakota Raynes's timely study of how citizens in Oklahoma responded to the increasingly visible risks of induced seismicity and earthquakes caused by wastewater injection from oil fracking. While state-industry representatives could more easily discount the relatively hidden threats posed by fracking to air and water quality, public health, climate change, and economic sustainability, it became increasingly difficult for officials to convince Oklahoma citizens that their five consecutive years of feeling the earth move, their houses shake, and witnessing their walls and foundations crumble were the result of "natural" geological causes. When an increasing number of peer-reviewed scientific studies, along with other environmental claims, began to publicly acknowledge a causal link between the increase in wastewater injection rates and the dramatic increase in hundreds of seismic events each year, the perceptions of a critical mass of citizens began to turn from curiosity and fear to anger, recreancy, and action. As revealed today in myriad natural resource and fracking conflicts across the land, states like Oklahoma represent one of the many available social laboratories for studying the linkages between technological risk, "normal accidents," institutional blame, and citizen mistrust of corporate power and state authority (Perrow 1999).

Despite the hundreds of community bans on fracking and energy infrastructure in the United States, many of the case studies featured in *Fractured Communities* also point to the industry backlash that has emerged and is growing in response. Clearly, of all the larger institutional structures that have facilitated the shale energy/fracking revolution today, perhaps none is more important than the rise of the neoliberal state over the last 30 years and its agenda to promote economic growth, capital accumulation, market deregulation, the privatization of resources, minimal public sector ownership, and the rollback of state power (Perrow 2015; Tierney 2015). Following a path paved by logging, mining, and other resource extraction industries historically, neoliberal policies seek to achieve unrestrained and deregulated oil and gas development wherever possible so that lucrative exchange values can be realized and profits maximized for corporate owners, managers, and shareholders. Accordingly, shale formations must be enclosed and intensive fracking encouraged on private and public lands through inequitable leasing agreements, declarations of eminent domain, legal end runs around existing local zoning laws, and the privileging of mineral rights over surface rights (Malin 2014; Malin et al. 2017).

Yet, as Sherry Cable's research in the Utica Shale region argues, the neoliberal state has increasingly turned its gaze in recent years to the leasing of public lands for energy development and natural resource exploitation, where profits can be more easily privatized and production costs more covertly socialized. States like Ohio have moved to permit fracking operations in and around state parks, forests, recreational lands, and wildlife areas, thereby creating significant profits for public-private sector alliances and allies, while violating public trust doctrines on which the protection of citizens' resources rests. In turn, state legislatures and supreme courts have acted to systematically overturn Home Rule statutes and local structures of democratic governance in an attempt to render citizens powerless to stop the oil and gas industry from turning their community into the next sacrifice zone. These governance impacts are acknowledged in the chapters by Vasi (Texas), Price and Maples (Ohio), Mix and Raynes (Oklahoma), Ladd (Louisiana), and Malin, Ryder, and Hall (Colorado). As one journalist (Healy 2015: Y-11) accurately summarizes the situation:

> In an aggressive response to a wave of citizen-led drilling bans, state officials, energy companies and industry groups are taking ... municipalities to court, forcing local governments into what critics say are expensive, long-shot efforts to defend the measures. While the details vary—some municipalities have voted for outright bans, and others for multiyear suspensions of fracking—energy companies in city after city argue that they have a right to extract underground materials, and that the drilling bans amount to voter-approved theft. They also say that state agencies, not individual communities, are the ones with the power to set oil and gas rules.

In a growing number of shale regions, many citizens are angry over what they see as the denial of their democratic liberties and private property rights. This is particularly the case when "forced pooling" laws—legal in 39 states—compel resistant landowners to join gas-leasing agreements with their neighbors, even if they strongly oppose drilling (Baca 2016). As one activist put it: "Local people should have a say in whether these industries, which damage our natural resources and quality of life, can come into our communities" (Hauter 2016: 257).

The ascendancy of the neoliberal state has also meant billions of dollars annually in massive economic subsidies to the oil and gas industry, which, depending on how they are calculated, have been estimated to run as high as $450 billion a year (Schwartz 2015). To prop up their profits and stock values in a period of low energy prices and diminished demand, both state and federal government agencies have worked behind the scenes with energy companies to expand production into new locales and environments previously off limits to fracking, including national parks and recreational areas, as well as the Pacific waters off the coast of California (Ladd 2014). More recently, it was reported that in the Gulf of Mexico, federal officials approved more than 1,200 offshore fracks in 630 different wells off the coasts of Texas, Louisiana, Mississippi, and Alabama between 2010 and 2014. During that time, no public involvement or site-specific tests were undertaken beforehand and some 76 billion gallons of drilling wastes were dumped into the Gulf in 2014 alone (Rowell 2016).

Some Directions for Future Sociological Research

To date, the literature on unconventional energy development and the risks of fracking has represented a relatively undertheorized area of study in the field of sociology. While researchers have creatively applied a wide panache of conceptual frameworks to the general issues at hand, from neoliberalism (Malin 2014; Malin and DeMaster 2016; Malin et al. 2017), to political ecology (Poole and Hudgins 2014), to green criminology (Opsal and Shelley 2014) and popular discourse analysis (Evensen et al. 2014), among others, the socioenvironmental consequences, ecological disruptions, policy debates, and social movement dynamics surrounding fracking have yet to receive much substantive theoretical attention. Yet, the controversy over fracking offers fertile ground for environmental sociologists who are interested in issues concerning technological risk, disaster, political economy, treadmill of production, or environmental justice. Some of the many important research questions in need of sociological analysis might include, for example: What will be the long-term impacts of shale energy development on public health, agricultural production, and water quality/availability? Which demographic groups and minority communities will disproportionally bear the risks of unconventional oil and gas extraction? What will be the longitudinal impacts of fracking and wastewater injection practices on induced seismicity and earthquake frequency? How is fracking affecting aquifers, wildlife, livestock, ecosystem services, and climate change trends? What role will global and national energy markets play in expanding fracking into unexplored regions like the Monterey Shale in California? How will intensive fracking exacerbate existing patterns of social and economic inequality? How has the oil and gas industry sought to increase its institutional legitimacy through the (dark money) funding of think tanks and "Frackademia" research centers at major state universities, as well as its behind-the-scenes push into K–12 education?

In turn, social movement researchers similarly have a wealth of opportunities to apply some of the theoretical insights of their field (e.g., resource mobilization, political opportunity, collective identity, frame alignment, etc.) to a range of questions exploring the mobilization of power and protest on both sides of the shale fracking controversy. For instance: Has opposition to fracking in the United States evolved into a national social movement or is it still largely contained in local and site-specific grievances? What are the mechanisms of boundary construction for the movement and is there a process of clear identity formation for activists that can serve as a precondition for collective action? How do the movement's resource base, organizational structure, and framing strategies contribute to its likelihood of success? What are the potential kinds of protest opportunities and mobilization techniques that can attract adherents and galvanize them to resist unconventional shale development in their communities and beyond? How is the fossil fuel industry stepping up its social control strategies to thwart citizen resistance efforts and manipulate public grievance perceptions? What are the explicit mechanisms of domination used by state actors to forestall opposition to fracking in extractive communities and create quiescence in the cultural ideas and beliefs of civil society? Clearly, the study of unconventional energy development and fracking offers environmental and movement researchers a timely opportunity to better understand one of the leading sources of conflict today that is fracturing both the community and the ecosystem on which humans depend.

Building on some of the other themes in this volume, future sociological research should explore in more detail the risk perceptions and opportunity-threat impacts associated with local-level unconventional energy development and how they compare across different shale regions of the country. Additional studies should also examine the empirical linkages between citizen actions, leadership initiatives, media frames, state-corporate responses, and various political mobilization efforts. What role do these factors play in influencing larger attitudes about energy use, technological hazards, environmental degradation, or government regulation? At the same time, more longitudinal research is needed to track public responses to unconventional energy development as boomtown conditions rise and fall; hidden social, economic, and environmental impacts are exposed; and post-development stages of "adaption" and "recovery" are achieved (or not) for local residents and surrounding ecosystems.

Despite the relatively polarized views about fracking that recent public opinion polls and surveys have brought to light, environmental sociologists still know little about the beliefs of other key stakeholders in the wider controversy, including those of oil and gas workers, business owners, scientists, environmentalists, and regulatory officials, among others, and how these attitudes differ by demographic status and geographic region. As the debate over unconventional energy development grows, it is imperative that these differential perceptions be further studied and better understood if policy makers and elected officials are to adequately navigate the difficult political complexities and landmines associated with the social, environmental, and public health impacts of hydraulic fracking (Schafft, Borlu, and Glenna 2013). Above all, future sociological research must link the rising tide of community conflicts surrounding energy development to larger global markets and geopolitical realities that transcend local concerns.

A Turning Point in U.S. Energy Policy?

In the past year, Americans have witnessed a string of major oil- and gas-related events that serve to highlight the centrality of energy issues to the future social, economic, and ecological health of the United States, as well as the sustainability of earth's atmosphere and life-support systems. Not surprisingly, many of these episodes also help expose the growing risks and conflicts associated with fracking and its role in expanding the treadmill of fossil fuel production. Consider the following key news stories of 2016:

- A gas well leak in the upscale San Fernando Valley community of Porter Ranch, California was finally capped on February 18, 2016 after 110 days of emitting hundreds of thousands of pounds of methane into the air since October 23, 2015, the forced evacuation of more than 4,700 families, and a state of emergency being declared by Governor Jerry Brown on January 6, 2016. Both Los Angeles County and the South Coast Air Quality Management District, among other local, state, and federal agencies, filed suit against Southern California Gas Co., charging the company with negligence, as well as significant negative impacts on climate. The cases are still pending (Food & Water Watch 2016a; *Los Angeles Times*, 2016).

- In April 2016, the number of rigs exploring for oil and natural gas in the United States dropped to an all-time low of 450, down from 1,028 rigs just a year ago and 4,530 rigs in 1981. Due to a glut of fracked gas and oil in the marketplace, energy prices have fallen significantly, which in turn has led to a decline in state tax revenues, severe state budget deficits, and the laying off of thousands of oil and gas company workers across the nation. With the average cost of fracking a single well now between $7.5 to $8 million, many drillers are pulling out of their U.S. leases and heading for more lucrative foreign markets until demand rises (Adler 2016; *Times-Picayune* 2016).

- On Earth Day, April 22, 2016, New York governor Andrew Cuomo and the Department of Environmental Conservation rejected the proposed fracked natural gas Constitution Pipeline that would have impacted water sources, forests, and communities on a 124-mile route across the state. The decision joins Cuomo's veto of the Port Ambrose LNG facility, as well as the existing ban on high-volume fracking, in making New York one of the nation's leading states that has successfully resisted shale development and hydraulic fracturing (Frack Action 2016a).

- On May 2, 2016, the Colorado Supreme Court issued a decision invalidating the community fracking bans in Longmont and Fort Collins, Colorado that had passed in ballot referendums in 2015. Less than two weeks later, Grant Township in western Pennsylvania passed a first-in-the-nation law legalizing nonviolent civil disobedience against Pennsylvania General Energy Company's plan to convert a well for fracked wastewater injection disposal. The tiny community of 700 residents said their action legalizing direct action protest was a response to the ongoing problem of rural residents seeing their voices excluded from discussions with state governments and energy corporations on issues of local importance (*Stringer* 2016).

- On May 13, 2016, Shell spilled 90,000 gallons of crude oil into the Gulf of Mexico, marking yet another offshore drilling accident in the Gulf since the historic *Deepwater Horizon* oil disaster in 2010 (Brune 2016; Food & Water Watch 2016b).

- In June 2016, a longtime environmental watchdog group in North Carolina, NC WARN, charged the federal EPA with "a persistent and deliberate cover-up that has prevented the agency from requiring the natural gas industry to make widespread, urgently needed and achievable reductions in methane venting and leakage across the nation's expanding natural gas infrastructure." The group specifically alleged that the EPA's Science Advisory Board used oil and gas industry–funded research for policy and regulatory purposes that seriously underestimated methane emissions from fracking sites (Knight 2016).

- In mid-July, a massive explosion took place on a fracking site in the Mancos Shale region of New Mexico, igniting 36 oil tanks, temporarily closing nearby Highway 550, and forcing 55 local residents out of their homes (Chow 2016).

- On September 4, 2016, a 5.6 magnitude earthquake took place near Pawnee, Oklahoma, in the north-central region of the state where many fracking wastewater disposal wells operate. The state Corporation Commission directed some 32 wells to shut down until the suspected connection between wastewater injection methods and earthquakes can be further studied (Miller 2016).

As *Fractured Communities* goes to press in the aftermath of the 2016 U.S. presidential election, the contested terrain surrounding unconventional oil and gas development is expanding and pushing the issue of fracking deeper into the mainstream of the American political arena. Opposition to fracking became a national campaign issue for the first time in the 2016 Democratic presidential race between Senator Bernie Sanders and Secretary Hillary Clinton. At the March primary debate in Flint, Michigan, for instance, Sanders pledged to oppose fracking, while Clinton said she would be in opposition only if a locality or state were against it, releases of methane or water contamination were present, or fracking operators refused to report what chemicals they were using (Leber 2016). Clinton's stance signified a significant departure from her previous enthusiastic support for natural gas fracking as part of the Obama administration's "All of the Above" energy policy.

In July 2016, more than 100 families personally impacted by oil and gas fracking wrote to President Obama demanding that he, along with top EPA officials, meet with them ahead of the Democratic National Convention to hear their personal testimonies of the harm they have experienced at the hands of the oil and gas industry, as well as the inaction by local, state, and federal agencies (Fulton 2016). On the day before the Democratic National Convention, 10,000 citizens and activists—including tribal members from a number of prominent western indigenous nations—mobilized in Philadelphia to support the "March for a Clean Energy Revolution" through the streets of the city. The protest march called on political leaders to support a national ban on fracking, keep fossil fuels in the ground, ensure environmental justice for all, and work for a quick and lasting transition to a 100% clean energy economy (Frack Action 2016b). While the protest march did not persuade the Democratic Party to endorse a moratorium on fracking, the party platform did call for a greater reliance on clean energy sources and support for the Paris Climate Accords, among other measures.

To return to a theme emphasized in the book's introductory chapter: *energy matters*—arguably now more than at any time in recent U.S. history. With the election of Donald J. Trump to the White House, the Republican Party now controls the executive and congressional branches of government, as well as federal judicial appointments and the majority of state governments. Already, Americans are witnessing a wholesale abandonment of the bipartisan energy policies and regulatory regimes that have been in place since the 1970s. Indeed, the GOP platform passed at the Republican National Convention in Cleveland in July 2016 took aim at "environmental extremists," called the environmental movement "a self-serving elite," and promised to move responsibility for environmental regulation from "the federal bureaucracy"—particularly the EPA—to the states where few fiscal resources and commitments to government oversight exist (Mufson 2016: A9). On energy and climate change issues especially, the Trump administration has pledged—and is rapidly advancing—an agenda that represents a carbon copy of the wish-list of the fossil fuel industry, including:

- Constructing the Keystone XL and Dakota Access pipelines
- Abandoning the COP 21 Paris Climate Agreement
- Reducing/blocking funding for the Environmental Protection Agency (EPA), National Oceanic and Atmospheric Administration (NOAA), and climate research at the National Aeronautics and Space Administration (NASA)
- Scrapping the Clean Power Plan
- Rescinding the Obama administration's ban on drilling in the Alaskan Arctic, as well as regulations to limit methane leaks from wells and pipelines

- Increasing fracking and drilling on public lands
- Increasing coal production and "clean-coal" technology

Clearly, these renewed carbon-intensive policies cannot be understood in isolation from the neoliberal deregulatory trends of the past three decades, nor detached from the surge of campaign money unleashed by Citizens United into the coffers of congressional representatives in key energy states (Perrow 2015). The enactment of the larger Republican environmental agenda will mean an expansion of offshore and Arctic drilling, tar sands development, mountaintop coal removal, and other extreme energy policies, while rolling back existing programs that increase energy efficiency, as well as provide tax incentives and credits for solar, wind, and biomass production—all carried out, ironically, in the name of U.S. "energy independence." After a two-year downturn in oil and gas prices, U.S. production is on the rise again, with the number of U.S. drilling rigs growing to 602 as of March 2017 (Bias 2017: A19).

Surrounded by advisers and cabinet officials with deep ties to the fossil fuel industry, President Trump—who called climate change a Chinese hoax—may in fact be presiding over what one researcher called "the most anti-science administration in American history" (Proctor 2016: 4). Trump's inner circle on environment and energy issues, including Secretary of State (former ExxonMobil CEO) Rex Tillerson, Secretary of Interior (former Montana congressman) Ryan Zinke, Secretary of Energy (former Texas governor) Rick Perry, Secretary of Agriculture (former governor of Georgia) Sonny Perdue, and most importantly, Administrator of the Environmental Protection Agency (former attorney general of Oklahoma) Scott Pruitt, all represent longtime climate change deniers and fossil fuel industry advocates, as well as steadfast opponents of international environmental accords on climate and carbon emissions, federal protections for public lands, or independent scientific research that addresses atmospheric or ecological threats. These political figures are not just *indifferent* to environmental problems, but openly *hostile* to their remediation through government regulations and policy making. Indeed, not since the Reagan years of the 1980s and the Bush years of the early 2000s have the hydrocarbon and petrochemical interests of the energy-industrial complex had a potentially greater political ally in the White House (Ladd and York 2017).

At its roots, the controversy over fracking must be seen as a subset of the larger conflict over the future direction of American (and global) energy policy and the resource base and power brokers that will shape U.S. political and economic interests well into the 21st century. As observers have noted since the Arab oil embargos of the 1970s, the nation continues to stand at a critical historical crossroads defined by two mutually exclusive and environmentally distinct energy futures. The question remains the same. Will the U.S. fossil fuel industry and its allies be permitted to maintain its hegemonic control over an emerging "Third Carbon Era" defined by hydraulic fracking and other invasive forms of extreme energy production? Or will the democratic majority create the political mechanisms necessary to initiate the transition to a clean energy society defined by renewable resources, efficient technologies, and green economic policies over the next two decades? (Ladd 2016). As the research assembled in *Fractured Communities* helps illuminate, we are at a decisive turning point in U.S. energy policy and the road we choose may very well turn out to spell the difference between peril and promise.

References

Adler, Ben. 2016. "Wanna See What Happens When You Rely on the Fossil Fuel Sector and Slash Taxes? Check Out Louisiana." *Grist Magazine*, March 7. Accessed March 8, 2016. (http://grist.org/politics/wanna-see-what-happens-when-you-rely-on-the...a/?utm _medium=email&utm_source=newletter&utm_campaign=daily-horizon).

Baca, Marie C. 2016. "Forced Pooling: When Landowners Can't Say No to Drilling." *Pro-Publica*, July 7. Accessed July 7, 2016. (http://www.propublica.org/article/forced-pooling -when-landowners-cant-say-no-to-drilling).

Bamberger, Michelle, and Robert Oswald. 2014. *The Real Cost of Fracking: How America's Shale Gas Boom Is Threatening Our Families, Pets, and Food*. Boston: Beacon Press.

Beck, Ulrich. 1995. *Ecological Politics in an Age of Risk*. Cambridge: Polity Press.

Bias, Javier. 2017. "Shale Oil Industry Leaner, Fitter." *Times-Picayune*, March 5: A19.

Brune, Michael. 2016. "Another Major Spill in the Gulf of Mexico." *Sierra Club*, March 13. Accessed March 13, 2016.

Chow, Lorraine. 2016. "Massive Fracking Explosion in New Mexico, 36 Oil Tanks Catch Fire." *Energy News*, July 13. Accessed July 14, 2016. (http://www.ecowatch.com/massive -fracking-explosion-in-new-mexico-1919567359.html).

Eaton, Emily, and Abby J. Kinchy. 2016. "Quiet Voices in the Fracking Debate: Ambivalence, Nonmobilization, and Individual Activism in Two Extractive Communities." *Energy Research and Social Science* 20: 22–30.

Erikson, Kai. 1994. *A New Species of Trouble: The Human Experience of Modern Disasters*. New York: W. W. Norton.

Evensen, Darrick, Jeffrey B. Jacquet, Christopher E. Clarke, and Richard C. Stedman. 2014. "What's the 'Fracking' Problem? One Word Can't Say It All." *Extractive Industries and Society* 1: 130–136.

Finkel, Madelon L. (ed.). 2015. *The Human and Environmental Impact of Fracking: How Fracturing Shale for Gas Affects Us and Our World*. Santa Barbara, CA: Praeger.

Food & Water Watch. 2016a. "Out of the News, But Not out of Trouble." April 20. Accessed April 20, 2016. (act@fwwatch.org).

Food & Water Watch. 2016b. "Big Court Decision in Colorado." May 2. Accessed May 3, 2016. (act@fwwatch.org).

Frack Action. 2016a. "Huge Victory on the Constitution Pipeline!" Accessed April 27, 2016. (info@frackaction.com).

Frack Action. 2016b. "We Need a National Ban on Fracking!" July 12. Accessed July 12, 2016. (infro@frackaction.com).

Freudenburg, William R., and Robert Gramling. 1992. "Community Impacts of Technological Change: Toward a Longitudinal Perspective." *Social Forces* 70(4): 937–955.

Fulton, Deidre. 2016. "Fracking-Affected Families Plead with President Obama: 'We Need Help.'" *Common Dreams*, July 1. Access July 3, 2016. (http://commondreams.org.news /07/01/fracking-affected-families-p...elp?utm_campaign=shareaholic&utm_medium= email_thi&utm_source=email).

Gullion, Jessica Smartt. 2015. *Fracking the Neighborhood: Reluctant Activists and Natural Gas Drilling*. Cambridge, MA: MIT Press.

Hauter, Wenonah. 2016. *Frackopoly: The Battle for the Future of Energy and the Environment*. New York: New Press.

Healy, Jack. 2015. "Heavyweight Responses to Local Fracking Bans." *New York Times* (National section) January 4: Y11–13.

Jerolmack, Colin. 2014. "Please in My Backyard: Risk, Reward, and Inequality in Central Pennsylvania's 'Gas Rush.'" Paper presented at the American Sociological Association meetings, August 16–19, San Francisco, CA.

Knight, Nika. 2016. "Whistleblower: EPA Officials Covered Up Toxic Fracking Emissions for Years." *Common Dreams*, June 9. Accessed June 9, 2016. (http://www.commondreams .org/news/2016/06/09/whistlelblower-epa-offic...rs?utm_campaign=shareaholic&utm _medium=email_this&utm_source-email).

Ladd, Anthony E. 2013. "Stakeholder Perceptions of Socio-environmental Impacts from Unconventional Natural Gas Development and Hydraulic Fracturing in the Haynesville Shale." *Journal of Rural Social Sciences* 28(2): 56–89.

Ladd, Anthony E. 2014. "Environmental Disputes and Opportunity-Threat Impacts Surrounding Natural Gas Fracking in Louisiana." *Social Currents* 1(3): 293–312.

Ladd, Anthony E. 2016. "Meet the New Boss, Same as the Old Boss: The Continuing Hegemony of Fossil Fuels and Hydraulic Fracking in the Third Carbon Era." *Humanity and Society* (http://dx.doi.org/10.1177/0160597616628908).

Ladd, Anthony E., and Richard York. 2017. "Hydraulic Fracking, Shale Energy Development, and Climate Inaction: A New Landscape of Risk in the Trump Era." *Human Ecology Review* (forthcoming).

Leber, Rebecca. 2016. "Hillary Clinton Has a New Tune on Fracking." *Grist Magazine*, March 6. Accessed March 7, 2016. (http://grist.org/politics/hilary-clinton-has-a-new -tune-on-fracking/?utm_medium=email&utm_source=newsletter&utm_campaign= daily-horizon).

Los Angeles Times. 2016. "Porter Range Gas Leak Updates." February 18. Accessed May 23, 2016. (http://www.latimes.com/local/lanow/la-me-porter-ranch-gas-leak-live-htmlstory .html).

Malin, Stephanie. 2014. "There's No Real Choice But to Sign: Neoliberalization and Normalization of Hydraulic Fracturing on Pennsylvania Farmland." *Journal of Environmental Studies and Science* 4(1): 17–27.

Malin, Stephanie A., and Kathryn Teigen DeMaster. 2016. "A Devil's Bargain: Rural Environmental Injustices and Hydraulic Fracturing on Pennsylvania's Farms." *Journal of Rural Studies* 47: 278–290.

Malin, Stephanie A., Adam Mayer, Kelly Shreeve, Shawn K. Olson-Hazboun, and John Adgate. 2017. "Free Market Ideology and Deregulation in Colorado's Oil Fields: Evidence for Triple Movement Activism?" *Environmental Politics* (http://dx.doi.org/10.1080 .09644016.2017.1287627).

Miller, Ken. 2016. "Oklahoma Quake Felt as Far as Arizona." *Times-Picayune*, September 4: A4.

Mufson, Steven. 2016. "Energy Platform Would Reverse Decades of Policy." *Times-Picayune*, July 20: A9.

Opsal, Tara, and Tara O'Connor Shelley. 2014. "Energy Crime, Harm, and Problematic State Response in Colorado: A Case of the Fox Guarding the Hen House?" *Critical Criminology* (http://dx.doi.org/10.1007/s10612-014-9255-2).

Perrow, Charles. 1999 (2nd ed.). *Normal Accidents: Living with High Risk Technologies.* Princeton, NJ: Princeton University Press.

Perrow, Charles. 2015. "Cracks in the Regulatory State." *Social Currents* 2(3): 203–212.

Poole, Amanda, and Anastasia Hudgins. 2014. "'I Care More about This Place, because I Fought for It': Exploring the Political Ecology of Fracking in an Ethnographic Field School." *Journal of Environmental Studies and Science* 4(1): 37–46.

Proctor, Robert N. 2016. "Climate Change in the Age of Ignorance." *New York Times*, November 20: (Sunday Review) p. 4.

Rowell, Andy. 2016. "Hundreds of Offshore Fracking Wells Dump Billions of Gallons of Oil Waste into Gulf." *Ecowatch*, July 11. Accessed July 14, 2016. (http://www.ecowatch.com/offshore-fracking-wells-gulf-1915520170.html).

Schafft, Kai A., Yetkin Borlu, and Leland Glenna. 2013. "The Relationship between Marcellus Shale Gas Development in Pennsylvania and Local Perceptions of Risk and Opportunity." *Rural Sociology* 78(2): 143–166.

Schwartz, John. 2015. "Subsidies Dwarf Spending on Climate Change." *Times-Picayune*, December 6: A-11.

Steingraber, Sandra. 2012. "The Whole Fracking Enchilada." Pp. 175–78 in *The Energy Reader: Overdevelopment and the Delusion of Endless Growth*, edited by Tom Butler, Daniel Lerch, and George Wuerthner. Santa Rosa, CA: Post Carbon Institute.

Stringer, Kate. 2016. "Faced with a Fracking Giant, This Small Town Just Legalized Civil Disobedience." *Yes Magazine*, May 14. Accessed May 14, 2016. (http://www.yesmagazine.org/planet/faced-with-a-fracking-giant-this-small-town-just-legalized-civil=disobedience-20160513).

Tierney, Kathleen. 2015. "Resilience and the Neoliberal Project: Discourses, Critiques, Practices—And Katrina." *American Behavioral Scientist* 59(10): 1327–1342.

Times-Picayune. 2016. "Rig Count at All-Time Low." April 3: A11.

Wilber, Tom. 2015. *Under the Surface: Fracking, Fortunes, and the Fate of the Marcellus Shale.* Ithaca, NY: Cornell University Press.

Acknowledgments

As is the case with any academic work, there are many individuals who contributed directly and indirectly to the completion of this edited volume that I would like to thank. First and foremost, I am extremely grateful to Peter Mickulas and Scott Frickel of Rutgers University Press for their guidance, support, and faith in my ability to deliver a book that lived up to the promise of what they saw in my original book proposal. Their critiques, suggestions, and insights on the chapter drafts were always valuable and I am especially proud to be a part of the Nature, Society, and Culture series in Environmental Sociology. I also wish to extend my thanks to Tom Shriver, Brett Clark, and Rob Benson for their helpful comments on the proposal and project. Indeed, the completion and publication of this book is the fruition of a dream come true since I began studying environmental controversies, issues, and disasters some 35 years ago. It is certainly my hope that the research assembled here will make a worthy contribution to the sociological literature on unconventional energy development and hydraulic fracturing and motivate other sociologists, particularly those just coming out of graduate programs, to build on some of the baseline research generated here. It is also my hope that nonsociologists will read this book and become motivated as citizens to support the global fossil fuel resistance movement today (what Naomi Klein calls "Blockadia") that is working diligently to bring about a more sustainable, renewable, and clean energy future.

In thinking about the people in my life who played a role in making this book possible, I have been reminded of the enormous intellectual debt I owe my graduate school mentor in sociology at the University of Tennessee, Professor Don Clelland. It was in Don's social stratification courses that I first became interested in the political economy of the nuclear power industry and, more importantly, the inherent risks that nuclear energy posed for humans and the environment. Thanks to Don, I started reading books by perceptive energy critics like John Goffman, Alice Stewart, Helen Caldicott, E. F. Schumacher, Hazel Henderson, and Amory Lovins, authors who not only helped me understand the myth of the "peaceful atom," but the promise of the solar age and how renewable energy technologies could provide the seedbed for a better world. Armed with this new worldview, I joined a student antinuclear group on the UT campus, become active in antinuclear activities, and eventually wrote my Ph.D. dissertation on the May 6th March on Washington, an antinuclear protest in the nation's capital that drew over 125,000 protesters in the wake of the 1979 Three Mile Island accident in Pennsylvania. As a result of Don's wisdom, support, and friendship, as well as the growing "energy crisis" of the time, I became interested in the field of environmental sociology and the role of energy production in creating many (if not most) of our ecological problems.

I also want to acknowledge some of the many environmental sociologists I have had the pleasure of knowing and sometimes collaborating with over the years who have consistently inspired me to work harder and think more substantively about the environmental risks and disasters that surround us. Thanks to colleagues and friends like Duane Gill, Liesel Ritchie, Steve Picou, Tom Shriver, Sherry Cable, Bob Edwards, Shirley Laska, Steve Kroll-Smith, Pam Jenkins, and the late Brent Marshall and Bob Gramling, among others, I have been most fortunate to know a talented network of like-minded scholars whom I could always learn from, bounce ideas off, and get their advice on projects when I asked for it. Then too, I would be remiss if I did not pay tribute to the work of people like Rachel Carson, Paul Ehrlich, Barry Commoner, Jane Goodall, William Catton, Amory Lovins, Riley Dunlap, Charles Anderson, Kai Erickson, Charles Perrow, Bill Freudenburg, Bill McKibben, Dave Foreman, Josh Fox, Naomi Klein, and many, many more whose intellectual insights, books, articles, and personal activism have inspired me every day to get out of bed and—as Gandhi said—*be the change I want to see in the world*. Through their example, they have helped me teach my students to be better caretakers and citizens of the planet, rather than just resource predators and mindless consumers.

Finally, this book would not have been possible were it not for the love of my family and friends, especially the influence of my late grandmother, Edna Hernley, and mother, Margaret Ladd, whose never-ending support and belief in my educational aspirations and professional choices helped make me the sociologist I am today. Indeed, these important maternal figures in my life taught me how to be a feminist long before I knew what the word meant. From both of these strong and independent women I learned many things, including the value of being informed about current events, the joys of daily newspapers, good books, and great movies, the importance of listening to others over talking about one's self, as well as how to treat other people the way I wanted to be treated. I am so lucky to have grown up knowing you both as I did and may your spirit, generosity, and lessons always guide me.

Lastly, I want to thank my amazing and talented wife—Dr. Kathleen Fitzgerald—for her constant support and patience throughout the entirety of this book project. More than just my best friend, colleague, fellow sociologist, and love of my life, Kathleen was a constant source of encouragement, good humor, and sound professional advice (having published four books herself) on how to navigate through some of the challenges and frustrations I routinely encountered. Thank you, Kathleen, for your scholarly example, comradeship, and dedication to the craft of sociology. I love you madly, need you badly, and—as you promised me in your wedding vows—you still make me laugh every day of my life. Best of all, your beautiful and shining Irish eyes always go straight to my heart whenever you look my way.

Notes on Contributors

HILARY BOUDET is an assistant professor in the School of Public Policy at Oregon State University. Her research interests include environmental and energy policy, social movements, and public participation in energy and environmental decision making. She coauthored, with Doug McAdam, *Putting Social Movements in Their Place: Explaining Opposition to Energy Projects in the United States, 2000–2005*. Her recent work focuses on public perceptions of energy development and community-based interventions designed to encourage sustainable behavior.

SHERRY CABLE is a professor of sociology at the University of Tennessee–Knoxville whose primary research interests are in environmental conflict, environmental inequalities, and environmental policy. Her most recent book is *Sustainable Failures: Environmental Policy and Democracy in a Petro-dependent World*. Recent articles include "Risk Society and Contested Illness: The Case of Nuclear Weapons Workers" (*American Sociological Review*) with Tom Shriver and Tamara Mix, for which they received the 2011 Allan Schnaiberg Outstanding Publication Award from the American Sociological Association; and a forthcoming chapter in an edited volume coauthored with doctoral student Kayla Stover, "Women's Environmental Activism: Motivations, Experiences, and Transformations."

BRITTANY GAUSTAD is a research analyst at the Institute for Transportation Research and Education at North Carolina State University. She is currently analyzing the economic impacts of historical transportation investments, as well as several other transportation-related research projects. She holds a Master in Public Policy degree from Oregon State University, where she completed research on community mobilization and public participation surrounding the siting of liquefied natural gas facilities in Astoria and Coos Bay, Oregon, as well as sustainable behavior change.

PETER M. HALL is an affiliate professor of sociology, Colorado State University and professor emeritus of sociology, University of Missouri. He is currently co-editor of *The Sociological Quarterly* and has received the George Herbert Mead Award for career contributions from the Society for the Study of Symbolic Interaction.

ANTHONY E. LADD is a professor of sociology in the Department of Sociology and the Environment Program and a former chair of the Department of Sociology and the Environmental Studies Program at Loyola University New Orleans. His major area of research concerns the study of environmental controversies, technological disasters, and the mobilization of protest in response to ecological degradation. In addition to *Fractured Communities*, he has published over 50 articles, chapters, and reviews in such journals as *Sociological Inquiry*, *Social Currents*, *Sociological Spectrum*, *American Behavioral Scientist*, *Human Ecology Review*, *Journal of Rural Social Sciences*, *Humanity and Society*, *Journal of Public Management and Social Policy*, and *Social Justice*. His most recent published research analyzes the environmental disputes and differential impacts of natural gas fracking in the Haynesville Shale region of Louisiana, as well as the growing socioenvironmental threats posed by our continued reliance on fossil fuels and unconventional energy development. He is currently serving on the advisory board for an NSF grant studying wastewater-induced seismicity in Colorado and Oklahoma, as well as researching oil and gas industry funding of "Frackademia" think tanks in U.S. universities.

STEPHANIE A. MALIN is an assistant professor of sociology at Colorado State University. Her major interests include environmental justice, environmental health, globalization, social mobilization, poverty, and the political economy of energy development. Her work has been published in such journals as the *Journal of Rural Studies*, *Society and Natural Resources*, and *Land Use Policy*, and her book, *The Price of Nuclear Power: Uranium Communities and Environmental Justice* (Rutgers University Press). Malin's current research centers around energy development's socioenvironmental impacts and includes a three-year study, funded by the National Institutes of Health, which examines the relationships between unconventional oil and gas development, quality of life, and stress levels.

JAMES N. MAPLES is an assistant professor of sociology at Eastern Kentucky University. He received his Ph.D. from the University of Tennessee. His major research interests focus on the political economy of tourism, including the economic impact of rock climbing and ultra-marathons in Kentucky, beer tourism in Appalachia, and damage to historic mountain cemeteries in West Virginia by mountaintop removal coal mining. He is also writing a book framing rock climbing as a renewable economic resource in Appalachia.

TAMARA L. MIX is an associate professor of sociology at Oklahoma State University with research interests in environmental justice, social movements, and race, class, and gender inequality. A community-engaged scholar, she has worked on projects involving a diverse range of stakeholders to address issues of community contamination, water access and quality, food justice, and the community dimensions of resource extraction and production. Her current projects focus on environmental justice and community dimensions of energy and resource extraction, with an emphasis on unconventional resource technology and induced seismicity in Oklahoma, as well as the implications of food inequality, food justice, and local food production on underserved communities.

CARMEL E. PRICE is an assistant professor of sociology at the University of Michigan–Dearborn. She earned her Ph.D. in environmental sociology from the University of Tennessee, her M.S.W. from Tulane University, and her B.A. in Education and Psychology from the University of North Carolina, Chapel Hill. Her work centers on themes of social justice and ecological public health, and she is currently engaged in research examining the environmental impact of migration and food insecurity among college students.

DAKOTA K. T. RAYNES is currently a Ph.D. candidate in the Department of Sociology at Oklahoma State University. Their teaching and research interests highlight issues of inequality and how social movements can work toward meaningful social change. Raynes's love of nature and deep commitment to community engagement led them to spend the last two and a half years on the front lines of the fracking controversy in Oklahoma as both a scholar and activist.

STACIA S. RYDER is a Ph.D. candidate in the Department of Sociology at Colorado State University and a 2016–2017 School of Global Environmental Sustainability Leadership Fellow. She received her M.A. in Sociology and a graduate certificate in Women's Studies from Colorado State University. Currently, she serves as the assistant editor for *Society & Natural Resources*, is a principal investigator for the Environmental Justice CSU Global Challenges Research team, and is a graduate research assistant for the Center for Disaster and Risk Analysis. Her work focuses on unconventional oil and gas development and incorporating intersectionality as a theoretical framework for environmental justice research. She has published her work in *Social Thought and Research* and is co-editing an upcoming special issue of *Environmental Sociology* with Stephanie Malin entitled "Environmental Justice and Deep Intersectionality."

SUZANNE STAGGENBORG is a professor of sociology at the University of Pittsburgh. She is the author of numerous articles and books on the dynamics of social movements, including *The Pro-Choice Movement: Organization and Activism in the Abortion Conflict*, *Methods of Social Movement Research* (co-edited with Bert Klandermans), and *Social Movements*. She is currently researching grassroots environmental groups and campaigns.

TRANG TRAN is a researcher at the Institute of Social and Economic Research (ISER), University of Alaska, Anchorage. At ISER, she conducts research on a wide variety of topics, including fuel costs for space heating, the economic impacts of teacher turnover, and rural and native education program evaluation. She holds a Master of Public Policy degree from Oregon State University, where she completed research on community mobilization and public participation surrounding the siting of liquefied natural gas facilities in Astoria and Coos Bay, Oregon, as well as community-based projects to address local challenges relating to education, energy, transportation, and gender equality.

ION BOGDAN VASI is an associate professor in the Department of Sociology and the College of Business at the University of Iowa. His research examines how social movements contribute to organizational change, industry creation, and policy-making. He also studies the social processes that shape the diffusion of technological innovations. He is the author of *Winds of Change: The Environmental Movement and the Global Development of the Wind Energy Industry*. His work has been published in the *American Sociological Review*, *American Journal of Sociology*, *Social Forces*, *Mobilization*, and other journals.

CAMERON THOMAS WHITLEY is an environmental sociologist with a particular interest in issues of justice and equality. He completed his graduate work at Michigan State University with specializations in Environmental Science and Policy (ESPP), Gender, Justice and Environmental Change (GJEC), and Animal Studies. His publications include a 2013 International Book Award–winning book, several book chapters, and a number of journal articles that have appeared in *American Emergency Medicine*, *Teaching Sociology*, *Sociological Perspectives*, *International Journal of Sociology*, *Qualitative Sociology Review*, and the *Proceedings of the National Academies of Science*. In addition, his work has also been featured in news outlets like the *Los Angeles Times*, *New York Times*, and *Washington Post*. He is currently engaged in research examining the formation of risk perceptions surrounding new energy technologies like hydraulic fracturing.

PATRICIA WIDENER is an associate professor of sociology at Florida Atlantic University. Her research interests include the political economy of oil, climate change, and transnational environmental activism. Currently, she is exploring how activists in Aotearoa, New Zealand are mobilizing against deep-sea oil drilling in the context of anthropogenic climate change and known oil disasters. She is the author of *Oil Injustice: Resisting and Conceding a Pipeline in Ecuador*.

Index

acid fracking, 229–31, 241, 274
acid stimulation, 237
activism: activists, 4, 45, 155, 158, 209–10, 240, 282; in Colorado, 213–17; diversity of, 114; in Oklahoma 185–87; in Pittsburgh, 109–13; place-based, 185; range of activism, 275–77; resident activists, 238, 241; in South Florida, 231–34; use of documentary films among, 61–62, 65. *See also* fractivists
agriculture, agricultural production, 235, 237, 278
Alabama, 277
Alaska, 282
Alberta tar sands, 177
Albright, Len, 202
Algeria, 15
Allegheny Armada, 120
animals: animal testing, 130; and energy development, 128–29, 272, 274; rights/ welfare organizations, 141–42
anthropogenic hazards, 178, 185
anti-fracking movement, 19, 27, 77–80, 150, 162–64, 175, 275; and documentary use, 65–73; organizations, 186–87; in Pittsburgh, 107–27
Antrim shale region, 12–13, 15
Apple, Benjamin E., 203
aquifers, 271; depletion of, 153, 158, 165. *See also* water
Arctic drilling, 283
Arkansas, 13, 14, 21, 70, 155, 179, 183

Argentina, 15
Australia, 20, 122, 150

Bakken Shale region, 12–13, 15, 26, 54, 200, 272
Ball Brothers Glass Manufacturing, ix
Ball State University, xi
Bamberger, Michelle, 18
bans: against fracking, 20, 73–77, 80, 108, 115, 200, 230, 240; Alaska Arctic drilling ban, 282; on bans, 186; differential outcomes on bans nationwide, 275–77, 280, 282; on fracking on public land, 47–53; overriding local bans in Colorado, 208–14; on using the term climate change, 238. *See also* moratorium
Barnett Shale region, 12–14, 19, 26, 54, 61–62, 153, 165, 272
Beck, Ulrich, 16
Benford, Robert D., 154
Big Cypress National Preserve, 230–32, 236
biodiversity, reduction of, 135, 141, 152
boom, 62, 151, 155–56, 174, 199, 205, 217; boomtown, 2, 15, 18, 62, 274, 279; and bust cycle of energy development, 164
Brady, William J., 201
bridge fuel, 23, 54, 150, 271
Bulgaria, 20, 150
Bureau of Land Management (BLM), 10, 42–44
Bush-Cheney administration, 62–63
Buttel, Frederick, 5